Integral Calculus

Integral Calculus

R.K. Pandey

ANMOL PUBLICATIONS PVT. LTD.
NEW DELHI - 110 002 (INDIA)

ANMOL PUBLICATIONS PVT. LTD.
H.O.: 4374/4B, Ansari Road, Daryaganj,
New Delhi-110 002 (India)
Ph.: 23278000, 23261597
B.O.: No. 1015, Ist Main Road, BSK IIIrd Stage
IIIrd Phase, IIIrd Block,
Bangalore - 560 085 (India)
Visit us at: www.anmolpublications.com

Integral Calculus

ISBN 978-81-261-3586-8

PRINTED IN INDIA

Printed at Mehra Offset Press, Delhi.

Contents

Preface

The present book *Integral Calculus* is a unique textbook on integration, aiming at providing a fairly complete account of the basic concepts required to build a strong foundation for a student endeavouring to study this subject. The analytical approach to the major concepts makes the book highly self-contained and comprehensive guide that succeeds in making the concepts easily understandable. All the elementary principles and fundamental concepts have been explained rigorously, leaving no scope for illusion or confusion. The focus throughout the text has been on presenting the subject matter in a well-knit manner and lucid style, so that even a student with average mathematical skill would find it accessible to himself. In addition, the book provides numerous well-graded solved examples, generally set in various university and competitive examinations, which will facilitate easy understanding besides acquainting the students with a variety of questions.

It is hoped that the book would be highly useful for the students and teachers of mathematics. Students aspiring to successfully accomplish engineering and also those preparing for various competitive examinations are likely to find this book of much help.

List of Symbols

±	⊄	⊂	⊆	∈	∉	ℜ	∂	≠	≈	∠	∇
∑	Π	Ω	ϕ	α	β	χ	ε	γ	η	λ	μ
ν	⊥	π	θ	ρ	σ	τ	ℜ	ω	ξ	ψ	ζ
ϒ		∃	∫	∬	∭	×	√	∵	≥	∮	⇒

Chapter 1

Introduction

The differential and integral calculus is based on two concepts of outstanding importance, apart from the concept of number, namely, the concepts of function and limit. While these concepts can be recognized here and there even in the mathematics of the ancients, it is only in modern mathematics that thfeir essential chfaracter and significance have been fully clarified. We shall attempt here to explain these concepts as simply and clearly as possible.

THE INDEFINITE INTEGRAL

Recall that if I is any interval and f is a function defined on I, then a function F on I is an *antiderivative* of f if $F'(x) = f(x)$ for all x I. A corollary of the Mean Value Theorem tells us that if G is any other antiderivative of f on I, then there is a constant C such that $G(x) = F(x) + C$.

Let I be an interval and f a function defined on I. Then the *indefinite integral* of f is defined to be the set of all antiderivatives of f on I. We denote this set by: $\int f(x)\,dx$

We call f the *integrand* and x the *variable of integration*.

Since all the antiderivatives of f differ only by a constant, it is customary to write:

$$\int f(x)\,dx = F(x) + C$$

where F is any particular antiderivative, and C is the *constant of integration* which implicitly takes on all real number values.

THE DEFINITE INTEGRAL

Let f be a function defined on a closed interval $[a, b]$. Given any n we let $\Delta x = (b-a)/n$, and we let $x_k = a + k\,\Delta x$. For each k, select *sample points* x_k^* $[x_{k-1}, x_k]$. We define the *definite integral* of f on $[a, b]$ to be

$$\int_a^b f(x)\,dx = \lim_{n\to\infty} \sum_{k=1}^{n} f(x_k^*)\Delta x$$

provided this limit exists and does not depend on the choice of sample points x_k^*.

If the limit depends upon the choice of sample points, or is undefined for any choice of sample points, then the integral does not exist.

We call a the *lower limit* of the integral, and b the *upper limit*. We call f the *integrand*.

We also make the defintions:

$$= -\int_a^b f(x)\,dx$$

$$\int_b^a f(x)\,dx = 0$$

The integral sign " $\int$ " is an archaic "S" and stands for "sum" (as does Σ, of course).

If it exists, the definite integral gives you a number as its result. In this regard, x is a "dummy variable" and the dx reminds us what variable we are integrating with respect to. We can just as easily replace x with t and write: $\int_a^b f(t)\,dt$ and the numerical value of the definite integral is unchanged.

Geometrically, if $f(x) \geq 0$ we interpret the integral as the area between the graph $y=f(x)$, the x-axis, and the lines $x=a$ and $x=b$. More generally, we think of the integral as a "signed" area, where the integral is equal to the area above the x-axis less the area below the x-axis.

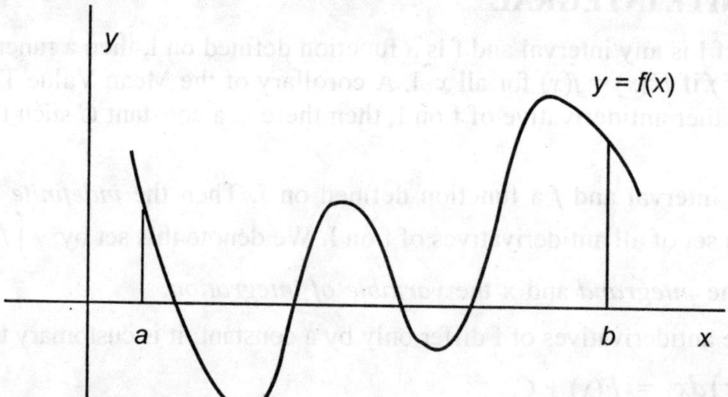

$$\int_a^b f(x)\,dx = A_+ - A_-$$

The sum: $\sum_{k=1}^{n} f(x_k^*)\Delta x$ in the definition of the definite integral is called a Riemann sum; the definite integral is sometimes called a Riemann integral (to distinguish it from other more general integrals used by mathematicians). Whenever you have a limit as n goes to infinity of a Riemman sum, you can replace it by an appropriate integral. This observation is critical in applications of integration.

For certain simple functions, you can calculate an integral directly using this definition. However, in general, you will want to use the fundamental theorem of calculus and the algebraic properties of integrals.

RULES OF INTEGRATION

Let f and g be functions and let a, b and c be constants, and assume that for each fact all the indicated definite integrals exist. Then the following are true:

1. Constants can be pulled out of integrals:

$$\int cf(x)dx = c\int f(x)dx$$

$$\int_a^b cf(x)dx = c\int_a^b f(x)dx$$

2. The integral of the sum of two functions equals the sum of the integrals of each function:

$$\int f(x)+g(x)dx = \int f(x)dx+\int g(x)dx$$

$$\int_a^b f(x)+g(x)dx = \int_a^b f(x)dx+\int_a^b g(x)dx$$

3. The integral of the difference of two functions equals the difference of the integrals of each function:

$$\int f(x)-g(x)dx = \int f(x)dx-\int g(x)dx$$

$$\int_a^b f(x)-g(x)dx = \int_a^b f(x)dx-\int_a^b g(x)dx$$

4. The integral from a to b of a function equals the integral from a to c plus the integral from c to b:

$$\int_a^b f(x)dx = \int_a^c f(x)dx+\int_a^b f(x)dx$$

Note that there are no general rule for integrals of products and quotients. Such integrals can sometimes, but not always, be calculated using substitution or integration by parts.

We can also give some facts about inequalities involving definite integrals.

1. If $f(x) \geq 0$ on $[a, b]$ then $\int_a^b f(x)dx \geq 0$
2. If $f(x) \geq g(x)$ on $[a, b]$ then $\int_a^b f(x)dx \geq \int_a^b g(x)dx$
3. The absolute value of an integral is less than the integral of the absolute value.

$$\left|\int_a^b f(x)dx\right| \leq \int_a^b |f(x)|dx$$

BASIC INTEGRALS

The following are some basic indefinite integrals. The table can also be used to find definite integrals using the fundamental theorem of calculus.

Integrals of Powers

$$\int k\,dx = kx + C$$

$$\int x^n\,dx = \frac{1}{n+1}x^{n+1} + C$$

$$\int \frac{1}{x}\,dx = \ln x + C$$

Integrals of Exponentials

$$\int e^x\,dx = e^x + C$$

$$\int a^x\,dx = \frac{1}{\ln a}\,a^x + C$$

Integrals of Trigonometric Functions

$$\int \sin x\,dx = -\cos x + C$$

$$\int \cos x\,dx = \sin x + C$$

$$\int \sec^2 x\,dx = \tan x + C$$

$$\int \sec x \tan x\,dx = \sec x + C$$

$$\int \csc x \cot x\,dx = -\csc x + C$$

$$\int \csc^2 x\,dx = -\cot x + C$$

Integrals Involving Inverse Trigonometric Functions

$$\int \frac{1}{\sqrt{1-x^2}}\,dx = \sin^{-1}x + C$$

$$\int \frac{1}{\sqrt{1+x^2}}\,dx = \tan^{-1}x + C$$

Integration By Parts

Recall that we derived the formula for integration by substitution by using the the Chain Rule and integrating it using the fundamental theorem of calculus. You might reasonably ask "What happens if we use one of the other rules of differentiation?"

This would be an exceedingly good question, and it even has an answer. If we take the product rule:

$$\frac{d}{dx}fg = f\frac{dg}{dx} + g\frac{df}{dx}$$

and integrate both sides, we get:

$$fg = \int fg'\,dx + \int gf'\,dx$$

This isn't particularly useful until we re-arrange and get:

$$\int f(x)g'(x)dx = f(x)g(x) - \int g(x)f'(x)dx$$

If you let $u = f(x)$ and $v = g(x)$, then we can abuse notation in the standard way and say that $du = f(x)$ and $dv = g'(x)$. Then the above is re-written as: $\int u\,dv = uv - \int v\,dv$. which may be easier to remember.

If we can split the integrand up into two bits:

- Something we can differentiate, ie. $f(x)$,
- And something we can integrate, ie. $g'(x)$,

then we can apply the formula. The hope is that the new integral we get, ie. $\int gf'\,dx$, will be easier to compute than the original integral.

Sometimes you will find that you need to apply the whole process again to the new integral, but if you do, be careful that you don't end up back where you started.

Unfortunately, choosing the correct way to split the integrand can be tricky to get right, and the only real way to get good at it is to practice. Initially you will make some missteps, but eventually you will sort out how to pick the right functions because you will be able to think ahead and guess what the final integral will look like.

Good ways to choose (although these are not written in stone) include:

- Polynomials make good choices for f, because they tend to get simpler as you differentiate them.
- log, arctan and arcsin are good choices for f if there are polynomials around, since you may get the chance to cancel.
- Exponential functions, sin and cos make good choices for g', because they tend not to get more complicated when you integrate.
- Since integration is harder than differentiation, try and make g' as complex as you can while still making sure you can integrate it.
- In general you should aim to be making the integrand that you get out "simpler" than the one you started with.
- If at first you don't succeed, try a different way of splitting the integrand.

Definite Integrals

Integration by parts works without much additional difficulty when doing definite integrals. The limits of the second integral are unchanged, and what you get is the following:

$$\int_a^b f(x)g'(x)dx = f(b)\,g(b) - f(a)\,g(a) - \int_a^b g(x)f'(x)dx$$

TRICKS OF THE TRADE

There are a handful of special tricks that employ integration by parts to solve otherwise difficult integration problems. Each of these is illustrated by a typical example which explains the technique.

Multiplying by 1

Sometimes you will be confronted by an integrand that you can't see how to split, but which you can see how to differentiate. The classic example of this is: $\int \log(x)dx$

The trick here is to think of this as: $\int 1 \log(x)dx$

and let $f(x)=\log x$, and $g'(x)=1$.

Getting the Same Integral After Integration by Parts

Sometimes you may find that after doing integration by parts (possibly multiple times), you find yourself back at the integral you started with. For example: $\int \cos(x)\, e^x\, dx$

can be differentiated by parts to get:

$$\int \cos(x)\, e^x dx = \cos(x)\, e^x - \int -\sin(x) e^x dx$$

and again to get:

$$\int \cos(x)\, e^x dx = \cos(x)\, e^x + \sin(x) e^x - \int \cos(x) e^x dx$$

The trick here is to let

$$I = \int \cos(x) e^x\, dx,$$

so that the expression becomes:

$$I = \cos(x)\, e^x + \sin(x)\, e^x - I$$

and we solve for I, to get:

$$I = \frac{\cos(x)e^x + \sin(x)e^x}{2} + C$$

This is perhaps the simplest possible case, but you can get more complex expressions involving your integral. The same trick will work, however. Be vary careful with this trick, since if you use it incorrectly you will end up with everything cancelling, which means that you chose the wrong u and dv in one of the integration by parts steps.

Reduction Formulas

Sometimes you will find that after you integrate by parts you have an almost identical integral, but one which differs from the original by some parameter. For example:

$$\int x^5 e^x \, dx = \int x^5 e^x - 5 \int x^4 e^x \, dx$$

Notice how the 5 has become a 4. In these cases it is often easiest to find a general rule, which is called a *reduction formula* , to help you evaluate the integral. In this example, the general rule is:

$$\int x^n e^x dx = \int x^n e^x - n \int x^{n-1} e^x \, dx$$

and by using this general rule repeatedly, you get:

$$\int x^5 e^x dx = \left(x^5 - 5x^4 + 20x^3 - 60x^2 + 120x - 120\right) e^x + c$$

Notice that the integral gets reduced down to $\int e^x \, dx$, which is an integral we can evaluate without needing integration by parts.

Sometimes you may need to combine this technique with the previous trick to get your reduction formula.

Integration By Parts

Recall that we derived the formula for integration by substitution by using the the Chain Rule and integrating it using the fundamental theorem of calculus. You might reasonably ask "What happens if we use one of the other rules of differentiation?"

This would be an exceedingly good question, and it even has an answer. If we take the product rule:

$$\frac{d}{dx} fg = f \frac{dg}{dx} + g \frac{df}{dx}$$

and integrate both sides, we get:

$$fg = \int fg' \, dx + \int g f' \, dx$$

This isn't particularly useful until we re-arrange and get:

$$\int f(x) g'(x) dx = f(x)\, g(x) - \int g(x) f'(x) dx$$

If you let $u = f(x)$ and $v = g(x)$, then we can abuse notation in the standard way and say that $du = f'(x)$ and $dv = g'(x)$. Then the above is re-written as:

$$\int u \, dv = uv - \int v \, du$$

which may be easier to remember.

THEORY OF INTEGRATION

We saw how the derivative can be used to approximate the anti-derivative, but how do we use it to give us the exact behavior of its anti-derivative?

Differentiation involves two main operations. The first is subtraction or letting a change in x, Δx get smaller. The second operation is division to calculate the relative change of the dependent variable, f with respect to the independent variable, x.

To begin our study, let us look at the definition of the derivative of a function, $f(x)$

$$\frac{dy}{dx} = \lim_{\Delta x \to 0} \frac{f(x+\Delta x)-f(x)}{\Delta x} = f'(x)$$

What this definition does is analyze a changing function over an infinitely small interval to calculate the rate of change of the function at that instant. By letting Δx go to zero, the terms interval and point become synonymous such that the derivative gives is the rate of change at any point x. If we multiply both sides of the equation by dx ,where $dx = \Delta x \to 0$ we get:

$$dy = f(x+dx) - f(x) = f'(x)dx$$

We can calculate the approximate change in the function, Δf by replacing the infinitesimal dx with a discrete Δx

$$\Delta f = f(x + \Delta x) - f(x) \approx f'(x)\Delta x$$

The answer is only an approximation because the equation assumes that the rate of change or derivative is constant over the interval Δx. For example consider the following linear function:

$$f(x) = 2x$$
$$f'(x) = 2$$

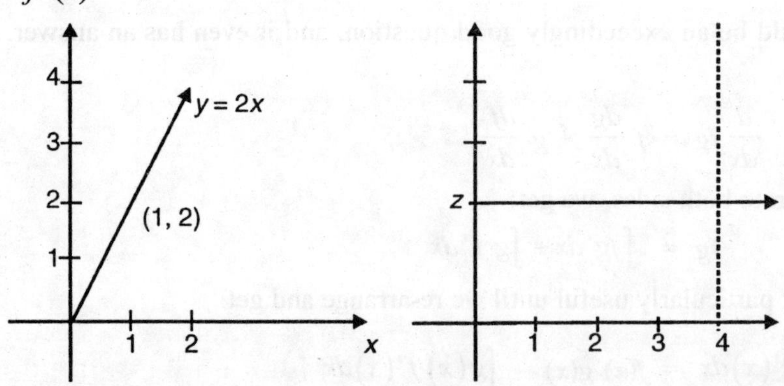

$\frac{dy}{dx} = 2$ or is a constant. At any point on the graph of $f(x) = 2x$, the rate of change is reflected by the constant steepness of the graph. Now let me pose a simple question. How much does the function, $f(x) = 2x$, change when x goes from 0 to 4? By evaluating $f(x) = 2x$ at 0 and 4 we get

$$\Delta f = \int_0^4 2x = 8 - 0 = 8$$

When x changes from 0 to 4, the dependent variable, *f*, changes by 8. How else could we arrive at the same answer? From the definition of the derivative we know that:

$$\Delta f = f(x+\Delta x) - f(x) \approx f'(x)\Delta x$$

Over an interval Δx, the change in f(x), can be approximated by evaluating $f'(x)\Delta x$ for the interval. Let us see how this applies to the function we were studying. We have shown that as x changes from 0 to 4, f changed from 0 to 8. Now let us find the same answer using the above equation:

$$\Delta f = f(x+\Delta x) - f(x) \approx f'(x)\Delta x$$

$$\Delta f = 8 \approx 2 \cdot 4$$

$$\Delta f = 8 = 8$$

The answer obtained from the derivative is exact because the rate of change of the function is constant over the discrete interval $\Delta x = 4$,

$$f'(x)\Delta x = f'(x+\Delta x) - f(x)$$

$$2.4 = f(4) - f(0)$$

$$8 = 8 - 0$$

But what about if the derivative is not a constant but varies with *x*? As we shall soon prove, the total change of a function *f*(x) from *x* = a to *x* = b is found by taking the infinite sum of $f'(x)dx$ from a to b where each $f'(x)dx$ gives an infinitely small change in the function, df, over the infinitely small interval, dx.

Let us look at a function whose derivative varies with x:

$$f(x) = \frac{x^2}{2}$$

$$\frac{df}{dx} = f'(x) = x$$

As *x* changes the rate of change of *f*(x) increases. We can begin by asking ourselves what is the change in *f*(x), when x changes from 0 to 6? In the graph this translates to:

For the function, $f(x) = \frac{x^2}{2}$, when Δ*x* goes from 0 to 8, Δ*f* is given by the equation:

$$\Delta f = f(x+\Delta x) - f(x) = \int_x^{x+\Delta x} f(x)$$

$$\Delta f \;=\; [_0^8 \frac{x^2}{2} = 36 - 0 - = 36$$

$$\Delta f \;=\; 36$$

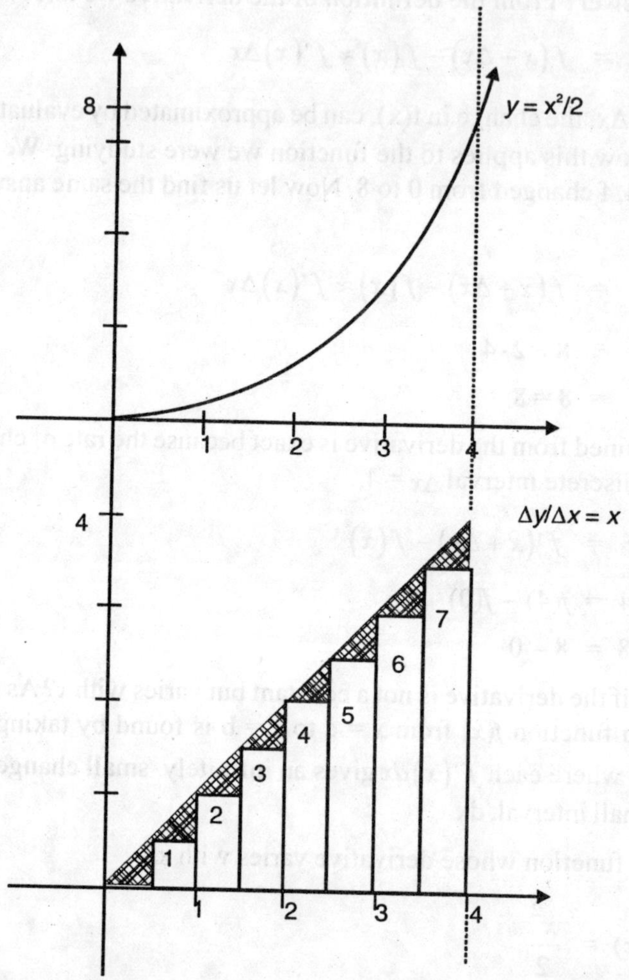

How else could we calculate the change in f? From the derivative we know that: $\Delta f \approx f'(x)\Delta x$

Since the derivative of the function changes with x, then we need to evaluate the above equation over small intervals of Δx to get an accurate answer. Remember the rate of change of $f(x)$ is changing and is not constant over a discrete interval, Δx.

By breaking up the derivative, $f'(x) = x$ into 8 intervals of $\Delta x = 1$, we can approximate the change in of $f(x) = \frac{x^2}{2}$ from $x = 0$ to $x = 4$ by using the derivative only. A small Δf of the graph is found by evaluating $\Delta f = f'(x) \cdot \Delta x = x \cdot \Delta x$ over each interval, $\Delta x = 1$

$$\Delta f_1 = f'(x_1)\cdot\Delta x = (0)\cdot 1 = 0$$
$$\Delta f_2 = f'(x_2)\cdot\Delta x = (1)\cdot 1 = 1$$
$$\Delta f_3 = f'(x_3)\cdot\Delta x = (2)\cdot 1 = 2$$
$$\Delta f_4 = f'(x_4)\cdot\Delta x = (3)\cdot 1 = 3$$
$$\Delta f_5 = f'(x_5)\cdot\Delta x = (4)\cdot 1 = 4$$
$$\Delta f_6 = f'(x_6)\cdot\Delta x = (5)\cdot 1 = 5$$
$$\Delta f_7 = f'(x_7)\cdot\Delta x = (6)\cdot 1 = 6$$
$$\Delta f_8 = f'(x_8)\cdot\Delta x = (7)\cdot 1 = 7$$

The net change in $f(x)$ is found by summing up the individual approximations of Δf_i, for each interval.

$$\Delta f \approx \sum_{i=1}^{8}\Delta f_i \ = \ \sum_{i=i}^{8} f'(x_i)\Delta x$$

Substituting known values in:

$$\Delta f \approx \sum_{i=1}^{8}\Delta f_i \ = \ \sum_{i=i}^{8} x_i\Delta x = 0+1+2+3+4+5+6+7$$

$$\Delta f \approx 28$$

Adding up the Δf_i, we get 28. This corresponds to an error of 8 since we calculated the actual Δf of $f(x)$ to be 8 for a Δx or:

$$\Delta f = [_0^8 \frac{x^2}{2} = 36 - 0 = 36$$

To approximate this $\Delta f = 36$ we evaluated the derivative, $f'(x) = x$ over small sub-intervals to calculate the net change in the function, $f'(x) = \frac{x^2}{2}$. Relating the instantaneous rate of change to the slope of the tangent is:

By evaluating the derivative over each interval Δx we got the approximate Δf of $f(x)$ *over that interval only*. How could we get a more accurate result? The solution is to further divide the interval into a smaller Δx, such that the derivative is assumed to be constant over the infinitely small interval. If we let equal 0.5, we will have sixteen intervals over which we can calculate the Δf_i

$$\Delta f \approx \sum_{i=1}^{16}\Delta f_i = \sum_{i=i}^{16} x_i\Delta x = 34$$

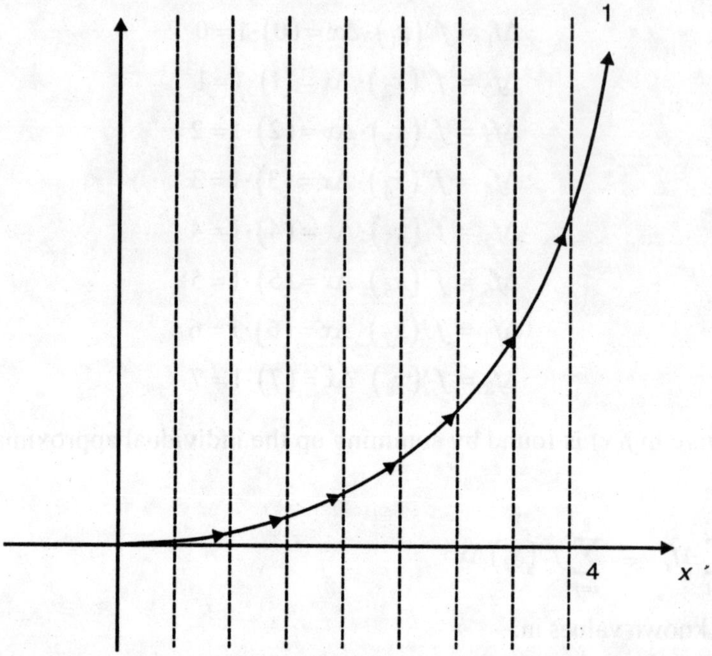

You can clearly see that by further reducing the interval, the error is difference between the approximate and actual Δf is very small. We can go on to conclude that as the interval, Δx, goes to zero, then the net change in $f(x)$, will correspond exactly to the infinite sum of the derivatives evaluated over each interval, Δx

$$\Delta f = \int_a^b f(x) = \sum_{i=1}^{\infty} f'(x_i)\cdot\Delta x$$

Thus the net change in a function from a to b is exactly equal to the infinite sum of its derivative evaluated over an infinitely small interval, $\Delta x \to 0$ One argument against such a conclusion is that as Δx decreases, the error gets smaller for each individual interval however the total error remains the same. This is because as $\Delta x \to 0$, we are adding up an infinite sum of small errors as opposed to a discrete sum of 8 or 10 for example. We have shown intuitively that this does not happen, but how can we prove it?

Returning to the definition of rate of change between two points on a graph, separated by a distance Δx

$$\frac{\Delta f}{\Delta x} = \frac{f(x+\Delta x)-f(x)}{\Delta x} = \frac{(x+\Delta x)^2 - x^2}{\Delta x}$$

$$\frac{\Delta f}{\Delta x} = \frac{x^2 + 2x\Delta x + \Delta x^2 - x^2}{\Delta x}$$

$$\frac{\Delta f}{\Delta x} = 2x + \Delta x$$

Notice that we are not taking any limit as $\Delta x \to 0$. The above equation holds true for any two points on the graph of $f'(x) = \frac{x^2}{2}$. Multiplying both sides by Δx gives us:

$$\Delta f \;=\; 2x\Delta x + \Delta x^2$$

$$\Delta f = \sum_{i=1}^{\infty} f'(x_i)\Delta x + \sum_{i=1}^{\infty}(\Delta x_i)^2$$

If we let $\Delta x \to 0$, then $2x$ is the derivative function such that:

$$\Delta f \;=\; f'(x)\Delta x + \Delta x^2$$

$$\Delta f \;=\; \sum_{i=1}^{\infty} f'(x_i)\Delta x + \sum_{i=1}^{\infty}(\Delta x_i)^2$$

Therefore Δx^2 represents the error associated with calculating Δf when we ignore it. The total sum of errors from $x = a$ to $x = b$, where $\Delta x = \lim_{x\to\infty}\left(\frac{b-a}{n}\right)$ is:

$$\sum_{i=1}^{\infty}(\Delta x_i)^2 \;=\; \lim_{x\to\infty} n\cdot\left(\frac{b-a}{n}\right)^2 = \frac{(b-a)^2}{n} = \frac{c}{n} = 0$$

Thus the total error for calculating the net change in $f'(x) = \frac{x^2}{2}$ is:

$$\Delta f \;=\; \sum_{i=1}^{\infty} f'(x_i)\Delta x + \sum_{i=1}^{\infty}(\Delta x_i)^2$$

$$\Delta f \;=\; \sum_{i=1}^{\infty} f'(x_i)\Delta x + 0$$

Remember that $dx = \Delta x \to 0$, therefore:

This important result confirms the fact that the sum of the derivatives evaluated over infinite times over infinitely small intervals of dx from $x = a$ to $x = b$ corresponds exactly to the net change in the anti-derivative, $f(x)$. We have proven this for the case of $f'(x) = \frac{x^2}{2}$, but how do we prove it for the general case for any $f(x)$ and its derivative? More specifically we want to prove the following fundamental theorem of Calculus.

$$\Big[f(x) \Big]_a^b \;=\; \sum_{i=1}^{\infty} f'(x_i)\cdot dx = \int_a^b f'(x)\cdot dx$$

Here, the integral sign, $\int$, replaces the summation sign, $\sum$, and represents the infinite sum of the derivative evaluated over an infinitely small interval, *dx*. To understand this let us take a more conceptual look at how the rate of change of a function is defined. The following formula give us the average rate of change of *f*(*x*), through any two points on the graph of *f*(*x*).

$$\frac{\Delta f}{\Delta x} = \frac{f(x+\Delta x)-f(x)}{\Delta x}$$

As we let $\Delta x \to 0$, the two points, (*x*) and $(x+\Delta x)$ come closer together. The average rate of change converges to the instantaneous rate of change of the function at that point, *x*.

$$\frac{df}{dx} = \lim_{\Delta x\to\infty} \frac{f(x+\Delta x)-f(x)}{\Delta x} = f'(x)$$

For any function, $f(x)=x^n$, as $\Delta x \to 0$ the rate of change converges to a constant value over an infinitely small interval, $dx = \Delta \dot{x} \to 0$. Since the instantaneous rate of change, given by $f'(x)=nx^{n-1}$ constant over the interval, *dx*, then an infinitely small change in *f*(*x*), *df*, is found by multiplying the definition of the derivative by *dx*, where $dx = \Delta x \to 0$

$$df = f(x+dx)-f(x) = f'(x)\cdot\Delta x$$

Thus the net change in *f*(*x*) is the infinite sum of the df's from $x = a$ to $x - b$ is:

$$\sum_{i=1}^{\infty} df_i = \sum_{i=1}^{\infty} f(x_i+dx)-f(x_i) = \sum_{i=1}^{\infty} f'(x_i)\cdot dx$$

The first term $\sum_{i=1}^{\infty} df_i$ converges to Δf, while the second term, $\sum_{i=1}^{\infty} f(x_i+dx)-f(x_i)$

represents the net change in the function, $f(x)$ from $x = a$ to $x = b$, where $x_1 = a$ and $x_\infty = b$. Finally the last term is the infinite sum of the derivative evaluate over each interval, *dx*. This infinite sum is represented by the integral sign, $\int$.

$$\Delta f = [_a^b f(x) = \int_a^b f'(x)\cdot dx$$

$$f(b)-f(a) = \int_a^b f'(x)\cdot dx$$

This the fundamental

UNDERSTANDING INTEGRATSION

The process of integration is an infinite summation of the product of a function of *x*, $f(x)$, and an infinitely small Δx. This infinite sum from $x = a$ to $x = b$ is the total change in the anti-derivative or integral of its function. The important point to understand was that a small change in a function *df* is given by its derivative multiplied by an infinitesimal *dx*.

Consider a mathematical function of the form.

$$F = f \,.\, x$$

Assuming *f* and *x* represent two different conditions, then the situation *F* is dependent on their product. In other words *f* and *x* are two constants whose product defines the dependent dimension, *F*.

$$F(f, x) = f \,.\, x$$

F is therefore, a function of two independent conditions, *f* and *x*. If we let *x* be a variable and *f* be a constant; we have:

$$F(x) = f \cdot x$$

$$F = f \cdot x$$

As *x* changes, *F* also changes, since its value is only dependent on *x* and the constant *f*. Thus:

$$\Delta F = f \cdot \Delta x$$

This should be rather obvious to you. Now what if the condition, *f*, is also a function of *x*? In other words, the value of *f* changes as *x* changes and is therefore, not a constant over an interval from *x* = *a* to *x* = *b*. This modifies the relationship to:

$$F(f(x), x) = f(x) \cdot x$$

Since *F* is a function of *f* and *x* and *f* is a function of *x*, then *F* is also some function of *x* only, such that;

$$F(x) = f(x) \cdot x$$

As the condition, *f*, is a function of *x*, then its value changes as *x* goes from *a* to *b*. Therefore $f(x)$ can be assumed to be constant only over some infinitely small interval, *dx*. A small change in $F(x)$ *df*, is proportional to the value of $f(x)$ multiplied by the infinitely small change in x, *dx*.

$$dF = f(x) \cdot x$$

Dividing both sides by *dx*:

$$\frac{dF}{dx} = f(x)$$

This extremely important result defines the instantaneous rate change of $F(x)$ with respect to *x* to be equal to $f(x)$. The relationship of $F(x)$ to $f(x)$ is the same as $f(x)$ is to $f'(x)$ i.e.

a function and its derivative. The only difference is the change in notation from f and f' to F and f. Multiplying both sides by dx again:

$$dF = f(x).dx$$

Integrating both sides results in:

$$\sum_{i=1}^{\infty} dF_i = \sum_{i=1}^{\infty} f(x_i)\cdot dx$$

$$[F(x)]_a^b = \int_a^b F(x)\cdot dx$$

The result is still a bit abstract and may not seem to be anything new to you. The important point to understand is that F is a function of two independent conditions, f and x. If f is also a function of x, then as x changes, f also changes such that its value is not constant over an interval from $x = a$ to $x = b$. Therefore an instantaneous change in F, df, needs to be defined over an infinitely small interval over which $f(x)$ is assumed to be constant.

$$\text{F} = f\cdot x$$

$$\text{F} = f(x)\cdot x$$

$$dF = f(x)\cdot dx$$

Integrating $dF = f(x) . dx$ from some value $x = a$ to $x = b$ gives us the net change in $F(x)$ over the interval. Remember that $F(x)$ is defined as the anti-derivative of $f(x)$

$$\sum_{i=1}^{\infty} dF = \sum_{i=1}^{\infty} f(x_i)\cdot dx$$

$$[F(x)]_a^b = \int_a^b f(x)\cdot dx$$

The result is the fundamental theory behind integral Calculus. Take some time to think deeply and abstractly about what is going on in this derivation. Mathematically it represents the same theory derived, but conceptually it explains integration in a new way. It allows us to analyze a changing situation in terms of infinitesimal changes which we can sum up to find the net change in the system as one variable changes.

THE CONTINUUM OF NUMBERS

We shall consider the numbers and start with the natural numbers 1, 2, 3, as given as well as the rules

$(a + b) + c = a + (b + c)$ – associative law of addition,

$a + b = b + a$ – commutative law of addition,

$(ab)c = a(bc)$ – associative law of multiplication,

$ab = ba$ – commutative law of multiplication,

$a(b + c) = ab + ac$ – distributive law of multiplication.

by which we calculate with them; we shall only briefly recall the way in which the concept of the positive integers (the natural numbers) has had to be extended.

The System of Rational Numbers and the Need for its Extension

In the domain of the natural numbers, the fundamental operations of addition and multiplication can always be performed without restriction, i.e., the sum and the product of two natural numbers are themselves always natural numbers. But the inverses of these operations, subtraction and division, cannot invariably be performed within the domain of natural numbers, whence mathematicians were long ago obliged to invent the number 0, the negative integers, and positive and negative fractions. The totality of all these numbers is usually called the class of rational numbers, since all of them are obtained from unity by means of the rational operations of calculation: Addition, multiplication, subtraction and division.

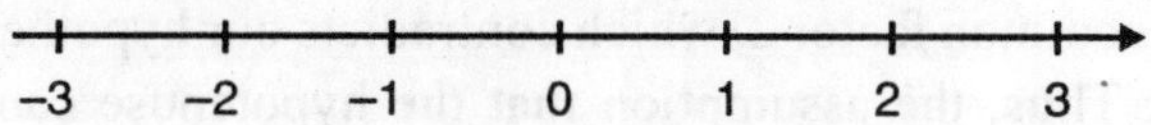

Fig. The Number axis

Numbers are usually represented graphically by means of the points on a straight line - the number axis - by taking an arbitrary point of the line as the origin or zero point and another arbitrary point as the point 1; the distance between these two points (the length of the unit interval) then serves as a scale by which we can assign a point on the line to every rational number, positive or negative. It is customary to mark off the positive numbers to the right and the negative numbers to the left of the origin. If, as is usually done, we define the absolute value (also called the numerical value or modulus) $|a|$ of a number a to be a itself when $a \geq 0$, and $-a$ when $a < 0$, then $|a|$ simply denotes the distance of the corresponding point on the number axis from the origin.

The symbol means that either the sign > or the sign = shall hold. A corresponding statement holds for the signs and $\mp$ which will be used later on.

The geometrical representation of the rational numbers by points on the number axis suggests an important property which can be stated as follows: The set of rational numbers is everywhere dense. This means that in every interval of the number axis, no matter how small, there are always rational numbers; in geometrical terms, in the segment of the number axis between any two rational points, however close together, there are points corresponding to rational numbers. This density of the rational numbers at once becomes clear if we start from the fact that the numbers $\frac{1}{2}, \frac{1}{2^2}, \frac{1}{2^3}, \ldots, \frac{1}{2^n}$, become steadily smaller and approach nearer and nearer to zero as n increases. If we now divide the number axis into equal parts of length $1/2^n$, beginning at the origin, the end-points $1/2^n, 2/2^n, 3/2^n$, of these intervals represent rational numbers of the form

$m/2^n$, where we still have the number n at our disposal. Now, if we are given a fixed interval of the number axis, no matter how small, we need only choose n so large that $1/2^n$ is less than the length of the interval; the intervals of the above subdivision are then small enough for us to be sure that at least one of the points of the sub-division $m/2^n$ lies in the interval.

Yet, in spite of this property of density, the rational numbers are not sufficient to represent every point on the number axis. Even the Greek mathematicians recognized that, if a given line segment of unit length has been chosen, there are intervals, the lengths of which cannot be represented by rational numbers; these are the so-called segments incommensurable with the unit. For example, the hypotenuse l of a right-angled, isosceles triangle with sides of unit length is not commensurable with the length unit, because, by Pythagoras' Theorem, the square of this length must equal 2. Therefore, if l were a rational number, and consequently equal to p/q, where p and q are non-zero integers, we should have $p = 2q$. We can assume that p and q have no common factors, for such common factors could be cancelled out to begin with. Since, according to the above equation, p is an even number, p itself must be even, say $p = 2p'$. Substituting this expression for p yields $4p' = 2q$ or $q == 2p'$, whence q is even, and so is q. Hence p and q have the common factor 2, which contradicts our hypothesis that p and q do not have a common factor. Thus, the assumption that the hypotenuse can be represented by a fraction p/q leads to contradiction and is therefore false.

The above reasoning - a characteristic example of an indirect proof - shows that the symbol cannot correspond to any rational number. Thus, if we insist that, after choice of a unit interval, every point of the number axis shall have a number corresponding to it, we are forced to extend the domain of rational numbers by the introduction of the new irrational numbers. This system of rational and irrational numbers, such that each point on the axis corresponds to just one number and each number corresponds to just one point on the axis, is called the system of real numbers.

Thus named to distinguish it from the system of complex numbers, obtained by yet another extension.

Real Numbers and Infinite Decimals

Our requirement that there shall correspond to each point of the axis one real number states nothing *a priori* about the possibility of calculating with these numbers in the same manner as with rational numbers. We establish our right to do this by showing that our requirement is equivalent to the following fact: The totality of all real numbers is represented by the totality of all finite and infinite decimals.

We first recall the fact, familiar from elementary mathematics, that every rational number can be represented by a terminating or by a recurring decimal; and conversely, that every such decimal represents a rational number. We shall now show that we can assign to every point of the number axis a uniquely determined decimal, so that we can represent the irrational points or irrational numbers by infinite decimals.

Let the points which correspond to the integers be marked on the number axis. By means of these points, the axis is subdivided into intervals or segments of length. In what follows, we shall say that a point of the line belongs to an interval, if it is an interior point or an end-point of the interval. Now let P be an arbitrary point of the number axis. Then this point belongs to one or, if it is a point of division, to two of the above intervals. If we agree that, in the second case, the right-hand point of the two intervals meeting at P is to be chosen, we have in all cases an interval with end-points g and $g+1$ to which P belongs, where g is an integer. We subdivide this interval into ten equal sub-intervals by means of the points corresponding to the numbers

$$g+\frac{1}{10}, g+\frac{2}{10}, \ldots, g+\frac{9}{10},$$

and we number these sub-intervals 0, 1, …. , 9 in their natural order from the left to the right. The sub-interval with the number a then has the end-points $g+a/10$ and $g+a/10 + 1/10$. The point P must be contained in one of these sub-intervals. (If P is one of the new points of division, it belongs to two consecutive intervals; as before, we choose the one on the right hand side.) Let the interval thus determined be associated with the number a_1. The end-points of this interval then correspond to the numbers $g+a_1/10$ and $g+a_1/10+1/10$. We again sub-divide this sub-interval into ten equal parts and determine that one to which P belongs; as before, if P belongs to two sub-intervals, we choose the one on the right hand side. Thus, we obtain an interval with the end-points $g+a_1/10+a_2/10$ and $g+a_1/10+a_2/10+1/10$, where a_2 is one of the digits 0, 1,…. , 9. We subdivide this sub-interval again and continue to repeat this process. After n steps, we arrive at a sub-interval, which contains P, has the length $1/10^n$ and end-points corresponding to the numbers

$$g+\frac{a_1}{10}+\frac{a_2}{10^2}+\ldots+\frac{a_n}{10^n} \text{ and } g+\frac{a_1}{10}+\frac{a_2}{10^2}+\ldots+\frac{a_n}{10^n}+\frac{1}{10^n},$$

where each a is one of the numbers 0, 1, … , 9, but

$$\frac{a_1}{10}+\frac{a_2}{10^2}+\ldots+\frac{a_n}{10^n}$$

is simply the decimal fraction $0.a_1a_2\ldots a_n$. Hence, the end-points of the interval may also be written in the form

$$g+0\cdot a_1a_2\ldots a_n \text{ and } g+0\cdot a_1a_2\ldots a_n+\frac{1}{10^n}.$$

If we consider the above process repeated indefinitely, we obtain an infinite decimal $0.a_1a_2\ldots$ which has the following meaning: If we break off this decimal at any place, say, the n-th, the point P will lie in the interval of length $1/10^n$ the end-points (approximating points) of which are

$$g+0\cdot a_1a_2\ldots a_n \text{ and } g+0\cdot a_1a_2\ldots a_n+\frac{1}{10^n}.$$

In particular, the point corresponding to the rational number $g + 0.a_1a_2 \ldots a_n$ will lie arbitrarily near to the point P if only n is large enough; for this reason, the points $g + 0.a_1a_2 \ldots a_n$ are called approximating points. We say that the infinite decimal $g + 0.a_1a_{2\ldots}$ is the real number corresponding to the point P.

Thus, we emphasize the fundamental assumption that we can calculate in the usual way with real numbers, and hence with decimals. It is possible to prove this using only the properties of the integers as a starting-point. But this is no light task and, rather than allowing it to bar our progress at this early stage,we regard the fact that the ordinary rules of calculation apply to the real numbers to be an axiom, on which we shall base all of the differential and integral calculus.

We insert here a remark concerning the possibility arising in certain cases of choosing in the above scheme of expansion the interval in two ways. It follows from our construction that the points of division, arising in our repeated process of sub-division, and such points only can be represented by finite decimals $g + 0.a_1a_{2\ldots}a_n$. Assume that such a point P first appears as a point of sub-division at the n-th stage of the sub-division. Then, according to the above process, we have chosen at the n-th stage the interval to the right of P. In the following stages, we must choose a sub-interval of this interval. But such an interval must have P as its left end-point. Therefore, in all further stages of the sub-division, we must choose the first sub-interval, which has the number 0. Thus, the infinite decimal corresponding to P is $g + 0.a_1a_{2\ldots}$ *an*. If, on the other hand, we had at the n-th stage chosen the left-hand interval containing P, then, at all later stages of sub-division, we should have had to choose the sub-interval furthest to the right, which has P as its right end-point. Such a sub-interval has the number 9. Thus, we should have obtained for P a decimal expansion in which all the digits from the $(n + 1)$-th onwards are nines. The double possibility of choice in our construction therefore corresponds to the fact that, for example, the number has the two decimal expansions 0.25000 and 0.24999.

Expression of Numbers in Scales other than that of 10

In our representation of the real numbers, we gave the number 10 a special role, because each interval was subdivided into ten equal parts. The only reason for this is the widely spread use of the decimal system. We could just as well have taken p equal sub-intervals, where p is an arbitrary integer greater than 1. We should then have obtained an expression of the form

$$g + \frac{b_1}{p} + \frac{b_2}{p^2} + \ldots,$$

where each b is one of the numbers $0, 1, \ldots, p - 1$. Here we find again that the rational numbers, and only the rational numbers, have recurring or terminating expansions of this kind. For theoretical purposes, it is often convenient to choose $p = 2$. We then obtain the binary expansion of the real numbers

$$g + \frac{b_1}{2} + \frac{b_2}{2^2} + \ldots,$$

where each b is either* 0 or 1.

Even for numerical calculations, the decimal system is not the best. The sexagesimal system, with which the Babylonians calculated, has the advantage that a comparatively large proportion of the rational. numbers, the decimal expansions of which do not terminate, possess terminating sexagesimal expansions.

For numerical calculations, it is customary to express the whole number g, which, for the sake of for simplicity, we assume here to be positive, in the decimal system, that is, in the form

$$a_m 10^m + a_{m-1} 10^{m-1} + \ldots + a_1 10 + a_0,$$

where each a_ν is one of the digits 0, 1, ..., 9. Then, for $g + 0.a_1 a_{2\ldots}$, we write simply

$$a_m a_{m-1} \ldots a_1 a_0 \cdot a_1 a_2 \ldots$$

Similarly, the positive whole number g can be written in one and only one way in the form

$$\beta_k p^k + \beta_{k-1} p^{k-1} + \ldots + \beta_1 p + \beta_0,$$

where each of the numbers β_ν is one of the numbers 0, 1, ... , $p-1$. Together with our previous expression, this yields: Every positive real number can be represented in the form

$$\beta_k p^k + \beta_{k-1} p^{k-1} + \ldots + \beta_1 p + \beta_0 + \frac{b_1}{p} + \frac{b_2}{p^2} + \ldots,$$

where β_ν and b_ν are whole numbers between 0 and p —1. Thus, for example, the binary expansion of the fraction 21/4 is

$$\frac{21}{4} = 1 \times 2^2 + 0 \times 2 + 1 + \frac{0}{2} + \frac{1}{2^{2*}}$$

Inequalities

Calculation with inequalities has a far larger role in higher than in elementary mathematics. We shall therefore briefly recall some of the simplest rules concerning them.

If $a > b$ and $c > d$, then $a + c > b + d$, but not $a - c > b - d$. Moreover, if $a>b$, it follows that $ac > bc$, provided c is positive. On multiplication by a negative number, the sense of the inequality is reversed. If $a>b>0$ and $c>$d>0, it follows that ac$>bd$.

For the absolute values of numbers, the following inequalities hold:

$$|a \pm b| \leqq |a| + |b|, |a \pm b| \geqq |a| - |b|.$$

The square of any real number is larger than or equal to zero, whence, if x and y are arbitrary real numbers

$$(x - y)^2 = x^2 + y^2 - 2xy \geqq 0,$$

or

$$2xy \leq x^2 + y^2.$$

Schwarz's Inequality: Let $a_1, a_2, \ldots, a_n$ and $b_1, b_2, \ldots, b_n$ be any real numbers. Substitute in the preceding inequality*

$$x = \frac{|a_i|}{\sqrt{(a_1^2 + a_2^2 + \ldots + a_n^2)}}, \quad y = \frac{|b_i|}{\sqrt{(b_1^2 + b_2^2 + \ldots + b_n^2)}}$$

for $i = 1, i = 2, \ldots\ i = n$ successively and add the resulting inequalities. We obtain on the right hand side the sum 2, because

$$\left(\frac{|a_1|}{\sqrt{(a_1^2 + \ldots + a_n^2)}}\right)^2 + \ldots + \left(\frac{|a_n|}{\sqrt{(a_1^2 + \ldots + a_n^2)}}\right)^2 = 1,$$

$$\left(\frac{|b_1|}{\sqrt{(b_1^2 + \ldots + b_n^2)}}\right)^2 + \ldots + \left(\frac{|b_n|}{\sqrt{(b_1^2 + \ldots + b_n^2)}}\right)^2 = 1.$$

If we divide both sides of the inequality by 2, we obtain

$$\frac{|a_1 b_1| + |a_2 b_2| + \ldots + |a_n b_n|}{\sqrt{(a_1^2 + \ldots + a_n^2)}\sqrt{(b_1^2 + \ldots + b_n^2)}} \leq 1,$$

or, finally,

$$|a_1 b_1| + |a_2 b_2| + \ldots + |a_n b_n| \leq \sqrt{(a_1^2 + \ldots + a_n^2)}\sqrt{(b_1^2 + \ldots + b_n^2)}.$$

* Here and hereafter, the symbol $\sqrt{x}$, where $x > 0$, denotes that positive number the square of which is x.

Since the expressions on both sides of this inequality are positive, we may take the square and then omit the modulus signs:

$$(a_1 b_1 + a_2 b_2 + \ldots + a_n b_n)^2 \leq (a_1^2 + \ldots + a_n^2)(b_1^2 + \ldots + b_n^2).$$

This is the Cauchy-Schwarz inequality.

THE CONCEPT OF FUNCTION

a. If an ideal gas is compressed in a vessel by means of a piston, the temperature being kept constant, the pressure p and the volume v are connected by the relation

$$pv = C,$$

where C is a constant. This formula - Boyle's Law - states nothing about the quantities v and p themselves; its meaning is: If p has a definite value, arbitrarily chosen in a certain range (the range being determined physically and not mathematically), then v can be determined, and conversely:

$$v = \frac{C}{p}, p = \frac{C}{v}.$$

We then say that v is a function of p or, in the converse case, that p is a function of v.

b. If we heat a metal rod, which at temperature 0 has the length l_0, to the temperature q, then its length l will be given, under the simplest physical assumptions, by the law where β - the coefficient of thermal expansion - is a constant. Again, we say that l is a function of q.

c. Let there be given in a triangle the lengths of two sides, say a and b. If we choose for the angle γ between these two sides any arbitrary value less than 180, the triangle is completely determined; in particular, the third side c is determined. In this case, we say that if a and b are given, c is a function of the angle γ. As we know from trigonometry, this function is represented by the formula

$$c = \sqrt{\left(a^2 + b^2 - 2ab\cos\gamma\right)}.$$

Formulation of the Concept of Function

In order to give a general definition of the mathematical concept of function, we fix upon a definite interval of our number scale, say, the interval between the numbers a and b, and consider the totality of numbers x which belong to this interval, that is, which satisfy the relation $a \leqq x \leqq b$.

If we consider the symbol x as denoting any of the numbers in this interval, we call it a continuous variable in the interval.

If now there corresponds to each value of x in this interval a single definite value y, where x and y are connected by any law whatsoever, we say that y is a function of x and write symbolically

$$y = f(x), y = F(x), y = g(x),$$

or some similar expression. We then call x the independent variable and y the dependent variable, or we call x the argument of the function y.

It should be noted that, for certain purposes, it makes a difference whether we include in the interval from a to b the end-points, as we have done above, or exclude them; in the latter case, the variable x is restricted by the inequalities

$a < x < b.$

In order to avoid a misunderstanding, we may call the first kind of interval - including its end-points - a closed interval, the second kind an open interval. If only one end-point and not the other is included, as, for example, in $a < x < b,$ we speak of an interval which is open at one end (in this case the end a). Finally, we may also consider open intervals which extend without bound in one direction or both. We then say that the variable x ranges over an infinite open interval and write symbolically

$a < x < \infty$ or $-\infty < x < b$ or $-\infty < x < \infty$.

In the general definition of a function, which is defined in an interval, nothing is said about the nature of the relation, by which the dependent variable is determined when the independent variable is given. This relation may be as complicated as we please and in theoretical investigations this wide generality is an advantage. But in applications and, in particular, in the differential and integral calculus, the functions with which we have to deal are not of the widest generality; on the contrary, the laws of correspondence by which a value of y is assigned to each x are subject to certain simplifying restrictions.

Graphical Representation. Continuity. Monotonic Function: Natural restrictions of the general function concept are suggested when we consider the connection with geometry. In fact, the fundamental idea of analytical geometry is one of giving a curve, defined by some geometrical property, a characteristic analytical representation by regarding one of the rectangular co-ordinates, say y, as a function $y = f(x)$ of the other co-ordinate x; for example, a parabola is represented by the function $y = x$, the circle with radius 1 about the origin by the two functions $y = (1 - x)$ and $y = -(1 - x)$.

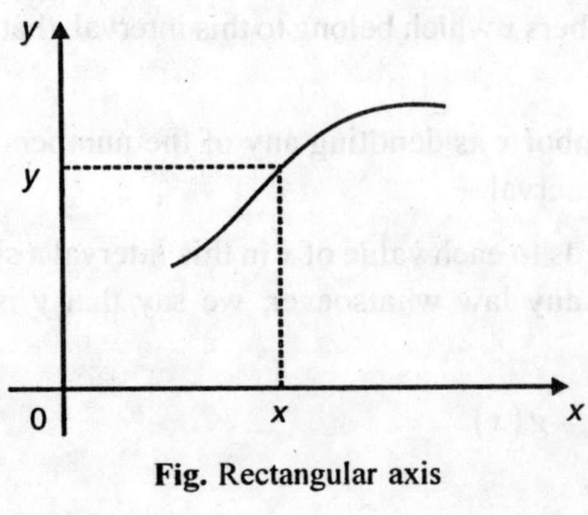

Fig. Rectangular axis

Conversely, if instead of starting with a curve which is determined geometrically, we consider a given function $y = f(x)$, we can represent the functional dependence of y on x graphically by making use of a rectangular co-ordinate system in the usual way.

If, for each abscissa x, we mark off the corresponding ordinate $y = f(x)$, we obtain the geometrical representation of the function. The restriction which we now wish to impose on the function concept is: The geometrical representation of the function shall take the form of a reasonable geometrical curve. It is true that this implies a vague general idea rather than a strict mathematical condition. But we shall soon formulate conditions, such as continuity, differentiability, etc., which will ensure that the graph of a function has the character of a curve capable of being visualized geometrically. At any rate, we shall exclude a function such as the following one: For every rational value of x, the function y has the value 1, for every irrational value of x, the value 0. This assigns a definite value of y to each x, but in every interval of x, no matter how small, the value of y jumps from 0 to 1 and back an infinite number of times.

Unless the contrary is expressly stated, it will always be assumed that the law, which assigns a value of the function to each value of x, assigns just one value of y to each value of x, as, f

or example, $y = x$ or $y = \sin x$. If we begin with a geometrically given curve, it may happen, as in the case of the circle $x + y = 1$, that the whole course of the curve is not given by one single (single-valued) function, but requires several functions - in the case of the circle, the two functions $y = (1 - x)$ and $y = -(1 - x)$. The same is true for the hyperbola $y - x = 1$, which is represented by the two functions $y = (1 + x)$ and $y = -(1 + x)$. Hence such curves do not determine the corresponding functions uniquely.

Consequently, it is sometimes said that the function corresponding to a curve is multi-valued. The separate functions representing a curve are then called the single-valued branches belonging to the curve. For the sake of clearness, we shall henceforth use the word function to mean a single-valued function. In conformity with this, the symbol x (for x 0) will always denote the non-negative number, the square of which is x.

If a curve is the geometrical representation of one function, it will be cut by any parallel to the y-axis in at most one point, since there corresponds to each point x in the interval of definition just one value of y. Otherwise, as, for example, in the case of the circle, represented by the two functions

$y = (1 - x)$ and $y = -(1 - x)$,

such parallels to the y-axis may intersect the curve in more than one point. The portions of a curve corresponding to different single-valued branches are sometimes so interlinked that the complete curve is a single figure which can be drawn with one stroke of the pen, for example, the circle on the other hand, the branches may be completely separated, for example, the hyperbola.

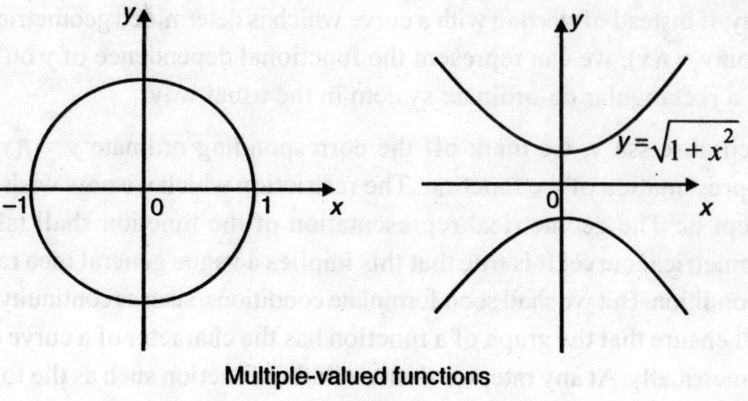

Multiple-valued functions

Here follow some more examples of the graphical representation of functions.

$y = ax$

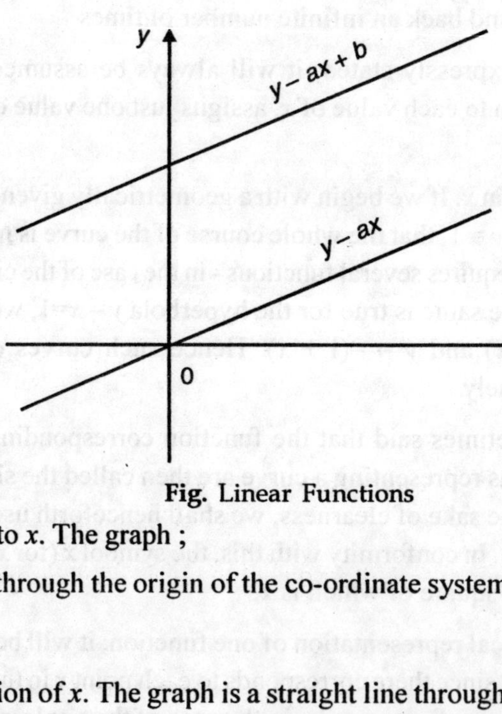

Fig. Linear Functions

y is proportional to x. The graph ;

is a straight line through the origin of the co-ordinate system.

$y = ax + b$

y is a linear function of x. The graph is a straight line through the point $x = 0$, $y = b$, which, if $a = 0$, also passes through the point $x = -b/a$, $y = 0$, and, if $a = 0$, runs horizontally.

$$y = \frac{a}{x}$$

y is inversely proportional to x. In particular, if $a = 1$, so that

$$y = \frac{1}{x},$$

we find, for example, that

$y = 1$ for $x = 1$, $y = 2$ for $x = \frac{1}{2}$, $y = \frac{1}{2}$ for $x = 2$.

The graph is a curve - a rectangular hyperbola, symmetrical with respect to the bi-sectors of the angles between the co-ordinate axes.

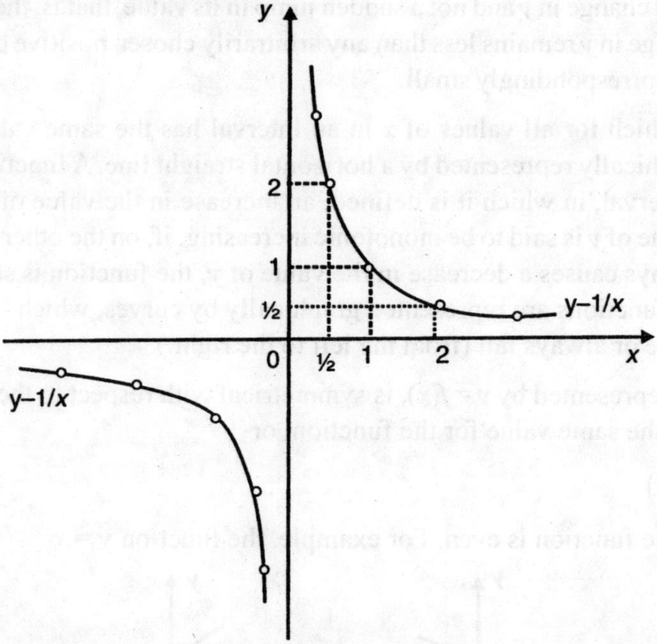

This last function is obviously not defined for the value $x = 0$, since division by zero has no meaning. The exceptional point $x = 0$, in the neighbourhood of which there occur arbitrarily large values of the function, both positive and negative, is the simplest example of an infinite discontinuity, a subject to which we shall return later.

$y = x^2$

As is well known, this function is represented by a parabola.

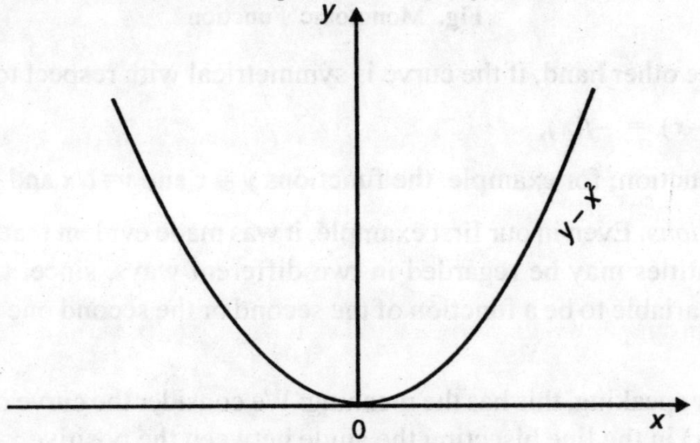

Fig. Parabola

Similarly, the function $y = x$ is represented by the so-called cubical parabola.

The Curves just considered and their graphs exhibit a property which is of the greatest importance in the discussion of functions, namely, the property of continuity. We shall later analyze this concept in more detail; intuitively speaking, it amounts to: A small change in x causes only a small change in y and not a sudden jump in its value, that is, the graph is not broken off or else the change in y remains less than any arbitrarily chosen positive bound, provided that the change in x is correspondingly small.

A function which for all values of x in an interval has the same value $y = a$ is called a constant; it is graphically represented by a horizontal straight line. A function $y = f(x)$ such that throughout the interval, in which it is defined, an increase in the value of x always causes an increase in the value of y is said to be monotonic increasing; if, on the other hand, an increase in the value of x always causes a decrease in the value of y, the function is said to be monotonic decreasing. Such functions are represented graphically by curves, which in the corresponding interval always rise or always fall (from the left to the right.).

If the curve, represented by $y = f(x)$, is symmetrical with respect to the y-axis, that is, if $x =$ - a and $x =$ a give the same value for the function, or

$$f(-x) = f(x)$$

we say that the function is even. For example, the function $y = x$

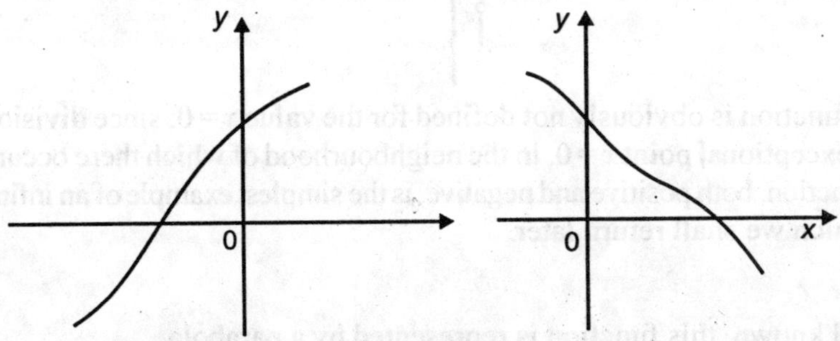

Fig. Monotonic Function

is even. On the other hand, if the curve is symmetrical with respect to the origin, that is, if

$$f(-x) = -f(x),$$

it is an odd function; for example, the functions $y = x$ and $y = 1/x$ and $y = 1/x$ are odd.

Inverse Functions: Even in our first example, it was made evident that a formal relationship between two quantities may be regarded in two different ways, since it is possible either to consider the first variable to be a function of the second or the second one a function of the first vatiable.

Geometrically speaking, this has the meaning: We consider the curve obtained by reflecting the graph of $y = f(x)$ in the line bisecting the angle between the positive x-axis and the positive *y*-axis*. This gives us at once a graphical representation of x as a function of y and thus represents the inverse function $x = \phi(y)$.

* Instead of reflecting the graph in this way, we could first rotate the co-ordinate axes and the curve $y = f(x)$ by a right angle and then reflect the graph in the *x*-axis.

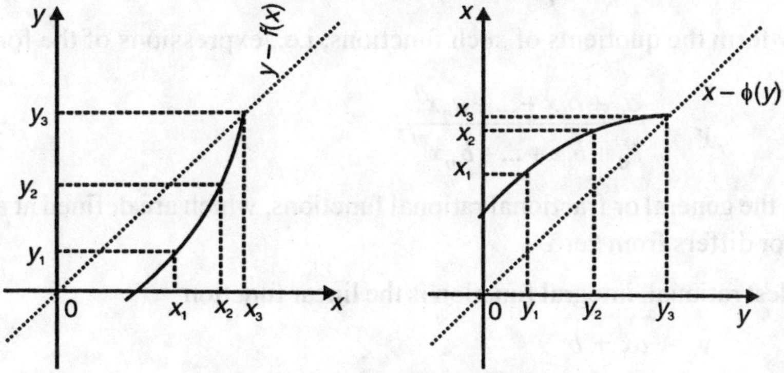

Fig. Inversion of a function

However, these geometrical ideas show at once that a function $y = f(x)$, defined in an interval, has not a single-valued inverse unless certain conditions are satisfied. If the graph of the function is cut in more than one point by a line $y=c$, parallel to the *x*-axis, the value $y = c$ will correspond to more than one value of x, so that the function cannot have a single-valued inverse. This case cannot occur if $y = f(x)$ is continuous and monotonic, because then Fig. shows us that there corresponds to each value of y in the interval $y_1 y y_3$ just one value of x in the interval x_1 x x_3, and we infer from the figure that a function which is continuous and monotonic in an interval always has a single-valued inverse, and this inverse function is also continuous and monotonic.

More Detailed Study of the Elementary Functions

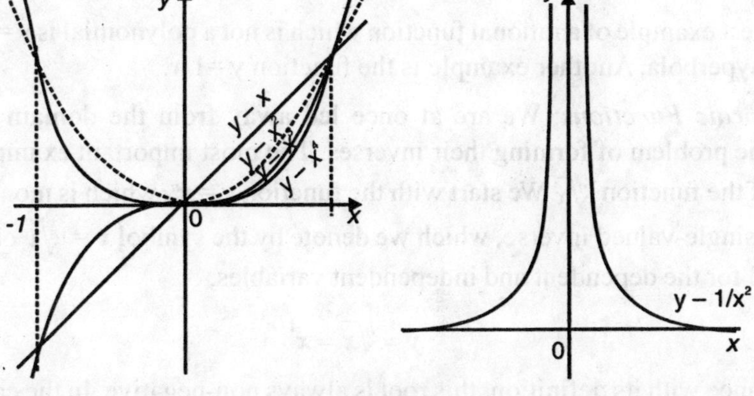

Fig. Power of x

The Rational Functions: We now continue with a brief review of the elementary functions which the reader has already encountered in his previous studies. The simplest types of function are obtained by repeated application of the elementary operations: Addition, multiplication, subtraction. If we apply these operations to an independent variable x and any real numbers, we obtain the rational, integral functions or polynomials:

$y = a_0 + a_1x + ... + a_nx^n$.

The polynomials are the simplest and, in a sense, the basic functions of analysis.

If we now form the quotients of such functions, i.e., expressions of the form

$$y = \frac{a_0 + a_1x + ... + a_nx^n}{b_0 + b_1x + ... + b_mx^m},$$

we obtain the general or fractional rational functions, which are defined at all points where the denominator differs from zero.

The simplest rational, integral function is the linear function

$$y = ax + b$$

It is represented graphically by a straight line. Every quadratic function of the form

$$y = ax^2 + bx + c$$

is represented by a parabola. The curves which represent rational integral functions of the third degree

$$y = ax^2 + bx^2 + cx + d,$$

are occasionally called parabolas of the third order or cubical parabolas, and so on.

As examples, we have in Fig. above the graphs of the function $y = x^n$ for the exponents. We see that for even values of n, the function $y = x^n$ satisfies the equation $f(-x) = f(x)$, whence it is an even function, while for odd values of n it satisfies the condition $f(-x) = -f(x\}$ and is an odd function.

The simplest example of a rational function which is not a polynomial is $y = 1/x$; its graph is a rectangular hyperbola. Another example is the function $y = 1/x$.

The Algebraic Functions: We are at once led away from the domain of the rational functions by the problem of forming their inverses. The most important example of this is the introduction of the function $\sqrt[n]{x}$ We start with the function $y = x^n$, which is monotonic for $x \rangle 0$. Hence it has a single-valued inverse, which we denote by the symbol $x = \sqrt[n]{y}$ or, interchanging the letters used for the dependent and independent variables,

$$y = \sqrt[n]{x} = x^{1/n}$$

In accordance with its definition, this root is always non-negative. In the case of odd n, the function x^n is monotonic for all values of x, including negative values, whence, for odd values of n, we can also define uniquely for all values of x; in this case, is negative for negative values of x.

More generally, we may consider

$$y = \sqrt[n]{R(x)},$$

where $R(x)$ is a rational function. We arrive at further functions of a similar type by applying rational operations to one or more of these special functions. Thus, for example, we may form the functions

$$y = \sqrt[m]{x} + \sqrt[m]{\left(x^2+1\right)}, y = x + \sqrt{\left(x^2+1\right)}$$

These functions are special cases of algebraic functions. (The general concept of an algebraic function cannot be defined here;

The Trigonometric Functions: While the rational and algebraic functions just considered are defined directly in terms of the elementary, computational operations, geometry is the source, from which we first draw our knowledge of the other functions, the so-called transcendental functions. We shall here consider the elementary transcendental functions - the trigonometric functions, the exponential function and the logarithm.

The word transcendental does not mean anything particularly deep or mysterious; it merely suggests the fact that the definition of these functions by means of the elementary operations of calculation is not possible, "quod algebrae vires transcendit" (Latin for what exceeds the forces of algebra).

In all higher analytical investigations, where there occur angles, it is customary to measure these angles not in degrees, minutes and seconds, but in radians. We place the angle to be measured with its vertex at the centre of a circle of radius 1 and measure the size of the angle by the length of the arc of the circumference which the angle cuts out. Thus, an angle of 180 is the same as an angle of π radians (has radian measure π), an angle of 90 has radian measure $\pi/2$, an angle of 45 a radian measure $\pi/4$), an angle of 360 a radian measure 2π. Conversely, an angle of 1 radian, expressed in degrees, is

$$\frac{180°}{\pi}, \text{ or approximately } 57° \ 17' \ 45''$$

From here on, whenever we speak of an angle x, we shall mean an angle the radian measure of which is x.

After these preliminary remarks, we may briefly remind the reader of the meanings of the trigonometric functions $\sin x$, $\cos x$, $\tan x$, $\cot x$*. These are shown in Fig. in which the angle x is measured from the arm OC (of length 1), angles being positive in the counter-clockwise direction.

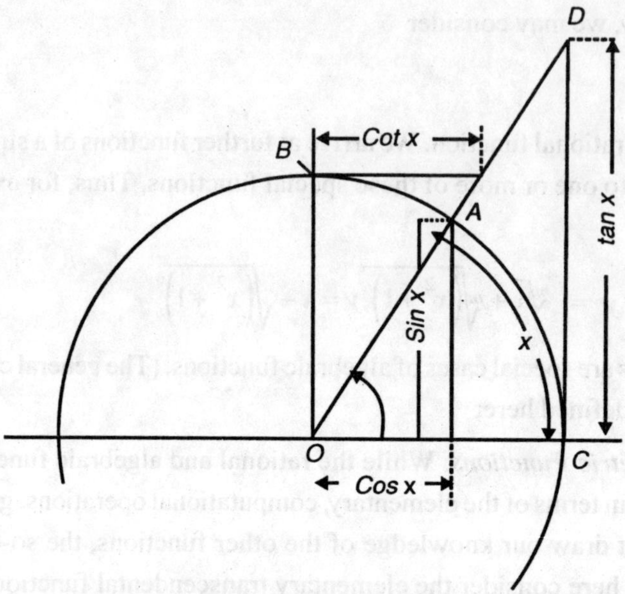

Fig. The trigonometric functions

At times, it is convenient to introduce the functions sec $x = 1/\cos x$, cosec $x = 1/\sin x$.

The rectangular co-ordinates of the point A yields at once the functions $\cos x$ and $\sin x$. The graphs of the functions $\sin x$, $\cos x$, $\tan x$, $\cot x$ are given in Figs.

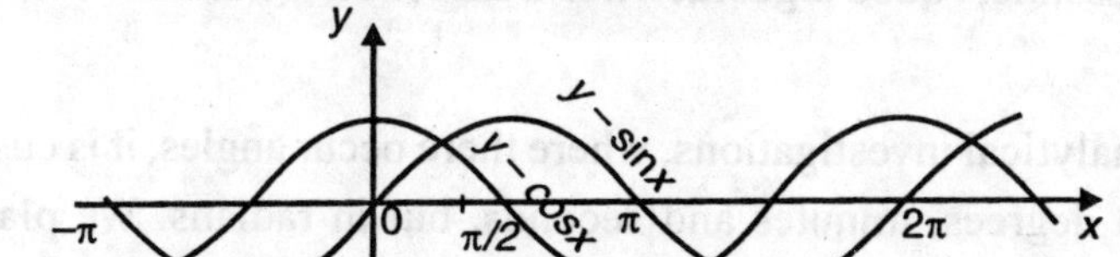

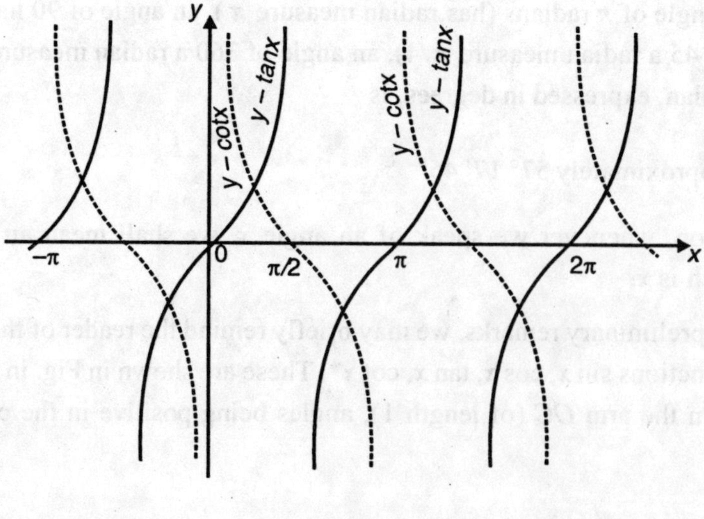

The Exponential Function and the Logarithm: In addition to the trigonometric functions, the exponential function with the positive base a,

$$y = a^x,$$

and its inverse, the logarithm to the base a,

$$x = \log_a y,$$

are also referred to as elementary transcendental functions. In elementary mathematics, it is customary to disregard certain inherent difficulties in the definition of these functions; we too shall postpone the exact discussion of these functions until we have better methods at our disposal (3.6 *et sequ.*). However, we will at least state here the basis of the definitions. If $x = p/q$ is a rational number (where p and q are positive integers), then - assuming the number a to be positive - we define a^x as $\sqrt[q]{a^p} = a^{p/q}$, where the root, by convention, is to be taken as positive. Since the rational values of x are everywhere dense, it is natural to extend this function a^x so as to make it into a continuous function, defined also for irrational values of x, giving a^x values when x is irrational, which are continuous with the values already defined when x is rational. This yields a continuous function $y = a^x$ - the exponential function - which for all rational values of x gives the value of a^x found above.

The function

$$x = \log_a y$$

can then be defined for $y > 0$ as the inverse of the exponential function.

Functions of an Integral Variable. Sequences of Numbers

Hitherto, we have considered the independent variable as a continuous variable, that is, as varying over a complete interval. However, there occur numerous cases in mathematics in which a quantity depends only on an integer, a number n which can take the values 1, 2, 3,... ; it is called a function of an integral variable. This idea will most easily be grasped by means of examples:

Chapter 2

Integral Properties

THE INTEGRAL

There are a number of different ways to approach the Integral Calculus. We will start with the least useful approach, which has the merit of being very simple to understand. We will then look at a more complicated approach which is actually much more useful.

Our first approach is going to be to say that Integration is just the opposite (or inverse operation) to Differentiation. If $f(x)$ differentiates to give $F(x)$ then, by definition, $F(x)$ integrates to give $f(x)$. We have been calling $F(x)$ the derivative of $f(x)$ and we will now call $f(x)$ an *Indefinite Integral* of $F(x)$. We use the notation

$$f(x) = \int F(x)\, dx$$

(This rather complicated notation derives from the other approach to integration that we will be considering. Note that both the $\int$ and the dx are part of the notation. If you like, the $\int$ says that we are dealing with an integral and the dx tells us the name of the variable under consideration.)

As a simple example, $\sin(x)$ differentiates to give $\cos(x)$, so $\cos(x)$ integrates to give $\sin(x)$. Or

$$\int \cos(x)\, dx = \sin(x)$$

Now we notice a slight problem. $\sin(x)$ differentiates to give $\cos(x)$, but so do $\sin(x) + 1$ and $\sin(x) -34.23$. In that case which of these is *the* integral of $\cos(x)$? We seem to have an ambiguity. This is true, and it is an ambiguity that will not go away.

How much ambiguity do we have to deal with? Well, suppose that $f_1(x)$ and $f_2(x)$ are both indefinite integrals of $F(x)$. That simply means that both $f_1(x)$ and $f_2(x)$ differentiate to give $F(x)$. But if $f_1'(x) = f_2'(x)$ then, by the rules of differentiation the derivative of $f_1(x)$ -

$f_2(x)$ must be zero. So $f_1(x) - f_2(x)$ must be a constant. So the *only* ambiguity is that we can add an arbitrary constant to our integral.

Many authors make this explicit by writing something like

$$\int \cos(x)\,dx \;=\; \sin(x) + C$$

where C stands for the arbitrary constant that we can add in. You may have been brought up to do this all the time. You are welcome to do it if that is what you are used to. I will tend to leave it out—simply because I know that it is always there (though there will come a time later in the course when I will have to be more careful).

Let me now gather together some terminology. Consider the formula

$$\int \cos(x)\,dx \;=\; \sin(x) + C$$

The function $\cos(x)$ is called the Integrand, the variable x is called the variable of integration, the constant C is called the Constant of Integration and $\sin(x)$ is called the Indefinite Integral or Anti-Derivative.

Doing Integrals

Now that we know what integrals are the next problem is to find out how to do them. This turns out to be a heavy problem. For differentiation we had a tidy set of rules that allowed us to work out the derivative of just about any function that we cared to write down—the procedure is basically mechanical and can be done quite well by computers. There is nothing like this for integration, in general. Even (apparently) simple integrals can be very difficult or downright impossible to work out. More of this later. Integration is more of a skill than a routine.

One approach that can work sometimes in simple cases is just 'spotting the answer'. If you can spot a function which differentiates to give your function then you have found an integral. Look at the following simple examples.

$$\frac{d}{dx}\left(\frac{1}{n+1}x^{n+1}\right) = x^n \quad \text{so} \quad \int x^n dx = \frac{1}{n+1}x^{n+1} \qquad (n \neq -1)$$

$$\frac{d}{dx}\sin x = \cos x \quad \text{so} \quad \int \cos x\,dx = \sin x$$

$$\frac{d}{dx}\cos x = -\sin x \quad \text{so} \quad \int \sin x\,dx = -\cos x$$

$$\frac{d}{dx}e^{ax} = a^{ax} \quad \text{so} \quad \int e^{ax}dx = \frac{1}{a}e^{ax}$$

$$\frac{d}{dx}\ln x = \frac{1}{x} \quad \text{so} \quad \int \frac{dx}{x} = \ln x$$

Let me do something a little more complicated. We have seen that $\int\cos(x)\,dx = \sin(x)$. What then is the value of $\int\cos(2x)\,dx$?

We want to know what function differentiates to give $\cos(2x)$. Well, if you differentiate $\sin(2x)$ you get $2\cos(2x)$, which is not quite right because of the factor 2. That is easily adjusted for and we see that the required integral is $\sin(2x)/2$.

Note that this simple approach does not work with an integral like $\int\cos(x^2)\,dx$. The obvious function to play with seems to be $\sin(x^2)$ but the derivative of this is $2x\cos(x^2)$ (damn the chain rule) which is nowhere near the required answer (and no, you can't go dividing by x!). Our simple approach has failed.

Your work on differentiation will allow you to check that the following short table is correct. These integrals should be known.

$f(x)$	$\int f(x)\,dx$
$x^a a \neq 1$	$\frac{1}{a+1} x^a + 1$
$\sin x$	$-\cos(x)$
$\cos(x)$	$\sin(x)$
e^x	e^x
$\frac{1}{x}$	$\ln(x)$
$\frac{1}{1+x^2}$	$\arctan x$
$\frac{1}{\sqrt{1-x^2}}$	$\arcsin x$

There is a table of integrals at the end of this booklet. There are books in the library that devote hundreds of pages to listing integrals.

Most of the 'techniques' for working out integrals are just methods for changing one integral into another in the hope that you will eventually come to an integral that you can 'spot' or find in the tables.

Let me finish this section by raising (quietly) yet another difficulty with the indefinite integral—it really is plagued with problems. Consider the statement

$$\int\frac{dx}{x} = \ln(x)$$

which is in the above table. This looks correct because we have already found that the derivative of $\ln(x)$ is $1/x$. But look a little closer. The integrand $(1/x)$ makes sense for any value of x other than 0. The integral $(\ln(x))$ makes no sense at all if x 0. What we have in this formula is really an integral for $1/x$ valid for $x > 0$. We have had to limit the 'domain of definition' of the integrand in order to make sense. This problem will have to be coped with later on.

SOME PROPERTIES OF THE INDEFINITE INTEGRAL

These are immediate consequences of the corresponding properties of derivatives. In each equation there is really an arbitrary constant of integration hanging around.

Let $f(x)$ and $g(x)$ be functions and λ a constant. Then

$$\int \lambda f(x)\,dx = \lambda \int f(x)\,dx$$

$$\int f(x)+g(x)\,dx = \int f(x)\,dx + \int g(x)\,dx$$

provided that the integrals exist; we have not yet shown that there necessarily exists a function $F(x)$ that differentiates to give a specified function $f(x)$.

These rules may allow us to reduce an integral to the point where we can spot the answer.

$$\int 2x + e^x dx = 2\int x\,dx + \int e^x dx = x^2 + e^x$$

$$\int \sin x + 2\cos x\,dx = \int \sin x\,dx + 2\int \cos x\,dx = -\cos x + 2\sin x$$

Impossible Integrals

This is an aside. Ignore it unless interested.

You will frequently be told that some integrals are 'impossible' they cannot be done. Common examples are such seemingly respectable integrals as $\int e^{x2}\,dx$, $\int \sin(x^2)\,dx$, $\int \frac{dx}{\sqrt{x^2+1}}\,dx$

This does *not* mean that, for example, there is no function that differentiates to give e^{x2}; there certainly is. The problem is that you do not know how to write it down. The answer cannot be expressed in terms of 'elementary functions'.

Let me give another example to make this clearer. We have seen that

$$\int \frac{dx}{1+x^2} = \arctan(x)$$

That is simple enough. But suppose I had not done the inverse trig functions. Then you would never have met arctan(x), so I would have to say to you that, though the integral could certainly be done, it could not be done in terms of functions known to you. The same kind of thing applies to the 'impossible' examples given above.

It is a distressing fact that the vast majority of integrals are, in this sense, impossible. When I write exercise sheets on differentiation I can get away with writing functions down at random, knowing full well that they can be differentiated. When writing exercises on integration I have to tread very carefully indeed and check everything out before I set anything.

In some ways it is *more* surprising that derivatives *can* be done. It just happens to be the case that the derivatives of all the elementary functions can be expressed in terms of elementary functions. There is no logical reason for this—it is pure luck. That's a lie actually, there are good historical reasons why our idea of 'elementary' for functions leads to simple derivatives.

The Definite Integral

Suppose that $F(x)$ is an indefinite integral of $f(x)$, i.e. $F'(x) = f(x)$. The Definite Integral

$$\int_a^b f(x)\, dx$$

where a and b are numbers, is defined to be the number $F(b) - F(a)$:

$$\int_a^b f(x)dx = \left[F(x)\right]_a^b = F(b) - F(a)$$

The choice of indefinite integral (choice of constant of integration) does not matter—the constant of integration cancels out.

a and b are called the Limits of Integration. a is called the Lower Limit and b is called the Upper Limit.

Note the 'square bracket' notation. $\left[f(x)\right]_a^b$ stands for $f(b) - f(a)$. Some books use $f(x)\Big|_a^b$, but this can be confusing in complicated expressions because it does not tell you where the expression starts.

This may all seem a strange idea at this stage. The justification for definite integrals will come quite soon.

Differentiating an Integral

Consider the function

$$g(x) = \int_a^x f(t)dt$$

where a is a constant and we are thinking of the upper limit of the integral as a variable. If $F(x)$ is an indefinite integral for $f(x)$ then

$$g(x) = F(x) - F(a)$$

So, by definition of $F(x)$,

$$g'(x) = F'(x) = f(x)$$

This gives the result

$$\frac{d}{dx}\int_a^x f(t)dt = f(x)$$

PROPERTIES OF THE DEFINITE INTEGRAL

If we assume that the function $f(x)$ has an integral then

$$\int_a^b f(x)\,dx = -\int_b^a f(x)dx \quad (F(b)-(F(a) = -(F(a)-(F(b))$$

$$\int_a^b f(x)dx + \int_a^c f(x)\,dx = -\int_a^c f(x)dx \quad ((F(b)-F(a)) + (F(c)-(F(b))$$

$$= (F(c)-(F(a))$$

Infinite Limits

Have a quick look at this section, but don't get worried by it.

We have seen what we mean by a definite integral where both limits of integration are *finite*. In many important applications of integration you will come across cases where one or more of the limits of integration is infinite. There are three possibilities:

$$\int_a^\infty f(x)\,dx \quad \int_{-\infty}^a f(x)\,dx \quad \int_{-\infty}^\infty f(x)\,dx$$

The interpretation of these is that they are to be taken as limiting cases of the corresponding integrals with finite limits. For example,

$$\int_a^\infty f(x)\,\mathrm{dx} = \lim_{b\to\infty}\int_a^b f(x)\,\mathrm{dx}$$

$$\int_{-\infty}^a f(x)\,\mathrm{dx} = \lim_{b\to-\infty}\int_b^a f(x)\,\mathrm{dx}$$

Here is a simple example: consider the integral $\int_1^\infty \frac{dx}{x^2}$

Now, for any finite positive value of b,

$$\int_1^b \frac{dx}{x^2} = \left[\frac{-1}{x}\right]_1^b = 1 - \frac{1}{b}$$

Now we look at the limiting behaviour as $b\to\infty$. In the limit $1/b\to 0$ and so the value of the integral tends to 1. Therefore

$$\int_1^\infty \frac{dx}{x^2} = 1$$

Of course, in most cases the limits will not exist, so the integral will not exist either. For example,

$$\int_0^\infty x\,\mathrm{dx} = \lim_{a\to\infty}\frac{1}{2}a^2$$

This limit does not exist, so the integral does not exist, though you might get away with saying that it is 'infinite'.

More seriously, consider the integral

$$\int_0^\infty \cos x \, dx = \lim_{a\to\infty} \sin a$$

Here the limit does not exist either, nor does it do anything so straightforward as going off to infinity—it just oscillates to and fro between –1 and 1.

Finally, let me note a small problem. The integral $\int_{-\infty}^{\infty} f(x)\,dx$ is the limiting case of $\int_a^b f(x)\,dx$ as $a \to \infty$ and $b \to \infty$. This need not be the same thing as $\lim_{a\to\infty} \int_{-a}^{a} f(x)dx$.

It is best to think of such doubly-infinite limits in two stages: first let the top limit tend to infinity to get $\int_a^\infty$ and then let the bottom limit tend to - ∞ to get the answer. The reason for such care is shown by the integral of sin(x). For any finite value of a we have $\int_{-a}^{a} \sin(x)\,dx = 0$ (check). If we took the limit $a \to \infty$ in this we would get

$$\int_{-\infty}^{\infty} \sin(x)\,dx = 0$$

but this is actually nonsense. If, on the other hand, you start by studying $\int_a^\infty \sin(x)\,dx$ you will see at once that no meaningful limiting value exists.

APPLICATIONS OF INTEGRATION

Integrals and Area

It is an odd fact that the basic ideas of integration predate those of differentiation by nearly 2000 years. Archimedes was quite good at 'doing integrals', albeit in a rather different context. He was interested in the problem of calculating the areas of curved regions. The ancient Egyptians and Babylonians were quite good at finding the areas of straight-sided regions (basic surveying—much needed in Egypt because of the regular flooding of the Nile). Curved regions are an altogether different problem and the ancient Ancients knew little more than that the area of a circle is r^2. More complicated regions need much more complicated ideas. It is not enough just to know the area of a triangle.

Archimedes, in an remarkable anticipation of the methods of the Calculus, tried to adopt a 'limiting values' approach to calculating areas. His idea was that you could approximate a curved region arbitrarily closely by straight-sided regions and could therefore hope to obtain the area of the curved region as a limiting case of the (calculable) areas of the approximations. This has since become one of the most basic ways of defining integrals because, as we will see in this section, there is a close connection between the idea of an integral (which we just introduced as the opposite of differentiation) and the idea of an area.

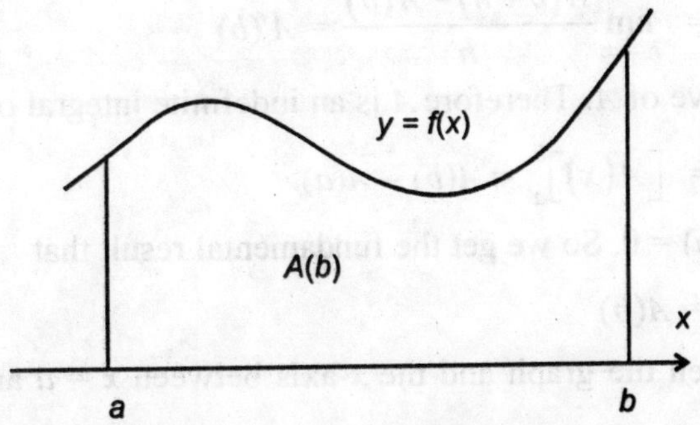

Fig. The area between a curve and the *x*-axis.

Consider the graph of $y = f(x)$ between $x = a$ and $x = b$. We want to calculate the area between this curve and the *x*-axis shown in Fig. 2.1.

In what follows I am going to regard a as fixed and denote the area between a and b by $A(b)$.

Now consider a point slightly to the right of b with coordinate $b + h$.

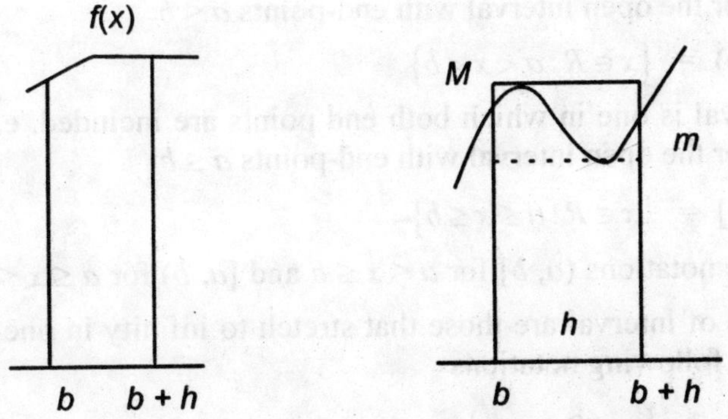

Fig. : Bounding an area between two rectangles

Then

$$A(b + h) = A(b) + \text{shaded area}$$

Let me draw a bigger picture of the shaded area. M is the highest point on the graph on this interval and m is the lowest. By looking at the two rectangles in the diagram it is clear that the shaded area has value between hm and hM. So, by above, $mh \leq A(b + h) - A(b) \leq Mh$ or $m \leq \dfrac{A(b+h)-A(b)}{h} \leq M$

(we have $h > 0$). Note that the central term is a Newton Quotient.

Now let h shrink down to zero. If f is continuous it is clear that as $h \to 0$ both m and M tend to $f(b)$. The Newton Quotient is sandwiched between them, so

$$f(b) = \lim_{h\to\infty} \frac{A(b+h)-A(b)}{h} = A'(b)$$

So f is the derivative of A. Therefore A is an indefinite integral of f. So

$$\int_a^b f(x)\,dx = \left[A(x)\right]_a^b = A(b) - A(a)$$

but, obviously, $A(a) = 0$. So we get the fundamental result that

$$\int_a^b f(x)dx = A(b)$$

So the area between the graph and the x-axis between $x = a$ and $x = b$ is given by the formula

$$\text{area} = \int_a^b f(x)dx$$

Intervals

Before going on to the next section it is worth introducing some standard notation for intervals. An Interval on the real line is all the points between two points (apart from a special case to be done in a moment). These intervals come in a number of types, depending on whether the end-points are included in the range or not.

An Open Interval is one in which the end-points are not included. e.g. $0 < x < 1$. We use the notation (a, b) for the open interval with end-points $a < b$

$$(a, b) = \{x \in R : a < x < b\}$$

A Closed Interval is one in which both end points are included. e.g. $0 \le x \le 1$. We use the notation $[a, b]$ for the open interval with end-points $a \le b$

$$[a, b] = \{x \in R : a \le x \le b\}$$

We also use the notations $(a, b]$ for $a < x \le b$ and $[a, b)$ for $a \le x < b$.

The other types of interval are those that stretch to infinity in one direction or another. For these we use the following notations

$$(-\infty, a) = \{x \in R : x < a\}$$

$$(-\infty, a] = \{x \in R : x \le a\}$$

$$(a, \infty) = \{x \in R : a < x\}$$

$$[a, \infty) = \{x \in R : a \le x\}$$

$$(-\infty, \infty) = R$$

Note that things like $[-\infty, a)$ do not make sense. 'Infinity' is not a number.

FURTHER PROPERTIES OF THE DEFINITE INTEGRAL

The 'area' interpretation of the definite integral gives us the following extra properties:

If $f(x) \le 0$ on $[a, b]$ then $\int_a^b f(x)dx \ge 0$

(If the graph does not go below the x-axis the integral is not going to be negative.)

$$\text{If } f(x) \le g(x) \text{ on } [a, b] \text{ then } \int_a^b f(x)\,dx \ge \int_a^b g(x)\,dx$$

(If the graph of f never goes below the graph of g then the area under the f graph is greater than that under the g graph. That explanation makes sense if both graphs are above the x-axis, but the result is true in general as you can see by using the first property on the function $h(x) = f(x) - g(x)$.)

$$\text{If } m \le f(x) \le M \text{ on } [a, b] \text{ then } m(b-a) \le \int_a^b f(x)\,dx \le M(b-a)$$

(Think of m as the constant function $g(x) = m$ and use the earlier results, together with $\int_a^b m\,dx = m(b-a)$. Similarly for M.)

Finally, a property that is often quite useful $\left|\int_a^b f(x)\,dx\right| \le \int_a^b |f(x)|\,dx$.

ANOTHER VIEW OF THE DEFINITE INTEGRAL

Consider the interval $[a, b]$. Let $f(x)$ be a continuous function defined on this interval. Let us think once more of the problem of finding the area under the graph of $f(x)$ between $x = a$ and $x = b$.

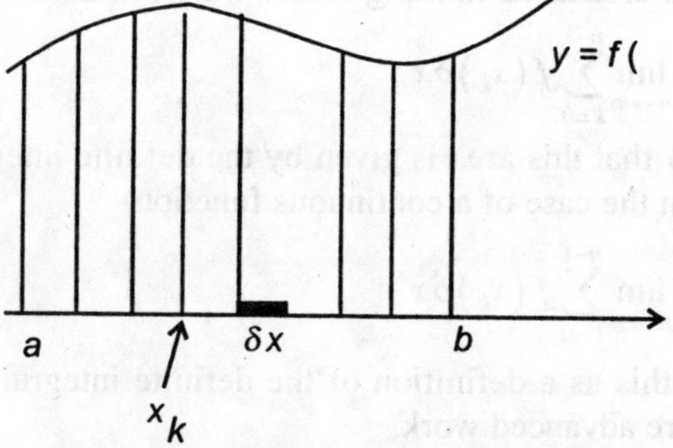

Fig. Dividing up the area under a curve

We will adopt an approach that is much more elementary (and much older) than our previous method. Divide the interval $[a, b]$ up into a large number of small parts. For convenience we will take them all to be of the same width, but that is not very important. Now use this subdivision to break up the area into *thin strips* as shown.

Denote the subdivision points by $x_0, x_1, \ldots, x_n$ where $x_k = a + k\delta x$ and δx is the width of each strip $\delta x = (b-a)/n$.

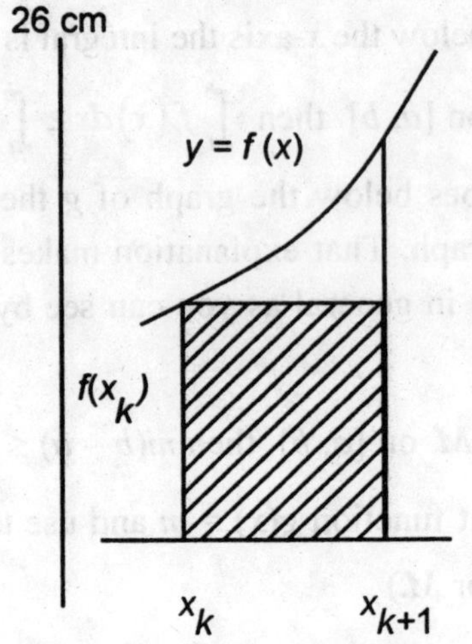

Fig. One of the rectangles

We can get an approximation to the area under the graph by adding up the areas of the n rectangles shown in the picture. The rectangle on the base $[x_k, x_{k+1}]$ has area $f(x_k)\delta x$. So

$$\text{approximate area} = \sum_{k=0}^{n-1} f(x_k)\delta x$$

Now we use the same kind of argument that we used when inventing the derivative. As n gets bigger and bigger we expect the sum of the areas of the rectangles to get closer and closer to the true area under the graph. We would hope that if we took the limit as $n \to \infty$ the sum would tend to the true area as its limiting value. We will assume that this is true. So

$$\text{area under graph} = \lim_{n\to\infty} \sum_{k=0}^{n-1} f(x_k)\delta x$$

But we already know that this area is given by the definite integral. So, putting our two results together, we get (in the case of a continuous function)

$$\int_a^b f(x)\,dx = \lim_{n\to\infty} \sum_{k=0}^{n-1} f(x_k)\delta x$$

We can in fact take this as a definition of the definite integral (the Riemann Integral), and normally do so in more advanced work.

This interpretation of the definite integral is the one that is most useful in applications, as we will soon see.

FURTHER APPLICATIONS OF THE DEFINITE INTEGRAL

Our new interpretation of the definite integral is very useful. It shows us how to turn a lot of different problems into integration problems. We introduced it in the context of finding areas, but it is vastly more general than that. Almost anything that comes into the category of 'break it into small bits, approximate on the bits and then add up the bits' yields a definite integral of some form.

In this section I will look at a few applications. You may meet others in other subjects.

The Length of a Curve

Suppose we want to calculate the length of the graph of $y = f(x)$ between $x = a$ and $x = b$.

Subdivide $[a, b]$ as before into n small parts. Look at the graph on one of these parts. The idea is to get an approximation to the length of the graph on this part, in the same way that we used the rectangle approximation when finding the area. Then we add up the approximations and take the limit as n so as to produce an integral which gives the true length.

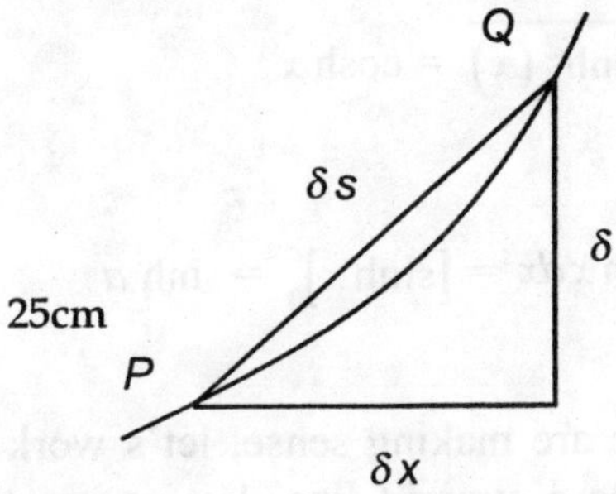

Fig. 2.8: Straight line approximation

What approximation do we use?

The obvious approach is to use the length of the chord PQ as an approximation to the length of the graph between P and Q. In the notation of Fig. 2.8 this length is

$$\delta s = \sqrt{\delta x^2 + \delta y^2}$$

which we can write more conveniently as

$$\delta s = \sqrt{1 + \left(\frac{\delta y}{\delta x}\right)^2}\ \delta x$$

We now have the approximation to the length which I will write crudely as

$$\sum \sqrt{1 + \left(\frac{\delta y}{\delta x}\right)^2}\ \delta x$$

Our new interpretation of the definite integral tells us that, as n this tends to the value of the definite integral $\int_a^b \sqrt{1 + \left(\frac{dy}{dx}\right)^2}\ dx$

(the $\delta y/\delta x$ is really the Newton Quotient and tends to dy/dx in the limit.)

So we have obtained the formula

$$s = \int_a^b \sqrt{1 + \left(\frac{dy}{dx}\right)^2}\ dx$$

for the length of the graph of $y = f(x)$ between $x = a$ and $x = b$.

Note: if the curve is given parametrically by $x = x(t)$ and $y = y(t)$ then a very similar argument gives us the formula

$$s = \int_a^b \sqrt{\dot{x}^2 + \dot{y}^2}\, dt$$

for the length of the curve between $t = a$ and $t = b$.

Example 2.6 What is the length of the graph $y = \cosh(x)$ between $x = 0$ and $x = a$? In this case $dy/dx = \sinh(x)$ so

$$\sqrt{1+\left(\frac{dy}{dx}\right)^2} = \sqrt{1+\sinh^2(x)} = \cosh x$$

So the length is

$$s = \int_0^a \cosh x\, dx = [\sinh x]_0^a = \sinh a$$

Example 2.7

As a check on whether we are making sense, let's work out the length of a part of a straight line. Let $y = mx + c$ be a straight line. Let us use the formula to find its length between $x = a$ and $x = b$.

$$y' = m \quad \text{so} \quad \sqrt{1+\left(\frac{dy}{dx}\right)^2} = \sqrt{1+m^2}$$

So

$$s = \int_a^b \sqrt{1+m^2}\, dx = (b-a)\sqrt{1+m^2} = \frac{b-a}{\cos\phi}$$

where ϕ is the angle that the line makes with the x-axis ($m = \tan$). You can easily check that this is indeed the right answer.

Example 2.8 What is the circumference of a circle of radius r?

Let the circle be given by the equation $x^2 + y^2 = r^2$. We will get the circumference by working out the length of the top semicircle and doubling the result.

On the top half $y = \sqrt{r^2 - x^2}$. So

$$\frac{dy}{dx} = \frac{-x}{\sqrt{r^2-x^2}}$$

and

$$\sqrt{1+\left(\frac{dy}{dx}\right)^2} = \frac{r}{\sqrt{r^2-x^2}}$$

So the length of the semicircle is given by

$$l = \int_{-r}^{r} \frac{r\,dx}{\sqrt{r^2 - x^2}}$$

This is actually a fairly standard integral and you will find it in most tables.

The solution is $dx/ = \int dx/\sqrt{r^2 - x^2}$ arcsin(x/r). You should check this by differentiating the RHS.

$$l = \left[r\arcsin(x/r)\right]_{-r}^{r} = r(\arcsin(1) - \arcsin(-1)) = r\pi$$

So the total circumference of the circle is twice this, i.e. circumference = $2\pi r$

Volumes of Revolution

Take the graph of $y = f(x)$ on the interval $[a, b]$ and spin it round the x-axis so as to produce what is known as a solid of revolution as shown in Fig. 2.9. We want to get a formula for the volume of this solid.

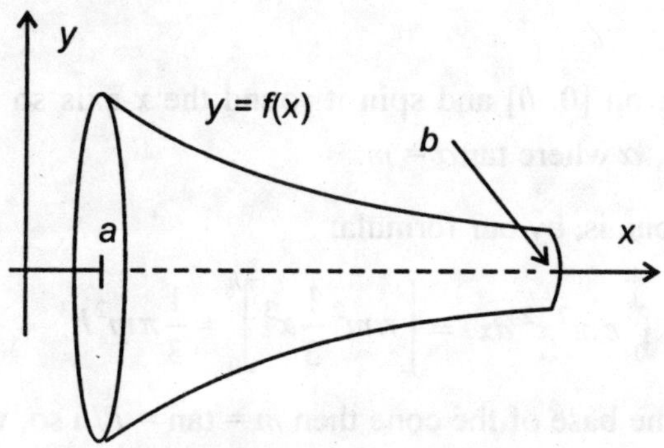

Fig. A volume of revolution

The method is almost exactly the same as in the previous examples. Think of the interval $[a, b]$ being subdivided into lots of little bits. Now look at one of the bits and try to get an approximation for the volume of the 'thin slice' of the solid obtained by rotating the piece of the graph on this interval.

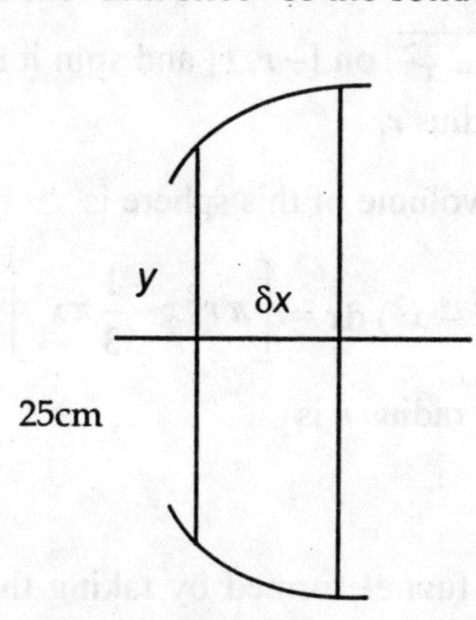

Fig. The volume of a slice

In the notation of the diagram, the thin slice of the solid is virtually a *cylinder* of radius y and thickness x. The volume of a cylinder is the product of its height and the area of its base. So we get the approximation

$$\delta V = \pi y^2 \, \delta x$$

for the volume of the slice.

The approximation to the total volume can then be written as

$$V \approx \sum \pi y^2 \, \delta x$$

Now take the limit as $n \to \infty$ and get

$$\text{Volume} = \int_a^b \pi y^2 \, dx$$

Example 2.9 Cone

Take the line $y = mx$ on $[0, h]$ and spin it round the x-axis so as to produce a cone of height h and 'semi-angle' , α where $\tan\alpha = m$.

The volume of this cone is, by our formula,

$$V = \int_0^h \pi m^2 x^2 \, dx = \left[\pi m^2 \frac{1}{3} x^3 \right]_0^h = \frac{1}{3} \pi m^2 h^3$$

If R is the radius of the base of the cone then $m = \tan = R/h$ so, with a bit of rearranging, we get

$$V = \frac{1}{3} \pi R^2 h = \frac{1}{3} \text{ base} \times \text{height}$$

Example 2.10 Sphere

Take the semicircle $y = \sqrt{r^2 - x^2}$ on $[-r, r]$ and spin it round the x-axis. We get, as our solid of revolution, a Sphere of radius r.

Our formula tells us that the volume of this sphere is

$$V = \int_{-r}^{r} \pi (r^2 - x^2) \, dx = \left[\pi r^2 x - \frac{1}{3} \pi x^3 \right]_{-r}^{r} = \frac{4}{3} \pi r^3$$

So the volume of a sphere of radius r is

$$V = \frac{4}{3} \pi r^3$$

Example 2.11 Consider the funnel formed by taking the curve $y = 1/x$ and rotating it round the x-axis on the interval $[1, a]$, as shown in Fig.

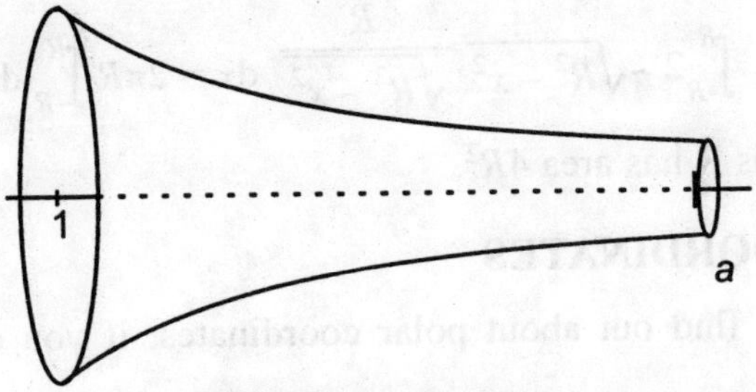

Fig. Volume of rotation.

The volume of this funnel is

$$V = \int_1^a \pi \frac{1}{x^2}\, dx = \left[-\frac{\pi}{x}\right]_1^a = \pi\left(1-\frac{1}{a}\right)$$

Now notice that, as $a \to \infty$, this volume tends to the finite value π(because $\pi/a \to 0$).

We write

$$\int_1^\infty \frac{\pi}{x^2}\, dx = \lim_{a\to\infty} \int_1^a \frac{\pi}{x^2}\, dx = \lim_{a\to\infty}\left(\pi - \frac{\pi}{a}\right) = \pi$$

Area of Surface of Revolution

I'm going to be brief here and just give you the formula.

Suppose we take the graph $y = f(x)$ on the range $[a, b]$ and rotate it around the x-axis as before. Then the *Surface Area* of the surface formed by this is given by

$$\text{area} = \int_a^b 2\pi y \sqrt{1+\left(\frac{dy}{dx}\right)^2}\, dx$$

Example 2.12 Sphere

As before, we obtain our sphere by rotating the semicircle

$$y = \sqrt{R^2 - x^2} \quad -R \le x \le R$$

around the x-axis.

For this curve

$$\left(\frac{dy}{dx}\right) = \frac{-x}{\sqrt{R^2 - x^2}}$$

So

$$\sqrt{1+\left(\frac{dy}{dx}\right)^2} = \sqrt{1+\frac{x^2}{R^2-x^2}} = \frac{R}{\sqrt{R^2-x^2}}$$

So the area is given by

$$\text{Area} = \int_{-R}^{R} 2\pi\sqrt{R^2 - x^2}\,\frac{R}{\sqrt{R^2 - x^2}}\,dx = 2\pi R\int_{-R}^{R} dx = 4\pi R^2$$

So a sphere of radius R has area $4R^2$.

AREA IN POLAR COORDINATES

Read this after you find out about polar coordinates, if you don't know about them already.

Suppose we have a curve given in polar coordinate by $r = f(\theta)$. How do we find the area of the region bounded by the curve and the radial lines$\theta = a$ and $\theta = b$?

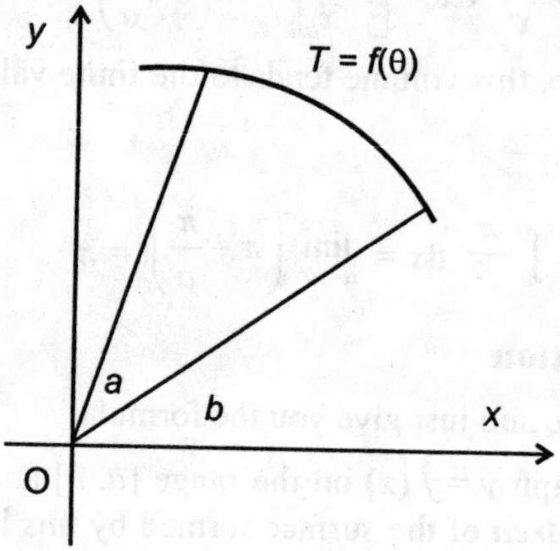

Fig. Describing an area in polar co-ordinates

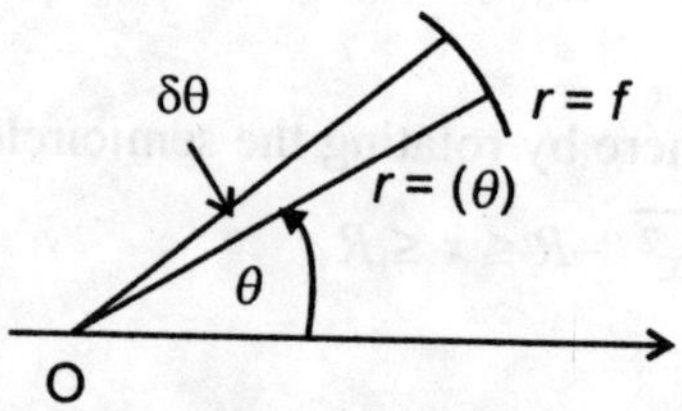

Fig. The area of a very small segment

This is another example of the 'thin slices' approach, though a bit different to previous ones. We imagine the θ range from a to b being subdivided into lots of small parts of width $\delta\theta$ and the area consequently being divided up into lots of thin pie slices. Any one of these slices, at angle θ, is approximately a triangle with sides $f(\theta)$ and included angle $\delta\theta$. The area of such a triangle is $\frac{1}{2}f(\theta)^2\sin(\delta\theta)$. But, for very small angles, $\sin(x)\approx x$. So we approximate the area of the pie-slice by $\frac{1}{2}f(\theta)^2(\delta\theta)$.

Now, 'adding up' these thin slice areas and passing over to the limit we get the formula

$$\text{area} = \frac{1}{2}\int_a^b f(\theta)^2 d\theta$$

for the area of the region. This is usually written more simply as

$$\text{area} = \frac{1}{2}\int_a^b r^2 d\theta$$

NUMERICAL APPROXIMATION TO DEFINITE INTEGRALS

If a and b are numbers then the value of $\int_a^b f(x)\,dx$ is a number. In many applications we are not interested in the precise value of this number—a good approximation will do. We should be grateful for this because, in practice, very few integrals can be evaluated explicitly so there is no hope of getting an exact answer.

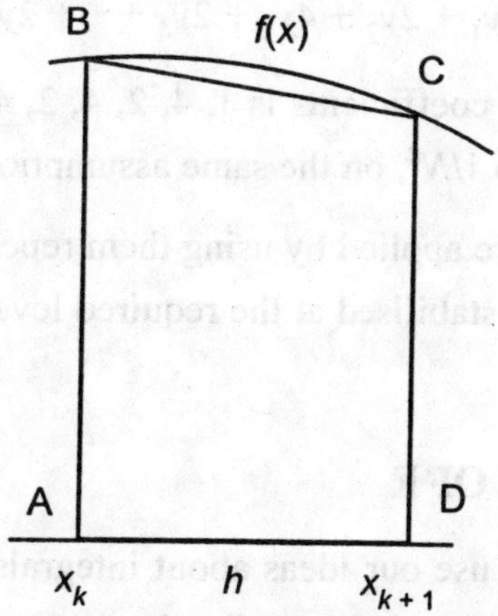

Fig. The area of below the curve is approximately the area of the trapezium

How do we approximate the value without being able to do the integral? The basic idea is our interpretation of the definite integral as an area (allowing for signs). You could get an approximation to the value of the integral simply by drawing the graph of $f(x)$ on graph paper and counting squares below the graph. This is crude and is a lot of work for a not very accurate result.

Our thin strips approach to areas gives us a better idea. Choose an integer N and subdivide the interval $[a, b]$ into N equal subdivisions, each of length $h = (b - a)/N$. Call the subdivision points $a = x_0, x_1, x_2,..., x_N = b$. Let the values of $f(x)$ at these points be $f(x_i) = y_i$.

If we look at one strip we can see that a reasonable approximation to its area is given by the area of the 'trapezium' ABCD. The smaller h becomes the better the approximation should be. This area is given by $\frac{h}{2}(y_k + y_{k+1})$. Adding up all these pieces we get

$$\int_a^b f(x)\,dx \approx \frac{h}{2}(y_0 + y_1) + \frac{h}{2}(y_1 + y_2) + \cdots + \frac{h}{2}(y_{N-1} + y_N)$$

This can be tidied up the produce what is known as the Trapezium Rule.

$$\int_a^b f(x)\,dx \approx \frac{h}{2}(y_0 + 2y_1 + 2y_2 + 2y_3 + \cdots + 2y_{N-1} + y_N)$$

The only real virtue of this method is that it is easy the derive (and is rather better than counting squares). Its accuracy turns out to be roughly proportional to $1/N^2$, at least if h is quite small and the graph is not too crazy. This means that doubling the number of strips should about quarter the error.

Let me write down without derivation a much better method known as Simpson's Rule. Same notation as before only this time the number of strips has to be *even*. We will take the number of strips to be $2N$, so $h = (b - a)/(2N)$. Simpson's Rule says that

$$\int_a^b f(x)dx = (y_0 + 4y_1 + 2y_2 + 4y_3 + 2y_4 + \cdots + 2y_{N-2} + 4y_{N-1} + y_N)$$

Note that the pattern of the coefficients is 1, 4, 2, 4, 2, 4, 2,..., 2, 4, 1. This method has accuracy roughly proportional to $1/N^4$, on the same assumptions as before.

In practice these methods are applied by using them repeatedly with increasing values of N until the results seem to have stabilised at the required level of accuracy. Very much a job for computers.

ESTIMATING THE VALUE OF E

This is a footnote. We can use our ideas about integrals and areas to get an idea about the value of the constant e. I quoted a value earlier, but did not deduce it.

Consider the graph of $y = e^x$ between $x = 0$ and $x = a > 0$ shown in Fig. 2.15. We know enough about e^x and its derivatives to be able to say that, on this interval, the graph lies above the tangent line at (0, 1) and below the chord ED.

This says that

$$\text{area ABCE} < \int_0^a e^x dx < \text{area ABDE}$$

Now ABCE has area $a + a^2/2$ and ABDE has area $a(1 + e^a)/2$. The integral has value $e^a - 1$. So $a + \frac{1}{2}a^2 < e^a - 1 < \frac{1}{2}a + \frac{1}{2}ae^a$

From these two inequalities we get, for $0 < a < 2$, $1 + a + \frac{1}{2}a^2 < e^a < \frac{2+a}{2-a}$

The outer values can be calculated. Put $a = 1$ and get $5/2 < e < 3$. True, but not very precise.

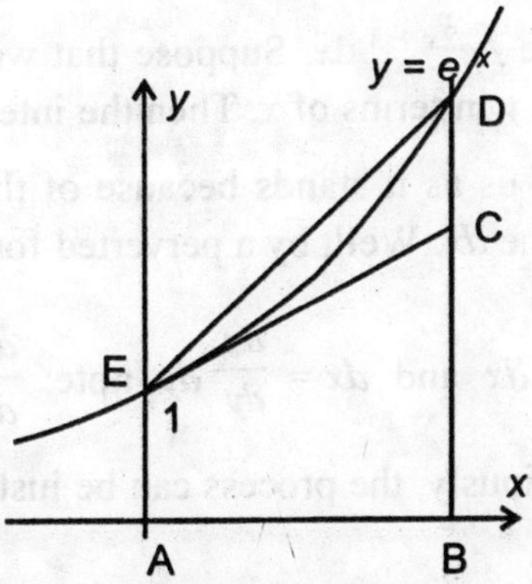

Fig. The graph of e^x lies between tangent and chord.

Now put $a = 1/n$ in the formula: $1 + \frac{1}{n} + \frac{1}{2n^2} < e^{1/n} < \frac{2n+1}{2n-1}$

Finally, raise everything to the nth power: $(1 + \frac{1}{n} + \frac{1}{2n^2})^n < e < \left(\frac{2n+1}{2n-1}\right)^n$

This is now a good result. The outer expressions can still, in principle, be evaluated by arithmetic. The larger a value we take for n a more precise result we get.

If $n = 10$ we get $2.7141 < e < 2.7206$ (getting close). If $n = 1000$ we get (with the help of a machine)

$2.71828138 < e < 2.7182821$

Larger values of n will produce correspondingly better results.

METHODS OF INTEGRATION

Introduction

So far our only way to work out $\int f(x)\,dx$ has been, in effect, to 'spot the answer'—i.e. notice the function $F(x)$ which, when differentiated, gives $f(x)$.

This method will not help us very much with integrals like $\int \frac{2x-3}{\sqrt{x^2+5x+7}}\,dx$

The purpose of this section is to introduce you to some techniques that can be used to evaluate *some* integrals. Most integrals cannot be evaluated in terms of functions that you know about.

The methods that I will show you are mostly techniques for converting one integral into another one in the hope that the new integral is more familiar.

Substitution

This is really the reverse process to the chain rule in differentiation.

Consider an integral like $\int e^{2x+3}\,dx$. Suppose that we make the *substitution* $y = 2x + 3$ which defines a new variable y in terms of x. Then the integral turns into $\int e^y\,dx$.

This is not much use to us as it stands because of the mixture of x's and y's. We still have to do something about the dx. Well, by a perverted form of the Chain Rule:

$$dy = \frac{dy}{dx}\,dx \text{ and } dx = \frac{dx}{dy}\,dy \text{ note: } \frac{dy}{dx} = \frac{1}{\frac{dx}{dy}}$$

(Don't take these too seriously, the process can be justified in other ways. I'm just telling you what to *do*.)

In our example, $dy = 2dx$, so $dx = \frac{1}{2}dy$ and our integral becomes $\int e^y \frac{1}{2}dy = \frac{1}{2}e^y$. Changing back to x's we get

$$\int e^{2x+3}\,dx = \quad e^{2x+3}$$

That is all there is to the method of substitution. The main problem in using it is in deciding what substitution to make. There are very few rules for this. It is mainly a matter of experience and common sense

Substitution in Definite Integrals

In one sense this is nothing new. If we want to work out $\int_a^b f(x)\,dx$ we work out the indefinite integral $F(x) = \int f(x)\,dx$ and then

$$\int_a^b f(x)\,dx = F(b) - F(a)$$

There is another approach to the problem which becomes more significant later. The idea is to use the substitution to change the definite integral itself. The process is very simple. We change the integrand and the dx just as before and we also change the limits of integration to their corresponding values in terms of the new variable of integration. (There are situations in which this needs a little care, but we will not worry too much about that at the moment.)

The definite integral $\int_a^b \frac{dx}{x}$

This needs care. As I pointed out, saying that

$$\int_a^b \frac{dx}{x} = \ln x$$

does not work unless $x > 0$

Now $\int_{-3}^{-2} \frac{dx}{x}$ is a perfectly sensible integral, as you will realise if you draw the graph, but

$$\int_{-3}^{-2} \frac{dx}{x} = \left[\ln(x)\right]_{-3}^{-2} = \ln(-2) - \ln(-3)$$

is nonsense, since you cannot take the log of a negative. (Unfortunately, if we carry on with our nonsense and write ln(- 2) - ln(- 3) = ln((- 2)/(- 3)) = ln(2/3) we end up with the right answer!)

The form that we really need for the definite integral is this:

$$\int_a^b \frac{dx}{x} = \ln(b/a)$$

provided that a and *b* have the same sign! If they don't then the integral is not defined.

It is a good exercise for you to check that this answer is actually correct.

Integration by Parts

This is a consequence of the product rule for differentiation.

$$(uv)' = u'v + uv'$$

Integrate this to get

$$\int (uv)'dx = \int u'v\,dx + \int uv'\,dx$$

But $\int (uv)'\,dx = uv$. So, rearranging things a bit,

$$\int u'v\,dx = uv - \int uv'\,dx$$

This is the Integration by Parts formula.

Again, this is a method for converting one integral into another one in the hope that the new one is easier to handle.

The method is usually appropriate in situations where the integrand is clearly the product of two, often rather different, terms, one of which can be integrated ($u' \to u$) and the other differentiated ($v \to v'$).

The Gamma Function

This is an interesting application of integration by parts. Let's have a look at the integral

$$I_n = \int_0^\infty x^n e^{-x}\,dx$$

where n is some non-negative whole number.

This seems to be a candidate for integration by parts. Differentiate the x^n and integrate the e^{-x}. This will give us

$$I_n = \left[-x^n e^{-x}\right]_0^\infty + \int_0^\infty x^{n-1} e^{-x}\,dx$$

It is an important property of the exponential function that, whatever the value of n, $x^n e^{-x} \to 0$ as $x \to \infty$(e^x grows far faster than any power of x). So, if $n > 0$ the first term on the RHS vanishes at both ends. The other term on the RHS involves an integral which is just the same as our original one with n replaced by $n-1$. So we have

$$I_n = n.I_{n-1} \quad \text{for} \quad n \geq 1$$

You grumble that this has not evaluated the integral—it has just turned it into another similar integral. True, but that does actually make it easy to evaluate the original integral. The point is that the above formula is valid for any positive integer *n*. So, climbing up the ladder, we have

$$\begin{aligned} I_1 &= 1.I_0 \\ I_2 &= 2.I_1 = 2.1.I_0 \\ I_3 &= 3.I_2 = 3.2.1.I_0 \\ I_4 &= 4.I_3 = 4.3.2.1.I_0 \end{aligned}$$

and so on. You will probably believe me if I say that the overall result is that $I_n = n!.I_0$.

This has reduced everything to the single problem of evaluating

$$I_0 = \int_0^\infty x^0 e^{-x} dx = 1$$

(as you should check).

So we have obtained the rather nice result that

$$\int_0^\infty x^n e^{-x}\, dx = n!$$

Now for a generalisation. We had a positive integer *n* in the integrand as a power. There was actually nothing in our integration procedure (the integration by parts) that relied on the fact that *n* was a whole number. We could just as well look at the integral

$$f(a) = \int_0^\infty x^a e^{-x}\, dx$$

where *a* is any positive value. Doing integration by parts on this will give

$$f(a) = a.f(a-1) \quad \text{if} \quad a > 1$$

as before. If *a* is a whole number then $f(a) = a!$, but $f(a)$ makes perfectly good sense for any positive value *a*. So we have produced a kind of generalisation of the factorial function. We can now talk about the factorial of $\frac{1}{2}$ or π. This may sound a rather silly thing to do, but it does actually have wide application.

Traditionally, we do not work with the function $f(a)$ but with the function $\Gamma(a) = f(a - 1)$. This is known as the Gamma Function,

$$\Gamma(a) = \int_0^\infty x^{a-1} e^{-x} dx, \quad \Gamma(a+1) = a.\Gamma(a)$$

for $a > 0$. Of course, we have $\Gamma(1) = \Gamma(2) = 1$.

To satisfy idle curiosity I can tell you that $\Gamma\left(\frac{1}{2}\right) = \sqrt{\pi}$

QUADRATICS AND TRIGONOMETRIC SUBSTITUTIONS

You often find terms like $\sqrt{x^2-3x+5}$ in integrals. This section suggests a way of dealing with them that might work.

The basic idea is to use variations on the formulas $1 - \sin^2 x = \cos^2 x$, $1 + \tan^2 x = \sec^2 x$, $\sec^2 x - 1 = \tan^2 x$, $\cosh^2 x - 1 = \sinh^2 x$ as suggestions for substitutions.

INTEGRATION OF RATIONAL FUNCTIONS AND PARTIAL FRACTIONS

We are now going to study integrals of the form $\int\left(\frac{p(x)}{q(x)}\right)\mathrm{d}x$

where p and q are polynomials.

We will not study the general case. Instead, we will start by assuming the following:

1. The denominator q can be factored as a product of *linear* factors $(ax + b)$, all different.
2. The degree of the numerator is less than that of the denominator.

As examples: $\int\frac{dx}{(x+1)(x+2)}$, $\int\frac{2x+3}{(3x-4)(1-2x)}dx$ $\int\frac{x^2}{(x+1)(x-1)(2x-3)}dx$

The following integrals do NOT satisfy the conditions

$\int\frac{x^2}{(x+1)(x+2)}\,\mathrm{d}x$ degrees wrong

$\int\frac{(x+1)}{(x-1)^2(x+2)}\,\mathrm{d}x$ square factor

$\int\frac{(x+1)}{x^2+x+1}dx$ cannot factorise denominator

To do this kind of integral we need the following result from algebra:

Any expression *fracp*$(x)(x+b_1)(x+b_2)\cdots(x+b_k)$

where $p(x)$ is a polynomial of degree less than k and all the factors in the denominator are different (in the above sense) can be expanded out in *Partial Fractions* as $\frac{A_1}{x+b_1}$ + $\frac{A_2}{x+b_2}$ + ... + $\frac{A_k}{x+b_k}$

where the As are constants.

That sounds complicated. What it means is shown by the following examples

$$\frac{1}{(x+1)(x+2)} = \frac{A}{x+1}+\frac{B}{x+2}$$

$$\frac{x+1}{(x-3)(2x+3)} = \frac{A}{x-1}+\frac{B}{2x+3}$$

$$\frac{x^2}{(x-1)(x+1)(x-3)} = \frac{A}{x-1}+\frac{B}{x+1}+\frac{C}{x-3}$$

If we have produced this kind of expansion then the integration is easy. All we have to know is that

$$\int\frac{dx}{ax+b} = \ln|ax+b|$$

So, using the above examples:

$$\int\frac{dx}{(x+1)(x+2)} = A\int\frac{dx}{x+1}+B\int\frac{dx}{x+2} = A\ln|x+1|+B\ln|x+2|$$

$$\int\frac{x^2dx}{(x-1)(x+1)(x-3)} = A\int\frac{dx}{x-1}+B\int\frac{dx}{x+1}+C\int\frac{dx}{x-3}$$

$$= A\ln|x-1|+B\ln|x+1|+C\ln|x-3|$$

The remaining problem is: how do we find the constants A, B etc.?

Consider the first example. We know, from the theorem, that

$$\frac{1}{(x+1)(x+2)} = \frac{A}{x+1}+\frac{B}{x+2} = \frac{A(x+2)+B(x+1)}{(x+1)(x+2)}$$

If this is going to work for *all* values of x then we must have

$$A(x+2)+B(x+1) = 1$$

for all values of x.

Let me now show you two methods for finding the values of A and B. The first is the most general but the second is usually the easiest in these cases.

1. Gather together powers of x and then compare coefficients on both sides of the equation: $(A+B)x+(2A+B) = 1$ so $A+B = 0$ and $2A+B = 1$. These simultaneous equations now solve to give $A = 1$ and $B = -1$.

2. $A(x+2)+B(x+1) = 1$ is true for all values of x, so it is true for any particular value of x that we put into it. If we put $x = -1$ we will kill off the B term and get$(-1+2)A = 1$, so $A = 1$. If we put $x = -2$ we kill off the A term and get$(-2+1)B = 1$, so $B = -1$.

So, either way,

$$\frac{1}{(x+1)(x+2)} = \frac{1}{x+1}-\frac{1}{x+2}$$

Example 3.23

$$\frac{x^2}{(x-1)(x+1)(x-3)} = \frac{A}{x-1} + \frac{B}{x+1} + \frac{C}{x-3}$$
$$= \frac{A(x+1)(x-3)+B(x-1)(x-3)+C(x-1)(x+1)}{(x-1)(x+1)(x-3)}$$

So we must have $A(x + 1)(x - 3) + B(x - 1)(x - 3) + C(x - 1)(x + 1) = x^2$ for all values of x.

Put $x = 1$ to kill off the B and C terms: $(1 + 1)(1 - 3)A = 1, A = -\frac{1}{4}$.

Put $x = -1$ to kill off the A and C terms: $8B = 1, B = \frac{1}{8}$.

Put $x = 3$ to kill off the A and B terms: $8C = 9, C = \frac{9}{8}$.

So

$$\frac{x^2}{(x-1)(x+1)(x-3)} = -\frac{1}{4}\frac{1}{x-1} + \frac{1}{8}\frac{1}{x+1} + \frac{9}{8}\frac{1}{x-1}$$

Now we can do the integral

$$\int\frac{x^2dx}{(x-1)(x+1)(x-3)} = -\frac{1}{4}\int\frac{dx}{x-1} + \frac{1}{8}\int\frac{dx}{x+1} + \frac{9}{8}\int\frac{dx}{x-1}$$

$$= -\frac{1}{4}\ln|x-1| + \frac{1}{8}\ln|x+1| + \frac{9}{8}\ln|x-3|$$

If the function to be integrated has a numerator of degree not lower than that of the denominator then you have to start by dividing the denominator into the numerator. For example

$$\frac{x3-1}{x(x+2)} = (x-2) + \frac{4x-1}{x(x+2)}$$

The first term on the RHS is easy and you can do partial fractions on the second.

Question is: how do you do the division? You may well have met some way of doing this in school. The basic method goes as follows. Suppose we are dividing $x^2 + x + 1$ into $3x^4 + x^2 + x + 1$. Look at the 'top' terms of the two polynomials: x^2 and $3x^4$. The first goes into the second $3x^2$ times. So multiply the first polynomial by $3x^2$ and subtract it from the second:

$$(3x^4 + x^2 + x + 1)$$
$$-\ 3x^2(x^2 + x + 1) = -3x^3 - 2x^2 + x + 1$$

Now start again. The top term x^2 of the divisor goes into the top term $-3x^3$ of our remainder $-3x$ times. So subtract $-3x$ times the divisor from the remainder:

$$(-\ 3x^3 - 2x^2 + x + 1)$$

$$-(-3x)(x^2 + x + 1) = x^2 + 4x + 1$$

Keep going. The top term x^2 of the divisor goes into the top term x^2 of the remainder 1 time. So subtract 1 times the divisor from the remainder:

$$(x^2 + 4x + 1) - 1.(x^2 + x + 1) = 3x$$

We have now got down to the point where the remainder has lower degree than the divisor, so we stop. The 'multipliers' that we have used make up the Quotient $3x^2 - 3x + 1$ and the final remainder is $3x$. So

$$\frac{3x^4 + x^4 + x + 1}{x^2 + x + 1} = 3x^2 - 3x + 1 + \frac{3x}{x^2 + x + 1}$$

Further Details

If the integrand has powers of factors in its denominator, for example a denominator like $x^3(x - 1)(x + 2)^2$, then you can still do partial fractions but you have to modify the expansion a little. Each square factor $(x + a)^2$ produces two terms in the expansion: $\frac{A}{x+a} + \frac{B}{(x+a)^2}$

Similarly, a third power like $(x + a)^3$ produces three terms in the expansion: $\frac{A}{x+a} + \frac{B}{(x+a)^2} + \frac{C}{(x+a)^3}$

The same pattern holds for higher powers. A term of power n, like $(x + a)^n$, in the denominator contributes n terms to the partial fraction expansion: $\frac{A_1}{x+a} + \frac{A_2}{(x+a)^2} + ... + \frac{A_n}{(x+a)^n}$

It may be the case that the polynomial in the denominator of the integrand cannot be factorised into linear factors. In theory, if not in convenient practice, it can always be factored into linear $(ax + b)$ and quadratic $(ax^2 + bx + c)$ factors. Single quadratic factors are handled in much the same way as linear factors with the difference that, instead of having a single constant in the numerator of the partial fractions term, you have a term $Ax + B$. For example:

$$\frac{2x+3}{(x^2+1)(x+2)} = \frac{Ax+B}{x^2+1} + \frac{C}{x+2}$$

$$\frac{x}{(x^2-x+3)(x+1)^2} = \frac{Ax+b}{x^2-x+3} + \frac{C}{x+1} + \frac{D}{(x+1)^2}$$

This adds greatly to the difficulty of integrating the resulting expansion because we now have to deal with integrals like $\int \frac{Ax+B}{Px^2+qx+r}dx$

These are 'easy', but can be tedious.

One final comment. I really ought to justify my earlier claim about 'comparing coefficients'. The claim was that if two polynomials in x have the same value for all values of x then they must be the same polynomial, i.e. have the same coefficients. This is a fact that is used quite frequently. How do we prove it? The easiest approach is to say that if polynomials $p(x)$ and $q(x)$ have the same values for all values of x then the polynomial $r(x) = p(x) - q(x)$ must be zero for all values of x. So we really have to prove that the only polynomial that is zero for all values of x is the 'zero polynomial' 0 (all coefficients are zero). There are lots of ways to do this. Since this is a calculus course I will suggest the following possible approach. Since $r(x)$ is always zero all its derivatives must always be zero as well. If you now evaluate these derivatives at $x = 0$ you will show that all the coefficients of $r(x)$ must be zero. As a simple example, if $r(x) = ax^2 + bx + c$ then $r'(x) = 2ax + b$ and $r''(x) = 2a$. So $r(0) = c = 0$, $r'(0) = b = 0$ and $r''(0) = 2a$, so $a = 0$. The same process works for any degree.

A final, final comment. I stated earlier that any polynomial can be factorized into a product of linear factors $(ax + b)$ and quadratic factors $(ax^2 + bx + c)$. This is not meant to be obvious. It is, in fact, a very profound theorem known as the *Fundamental Theorem of Algebra* first proved by Carl Gauss at the start of the Nineteenth Century (though the result had been accepted as true long before that). Knowing that such a factorisation is possible is a very different thing to finding the factorisation in a particular case. Indeed, it has been shown that there is unlikely to be any systematic method of doing this for polynomials of high degree (within certain constraints, it is not possible for degree higher than 4). Final exercise: $1 + x^4$ factorises into a product of two quadratic factors in a simple way; find them. When you have found them you can start feeling smug—it took mathematicians a surprisingly long time in the Eighteenth Century to solve this problem.

Chapter 3

The Fundamental Ideas of the Integral Calculus

Among the limiting processes of analysis, there are two processes with an especially important role, not only because they arise in many different connections, but chiefly due to their very close reciprocal relationship. Isolated examples of these two limiting processes, differentiation and integration, have even been considered in classical times; but it is the recognition of their complementary nature and the resulting development of a new and methodical mathematical procedure which marks the beginning of the real systematic differential and integral calculus. The credit of initiating this development belongs equally to the two great geniuses of the Seventeenth Century, Newton and Leibnitz, who, as we know to-day, made their discoveries independently of each other. While Newton, in his investigations, may have succeeded in stating his concepts more clearly, Leibnitz's notation and methods of calculation were more highly developed; even to-day, these formal portions of Leibnitz's work form an indispensable element in the theory.

THE DEFINITE INTEGRAL

We first encounter the integral in the problem of measuring the area of a plane region bounded by curved lines. Then, more refined considerations permit us to separate the notion of integral from the intuitive idea of area and to express it analytically in terms of the notion of number only. We shall find this analytical definition of the integral to be of great significance not only because it alone enables us to attain complete clarity in our concepts, but also because its applications extend far beyond the calculation of areas. We shall begin by considering the matter intuitively.

The integral as an Area: Let there be are given a function $f(x)$, which is continuous and positive in an interval, and two values a and b $(a < b)$ in that interval. We think of the function as being represented by a curve and consider the area of the region which is bounded above by the curve, at the sides by the straight lines $x = a$ and $x = b$ and below by the portion of the x-axis between the points a and b. That there is a definite meaning to speaking of the area of this region is an assumption inspired by intuition, which we state here

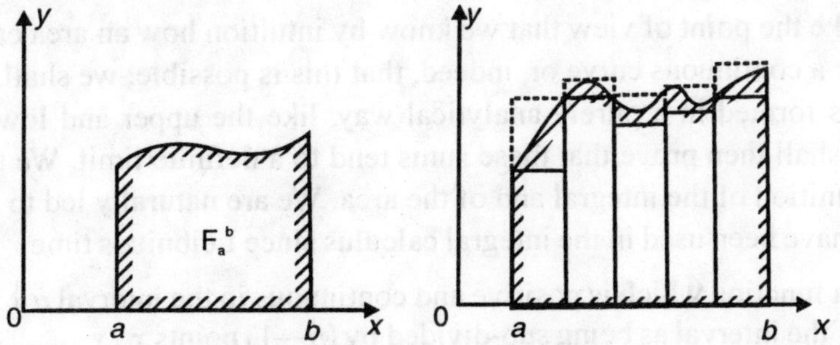

Fig. Upper sum and lower sum

expressly as a hypothesis. We call this area F_a^b the definite integral of the function $f(x)$ between the limits a and b. When we actually seek to assign a numerical value to this area, we find that we are, in general, unable to measure areas with curved boundaries, but we can measure polygons with straight sides by dividing them into rectangles and triangles. Such a sub-division of our area is usually impossible. It is, however, only a short step to conceive in the following manner the area as the limiting value of a sum of areas of rectangles. We subdivide the part of the x-axis between a and b into n equal parts and erect at each point of sub-division the ordinate up to the curve; the area is thus divided into n strips. We can no more calculate the area of such strips than we could that of the original surface; but if, we find first the least and then the greatest value of the function $f(x)$ in each sub-interval and then replace the corresponding strip (1) by a rectangle with height equal to the least value of the function, and by a rectangle with height equal to the greatest value of the function, we obtain two step-shaped figures. The first step-shaped figure obviously has an area which is at most equal to the area F_a^b which we are trying to determine; the second has an area which is at least as large as F_a^b. If we denote the sum of the areas of the first set of rectangles by $\underline{F_n}$ (lower sum) and the sum of the areas of the second set by $\overline{F_n}$ (upper sum), we find

$$\underline{F_n} \leq F_a^b \leq \overline{F}_n$$

If we now make the subdivision finer and finer, i.e., let n increase without limit, intuition tells us that the quantities $\overline{F_n}$ and $\underline{F_n}$ approach closer and closer to each other and tend to the same limit F_a^b. We may therefore consider our integral as the limiting value

$$F_a^b \;=\; \lim_{x\to\infty} \underline{F_n} = \lim_{x\to\infty} \overline{F_n}\,.$$

Intuition also tells us the possibility of an immediate generalization. It is by no means necessary that the n sub-intervals should all be of the same length. On the contrary, they may have different lengths provided only that, as n increases, the length of the longest sub-interval tends to 0.

The Analytical Definition of the Integral: In the above section, we have considered the definite integral as a number given by an area, hence to a certain extent as previously known, and have subsequently represented it as a limiting value. We shall now reverse this procedure.

We no longer take the point of view that we know by intuition how an area can be assigned to the region under a continuous curve or, indeed, that this is possible; we shall, on the contrary, begin with sums formed in a purely analytical way, like the upper and lower sums defined previously, and shall then prove that these sums tend to a definite limit. We take this limiting value as the definition of the integral and of the area. We are naturally led to adopt the formal symbols which have been used in the integral calculus since Leibnitz's time.

Let $f(x)$ be a function which is positive and continuous in the interval $a \leq x \leq b$ (of length $b - a$). We think of the interval as being sub-divided by $(n—1)$ points $x_1, x_2, \ldots\ldots, x_{n-1}$ into n equal or unequal sub-intervals and, in addition, we let $x_0=a$, $x_n=b$. In each interval, we choose a perfectly arbitrary point, which may be within the interval or at either end; suppose that in the first interval we choose the point ξ_1, in the second one the point ξ_2, and in the n-th interval the point ξ_n. Instead of the continuous function $f(x)$, we now consider a discontinuous function (step-function) which has the constant value $f(\xi_1)$ in the first sub-interval, the constant value $f(\xi_2)$ in the second sub-interval, the constant value $f(\xi_n)$ in the n-th sub-interval. As is shown in Fig., the graph of

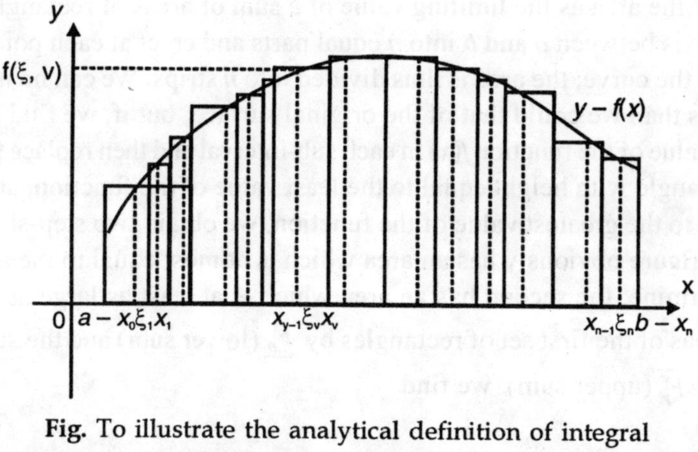

Fig. To illustrate the analytical definition of integral

this step-function defines a series of rectangles, the sum of the areas of which is given by

$$f_n = (x_1 - x_0) f(\xi_1) + (x_2 - x_1) f(\xi_2) + \ldots + (x_n - x_{n-1}) f(\xi_n)$$

This expression is usually abbreviated by means of the summation sign Σ:

$$f_n = \sum_{y=1}^{n} (x_v - x_{v-1}) f(\xi_v);$$

by introducing the symbol

$$\Delta x_v = x_v - x_{v-1}$$

we can simplify this formula to

$$F_n = \sum_{v=1}^{n} f(\xi_v) \Delta x_v$$

(Here the symbol Δ is not a factor, but denotes the word difference. By definition, the entire inseparable symbol Δx_ν means the length of the ν-th sub-interval.) Our basic assertion may now be stated as follows:

If we let the number of points of sub-division increase without limit and at the same time let the length of the longest sub-interval tend to 0, then the above sum tends to a limit. This limit is independent of the particular manner in which the points of division $x_1, x_2, \ldots, x_{n-1}$ and the intermediate point $\xi_1, \xi_2, \ldots, \xi_{n-1}$ are. chosen.

We shall call this limiting value the definite integral of the function $f(x)$, the integrand, between the limits a and b; as we have already mentioned, we shall consider it as the definition* of the area under the curve $y = f(x)$ for $a \;\; x \;\; b$. Our basic assertion may then be re-worded: If $f(x)$ is continuous in $a \} x \} b$, its definite integral between the limits a and b exists.

* Of course, we may also define the notion of are in a purely geometrical way and then prove that such a definition is equivalent to the above limit=definition.

This theorem on the existence of the definite integral of a continuous function can be proved by purely analytical methods and without appeal to intuition. We shall nevertheless pass it over for the present and return to it in A2.1 after the use of the concept of integral has stimulated the reader's interest in constructing for it a firm foundation. For the moment, we shall content ourselves with the fact that the intuitive considerations above have made the theorem appear to be extremely plausible.

Extensions, Notation. Fundamental Rules: The above definition of the integral as the limit of a sum led Leibnitz to express the integral by the symbol

$$\int_a^b f(x)\,dx$$

The integral symbol is a modification of a summation sign which has the shape of a long S. The passage to the limit from a sub-division of the interval into finite portions Δx_ν is suggested by the use of the letter d in place of Δ. However, we must guard ourselves against thinking of dx as an infinitely small quantity or infinitesimal, or of the integral the sum of an infinite number of infinitely small quantities.

In the above figures, we have assumed: that the function $f(x)$ is positive throughout the interval, and that $b > a$. However, the formula which defines the integrai as the limit of a sum is independent of any such assumptions. For if $f(x)$ is negative in all or a part of our interval, the only effect is to make the corresponding factors $f(\xi_\nu)$ in our sum negative instead of positive. We shall naturally assign to the region bounded by the part of the curve below the x-axis a negative area, which is in agreement with the familiar convention of sign in analytical geometry. The total area bounded by a curve will thus, in general, be the sum of positive and negative terms, corresponding to the portions of the curve above and below the x-axis, receptively.

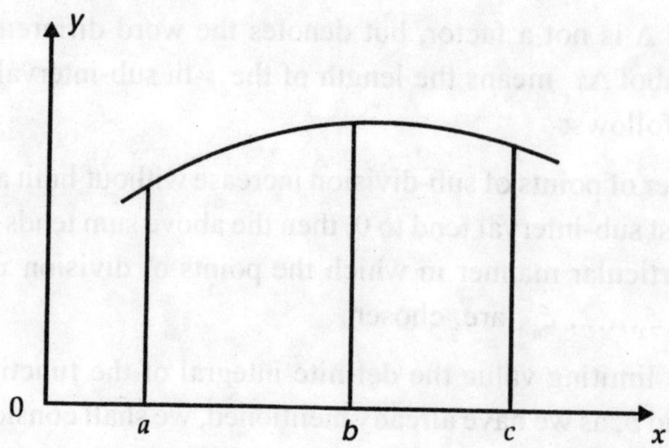

If we also omit the condition $a<b$ and assume that $a>b$, we can still retain our arithmetic definition of integral; the only change is that, if we traverse the interval from a to b, the differences Δx_ν are negative. We are thus led to the relation

$$\int_a^b f(x)dx = -\int_a^b f(x)dx$$

which holds for all values of a and b ($a \gtrless b$). Correspondingly, we define $\int_a^b f(x)dx$ as equal to zero.

Our definition immediately yields the basic relation :

$$\int_a^b f(x)dx + \int_b^c f(x)dx = \int_a^c f(x)dx$$

for $a<b<c$. By means of the preceding relations, we find at once that this equation is also true for any relative positions of the end-points a, b, x

We obtain a simple but important fundamental rule by considering the function $cf(x)$, where c is a constant. From the definition of the integral, we immediately obtain

$$\int_a^b cf(x)dx = c\int_a^b f(x)dx$$

Moreover, we assert the addition rule: If

$$f(x) = \phi(x)+\psi(x)$$

then

$$\int_a^b f(x)dx = \int_a^b \phi(x)dx + \int_a^b \psi(x)dx$$

The proof is quite simple.

We add a final remark about the variable of integration, which is perfectly obvious, but very important in applications. We have written our integral in the form $\int_a^b f(x)dx$. During evaluation of the integral, it does not matter whether we use the letter x or any other letter to denote the abscissae of the co-ordinate system, i.e., the independent variable. We use this particular symbol for the variable of integration and it is therefore a matter of complete indifference; instead of $\int_a^b f(x)dx$, we could equally well write $\int_a^b f(t)dt$ or $\int_a^b f(u)du$ or any similar expression.

Examples:

We are now in a position to carry out the limiting process prescribed by our definition of the integral and thus actually calculate the area in question in a number of special cases; we shall do this in a series of examples, where (except in No. 5) we shall employ only the upper or the lower sum.*

Integration of a Linear Function:

We first consider the function $f(x) = x^n$, where n is an integer greater than or equal to 0. For n = 0, i.e. for $f(x) = 1$, the result is so obvious that we simply write it down:

$$\int_a^b 1dx = \int_a^b dx = b - a$$

For the function $f(x) = x$, the integration is again a triviality from the geometrical point of view. Its integral

$$\int_a^b xdx$$

is simply the area of the trapezoid which by an elementary formula is

$$\frac{1}{2}(b-a)(b+a) = \frac{1}{2}\left(b^2 - a^2\right)$$

We shall now verify that our limiting process leads to exactly the same result. In calculating the limit, we can restrict ourselves to the discussion of upper sums or lower sums. We subdivide the interval from a to b into n equal parts by means of the points of sub-division

$$a+h, a+2h, \ldots, a+(n-1)h,$$

where $h = (b - a)/n$. The integral must then be the limit of the following sum, which is an upper sum, if $b < a$ and a lower sum if $b > a$:

$$h\{a+(a+h)+(a+2h)+\ldots+(a+(n-1)h)\} = h\{na+h+2h+\ldots+(n-1)h\}$$

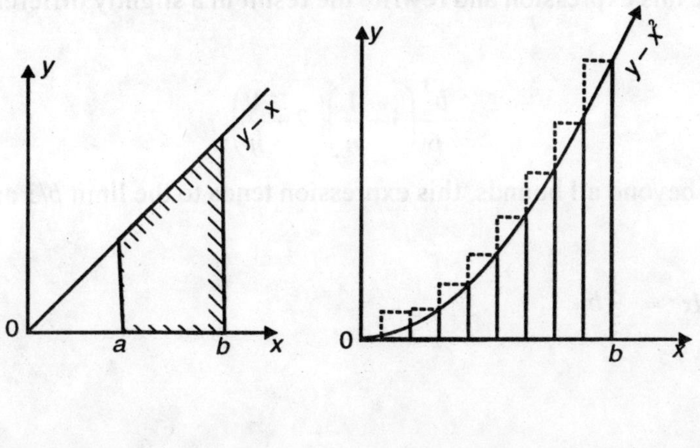

By an elementary formula, we have

$$1+2+...+(n-1)=\frac{1}{2}n(n-1)$$

and our expression may therefore be rewritten in the form

$$nh\left(a+h\frac{n-1}{2}\right) = (b-a)\left(a+\frac{b-a}{2}\frac{n-1}{n}\right)$$

As n increases, the right hand side obviously tends to the limit

$$(b-a)\left\{a+\frac{1}{2}(b-a)\right\}=\frac{1}{2}\left(b^2-a^2\right)$$

as was to be proved.

Integration of x:

A problem which is not quite as simple as that of the integration of the function $f(x) = x$, or, in geometrical language, of determining the area of the region bounded by a segment of a parabola, a segment of the x-axis and two ordinates. For example, consider the integral

$$\int_0^b x^2 dx$$

where b } 0 and sub-divide the interval 0 } x } b into n equal parts of length $h = b/n$; the area which we then wish to find is the limit of the expression (upper sum):

$$h\left(h^2+2^2h^2+3^2h^2+...+n^2h^2\right) = h^3\left(1^2+2^2+...+n^2\right)$$

$$= b^3\left(1^2+2^2+...+n^2\right)/n^3$$

However, the sum in brackets has already been found.

$$1^2+2^2+...+n^2 = \frac{1}{6}n(n+1)(2n+1)$$

If we substitute this expression and rewrite the result in a slightly different form, our sum becomes

$$\frac{b^3}{6}\left(1+\frac{1}{n}\right)\left(2+\frac{1}{n}\right)$$

As n increases beyond all bounds, this expression tends to the limit $b/3$ and we obtain the required integral

$$\int_0^b x^2 dx = \frac{1}{3}b^3$$

Using the general result above, we immediately derive the formula

$$\int_0^b x^2 dx \;=\; \int_0^b x^2 dx - \int_0^a x^2 dx = \frac{1}{3}\left(b^2 - a^3\right)$$

Integration of x^α, where α is any positive integer: As a third example, we consider the integration of the function

$$y \;=\; f(x) = x^n$$

where α is any positive integer. For the computation of the integral

$$\int_a^b x^n dx,$$

(where we assume $0 < \alpha < b$), it would be inconvenient to subdivide the interval into n equal parts.* However, the passage to the limit may be accomplished very easily, if we subdivide in geometric progression in the following manner. Let $\sqrt[n]{b/a} = q$ and subdivide the interval by the points

$$a, aq, aq^2, \ldots, aq^{n-1}, aq^n = b$$

* We should then be obliged to base the evaluation of the integral upon the calculation of the limit of

$$\frac{1}{n^{\alpha+1}}\left(1^\alpha + 2^\alpha + \ldots + n^\alpha\right)$$

as $n \to \infty$; the reader may work this result out as has been indicated in the footnote referred to above.

The required integral is then the limit of the sum:

$$a^\alpha (aq - a) + (aq)^\alpha \left(aq^2 - aq\right) + \left(aq^2\right)^\alpha \left(aq^3 - aq^2\right) + \ldots + \left(aq^{n-1}\right)^\alpha \left(aq^n - aq^{n-1}\right)$$

$$= a^{\alpha+1}(q-1)\left\{1 + q^{\alpha+1} + q^{2(\alpha+1)} + q^{3(\alpha+1)} + \ldots + q^{(n-1)(\alpha+1)}\right\}$$

The terms in the last bracket form a geometric progression with the common ratio $q^{\alpha+1} \neq 1$. If we sum this progression, we obtain for the entire expression the value

$$a^{\alpha+1}(q-1)\frac{q^{n(\alpha+1)} - 1}{q^{\alpha+1} - 1}$$

We now replace q by its value $(b/a)^{1/n}$; our sum then takes the form

$$\left(b^{\alpha+1} - a^{\alpha+1}\right)\frac{q-1}{q^{\alpha+1} - 1}$$

If we now let n increase without limit, the first factor retains its value. Since $q \neq 1$, we can use the formula for the sum of a geometric progression and write the second factor in the form

$$\frac{1}{q^{\alpha}+q^{\alpha-1}+\ldots+1};$$

as the equation $q = (b/a)^{1/n}$ shows that q tends to 1 as $n \to \infty$, the second factor will have the limit $1/(\alpha+1)$. Thus, finally, the value of our integral is given by

$$\int_a^b x^a\,dx = \frac{1}{\alpha+1}\left(b^{\alpha+1}-a^{\alpha+1}\right)$$

In principle,the above calculation is simple, but itsdetails are somewhat complicated. We shall later on discover that it can be entirely avoided once we have become better acquainted with integration theory.

Integration of x^{α}, where α is any Rational Number other than -1:

The result obtained above may be generalized considerably without essential complication of the method. Let $\alpha = r/s$ be a positive rational number, r and s positive integers; in the evaluation of the integral given above, nothing is changed except the evaluation of the limit $(q-1)/(q^{n+1}-1)$ as q approaches 1. This expression is now simply $(q-1)/(q^{(r+s)/s}-1)$. Let us set $q^{1/s} = \tau$ $(\tau \neq 1)$; then, as q tends to 1, τ will also tend to 1. We have therefore to find the limiting value of $(\tau^{r}-1)/(\tau^{r+s}-1)$ as τ approaches 1. If we divide both the numerator and denominator by $\tau - 1$ and transform them as before, the limit simply becomes

$$\lim_{\tau\to 1}\frac{\tau^{3-1}+\tau^{3-2}+\ldots+1}{\tau^{\tau+3-1}+\tau^{\tau+3-2}+\ldots+1}$$

Since both the numerator and denominator are continuous in τ, this limit is at once determined it we set $\tau = 1$. We thus arrive at the limit $s/(r+s) = 1/(\alpha+1)$ and obtain for every positive rational value of α

$$\int_a^b x^a\,dx = \frac{1}{\alpha+1}\left(b^{\alpha+1}-a^{\alpha+1}\right)$$

This formula remains valid for negative rational values of α, provided we exclude the value $\alpha = -1$ for which the formula used above for the sum of the geometric progression loses its meaning. We must now investigate the limit of the expression $(q-1)/(q^{n+1}-1)$ for negative values of α, say $\alpha = -r/s$. For this purpose, we set $q^{-1/s} = \tau$ and obtain

$$q = \tau^{-s}, q^{\alpha+1} = q^{-(r-s)/s} = \tau^{r-s}$$

Hence we seek to determine the limiting value of

$$\frac{\tau^{-s}-1}{\tau^{r-s}-1} = \frac{1-\tau^{s}}{\tau^{r}-\tau^{s}}$$

We leave it to the reader to prove that this limit is again equal to $1/(\alpha+1)$, that is, that we have the integral

$$\int_a^b x^a dx = \frac{1}{\alpha+1}\left(b^{\alpha+1} - a^{a+1}\right)$$

for the general case of rational values of either positive or negative α, with the exception of $\alpha = -1$.

The form of the right-hand side of this equation shows that the expression is not valid for $\alpha = -1$, since both the numerator and denominator would then be zero.

It is natural to assume that the range of validity of our last formula extends also to irrational values of α. We shall actually establish this in equation by a simple passage to the limit.

Integration of sin x and cos x: As a last example, consider the function $f(x) = \sin x$. Also in this case, we shall employ a special device. We express the integral

$$\int_a^b \sin x dx$$

as the limit of the sum:

$$S_n = h\{\sin(\alpha+h) + \sin(\alpha+2h) + \ldots + \sin(\alpha+nh)\}$$

where $h = (b-a)/n$. We multiply the right-hand bracket by 2 sin $h/2$ and recall the well-known trigonometrical formula

$$2\sin u \sin v = \cos(u-v) - \cos(u+v);$$

provided h is not a multiple of 2π, we obtain

$$\frac{h}{2\sin\frac{h}{2}}\left\{\begin{matrix}\cos\left(\alpha+\frac{h}{2}\right) - \cos\left(\alpha+\frac{3}{2}h\right) + \cos\left(\alpha+\frac{3}{2}h\right) - \cos\left(\alpha+\frac{5}{2}h\right) \\ + \ldots + \cos\left(\alpha+\frac{2n-1}{2}h\right) - \cos\left(\alpha+\frac{2n+1}{2}h\right)\end{matrix}\right\}$$

$$= \frac{h}{2\sin\frac{h}{2}}\left\{\cos\left(\alpha+\frac{h}{2}\right) - \cos\left(\alpha+\frac{2n+1}{2}h\right)\right\}$$

Since $a + nh = b$, the integral becomes the limit of

$$\frac{h}{2\sin\frac{h}{2}}\left\{\cos\left(\alpha+\frac{h}{2}\right) - \cos\left(b+\frac{h}{2}\right)\right\} \text{as } h \to 0$$

Now we know from that as h tends to 0, the expression $h/2/\sin h/2$ approaches the limit 1. The desired limit is then simply cos a - cos b and we thus arrive at the integral

$$\int_a^b \sin x dx = -(\cos b - \cos a)$$

In the same manner, the reader may verify the formula

$$\int_a^b \cos x dx = \sin b - \sin \alpha$$

Almost every one of these examples has been attacked by means of some special method or particular device. The essential point of the systematic Integral and Differential Calculus is the very fact that we utilize instead of such special devices considerations of a general character which lead us directly to the desired result. In order to arrive at these methods, we must now turn our attention to the other fundamental concept of higher analysis, the derivative.

The Derivative

The concept of the derivative, like that of the integral, has an intuitive origin. Its sources are:

1. The problem of construction of the tangent to a given curve at a given point and
2. The problem of finding a precise definition for the velocity during an arbitrary motion.

The Derivative and the Tangent: We shall first deal with the tangent problem. If P is a point on a given curve, we shall, in conformity with naive intuition, define the tangent to the curve at the point P by means of the following geometrical limiting process. In addition to the point P, we consider a second point P_1 on the curve. Through the two points P and P_1, we draw a straight line, a secant of the curve. If we now let the point P_1 move along the curve towards the point P, this secant will tend to a limiting position which is independent of the direction from which it approaches P. This limiting position of the secant is the tangent and the statement that such a limiting position of the secant exists is equivalent to the assumption that the curve has a definite tangent or a definite direction at the point P.

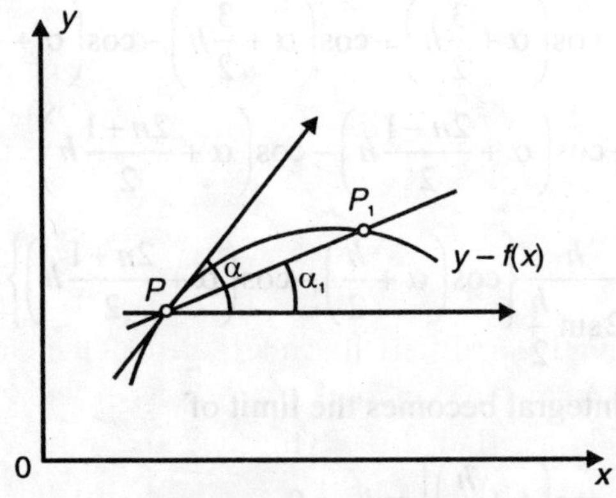

Fig. Chord and Tangent

Once we have represented our curve by means of a function $y = f(x)$, there arises the problem of representing our geometrical limiting process analytically, using the function $f(x)$. We take the angle, which a straight line l makes with the x-axis, as being the angle through which the positive x-axis must be turned in the positive direction* in order to be for the first time parallel to the line l. Let α_1 be the angle which the secant PP_1 forms with the positive x-axis and

α the angle which the tangent forms with the positive x-axis. Then, disregarding the case of a perpendicular tangent, we obviously have

$$\lim_{P_1 \to P} \alpha_1 = \alpha$$

where the meaning of the symbols is perfectly clear. If x, y ($= f(x)$) and x_1, y_1 ($= f(x_1)$) are the coordinates of the points P and P_1, respectively, we immediately find**

$$\tan \alpha_1 = \frac{y_1 - y}{x_1 - x} = \frac{f(x_1) - f(x)}{x_1 - x}$$

and thus our limiting process is represented by the equation

$$\lim_{x_1 \to x} \frac{f(x_1) - f(x)}{x_1 - x} = \tan \alpha$$

That is, in such a direction that a rotation of $\pi/2$ brings it into coincidence with the positive y-axis, in other words, counter-clockwise.

In order that this equation may have a meaning, we must assume that $0 < |x - x_1| < \delta$, δ being sufficiently small. In what follows, corresponding assumptions will often be made tacitly in the steps leading to limiting processes.

The Expression

$$\frac{f(x_1) - f(x)}{x_1 - x} = \frac{y_1 - y}{x_1 - x} = \frac{\Delta y}{\Delta x}$$

is called the difference quotient of the function $y = f(x)$, since the symbols Δy and Δx denote the differences of the function $y = f(x)$ and of the independent variable x, respectively. (As in 2.1.2, the symbol Δ is an abbreviation for the difference and is not a factor!) The tangent of α, the direction angle of the curve,* is therefore equal to the limit to which the difference quotient of our function tends when x_1 tends to x.

The slope or gradient of the curve is given by $\tan \alpha$, whence also the term gradient is used for the derivative of the function represented by the curve.

We call this limit the derivative or differential coefficient of the function $y = f(x)$ at the point x and, as Lagrange did, denote it by the symbol $y' = f'(x)$ or, as Leibnitz did, the symbol dy/dx or $df(x)/dx$ or $d/dx\, f(x)$.* On p. 100, we shall discuss the meaning of Leibnitz's notation in greater detail; here we just point out that the notation $f'(x)$ expresses the fact that the derivative is itself a function of x, since it has a definite value for each value of x in the interval under consideration. This fact is sometimes emphasized by the use of the terms derived function, derived curve. We again quote the definition of the derivative:

$$f'(x) = \lim_{x_1 \to x} \frac{f(x_1) - f(x)}{x_1 - x},$$

or

$$\frac{dy}{dx} = \frac{df(x)}{dx} = f'(x) = \lim_{x_1 \to x} \frac{f(x_1) - f(x)}{x_1 - x} = \lim_{\Delta x \to 0} \frac{\Delta y}{\Delta x}$$

$$= \lim_{h \to 0} \frac{f(x+h) - f(x)}{h},$$

where we have replaced in the last expression x_1 by $x + h$.

*Cauchy's notation $Df(x)$ if also occasionally found in the literature.

It is impossible to find the derivative merely by putting $x_1 = x$ in the expression for the difference quotient, for then the numerator and denominator would both be equal to 0 and we should then led to the meaningless expression 0/0. On the contrary, the actual performance of the passage to the limit in each individual case depends on certain preliminary steps (transformation of the difference quotient).

For example, for the function $f(x) = x$, we have

$$\frac{f(x_1) - f(x)}{x_1 - x} = \frac{x_1^2 - x^2}{x_1 - x} = x_1 + x.$$

This function $x_1 + x$ is not the same function as $\frac{x_1^2 - x^2}{x_1 - x}$, because the function $x_1 + x$ is defined at one point where the quotient $\frac{x_1^2 - x^2}{x_1 - x}$ is not defined, namely, at the point $x_1 = x$. For all other values of x_1, the two functions are equal to each other, whence in the above passage to the limit, in which we specifically required that $x_1 \neq x$, we obtain the same value $\lim_{x_1 \to x} \frac{x_1^2 - x^2}{x_1 - x}$ as for $\lim_{x_1 \to x} (x_1 + x)$ However, since the function $x_1 + x$ is defined and continuous at the point $x_1 = x$, we can do with it what we could not do with the quotient, namely, pass to the limit by simply putting $x_1 = x$. We then obtain for the derivative the expression

$$f'(x) = \frac{d(x^2)}{dx} = 2x$$

The performance of such a process, i.e., the actual formation of the derivative is called the differentiation of the function $f(x)$. We shall see later on how this process of differentiation can actually be carried out in all important cases.

Now, the fact that the problem of differentiating a given function has a definite meaning apart from the geometrical intuition of the tangent is of great significance. The reader will recall that, in the case of the integral, we freed ourselves from the geometrical intuition of area and, on the contrary, based the notion of area on the definition of the integral. Now, independently of the geometrical representation of a function $y = f(x)$ by means of a curve, we shall define the derivative of the function $y = f(x)$ as being the new function $y' = f'(x)$, given by the equation above, provided always that the limit of the difference quotient exists. If this limit exists, we say that the function $f(x)$ is differentiable. From now on, we shall assume that every function dealt under consideration is differentiable unless specific mention is made to the contrary.* It should be observed that, if the function $f(x)$ is to be differentiable at the point x, the limit as $h \to 0$ of the quotient $[f(x + h) - f(x)]/h$ must exist independently of the manner in which h tends to 0, whether through positive or through negative values or without restriction as to the sign.

Examples of cases in which this assumption is not satisfied will be given later (2.3.5).

Once we have found the derivative $f'(x)$, we take the direction which makes an angle α with the positive x-axis given by the equation $\tan \alpha = f'(x)$ as the direction of the tangent to the curve at the point (x, y). We thus avoid the difficulties which arise out of the indefiniteness of the geometrical view, since we base the geometrical definition on the analytical one and not *vice versa*.

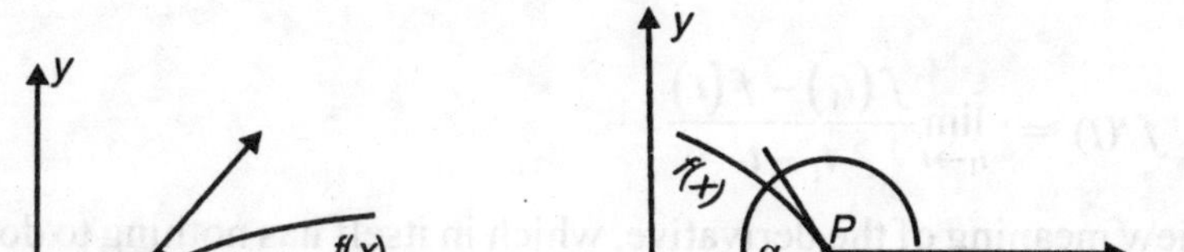

Nevertheless, the visualization of the derivative as the tangent to the curve is an important aid to understanding, even in purely analytical discussions. Accordingly, we shall at once accept the statement based on geometrical intuition: If $f'(x)$ is positive and the curve is traversed in the direction of increasing x, then the tangent slants upwards, and therefore at the point in question the curve rises as x increases; on the other hand, if $f'(x)$ is negative, the tangent slants downwards and the curve falls as x increases. Analytically, this follows from the remark that the limit of $[f(x + h) - f(x)]/h$ cannot be positive unless the function is increasing at the point x, by which we mean that for all values of h sufficiently close to 0 the value of $f(x + h)$ is greater or smaller than $f(x)$ according to whether h is positive or negative. We can, of course, make a corresponding statement for the case when $f'(x)$ is negative.

The Derivative as a Velocity: Just as naive intuition led us to the notion of the direction of the tangent to a curve, so it causes us to assign a velocity to a motion. The definition of velocity leads us once again to exactly the same limiting process which we have already called differentiation.

For example, consider the motion of a point along a straight line, the position of the point being determined by a single co-ordinate y, which is the distance, with its proper sign, of our moving point from a fixed point on the line. The motion is given, if we know y as a function of the time t, $y = f(t)$. If this function is a linear function $f(t) = ct + b$, we speak of uniform motion with the velocity c, and for every pair of values t and t_1, which are not equal to each other, we can write

$$c = \frac{f(t_1) - f(t)}{t_1 - t}$$

The velocity is therefore the difference quotient of the function $ct + b$, and this difference quotient is completely independent of the particular pair of instants considered. But what are we to understand by the velocity of motion at an instant t if the motion is no longer uniform?

In order to arrive at this definition, we consider the difference quotient $\frac{f(t_1) - f(t)}{t_1 - t}$ which we shall call the average velocity in the time interval between t_1 and t. If now this average velocity tends to a definite limit when we let the instant t_1 come closer and closer to t, we shall naturally define this limit as the velocity at the time t. In other words: The velocity at the time t is the derivative

$$f'(t) = \lim_{t_1 \to t} \frac{f(t_1) - f(t)}{t_1 - t}$$

From this new meaning of the derivative, which in itself has nothing to do with the tangent problem, we see that it really is appropriate to define the limiting process of differentiation as a purely analytical operation independently of geometrical intuitions. Here again, the differentiability of the position-function is an assumption which we shall always make tacitly and which, in fact, is absolutely necessary if the notion of velocity is to have any meaning.

As a simple example of the connection between motion and velocity, consider the case of a freely falling body. We begin with the experimentally established law that the distance traversed in time t by a freely falling body is proportional to t and therefore can be represented by a function of the form

$$y = f(t) = at^2$$

As before, we find immediately that the velocity is given by the expression $f'(t) = 2at$, which shows that the velocity of a freely falling body increases proportionally to the time.

Examples: We now proceed to work out a number of examples of the actual differentiation of functions and begin with the function $y = f(x)$, where c is a constant. It is then always true that $f(x + h) - f(x) = c - c = 0$, so that $\lim_{t_1 \to 0} \frac{f(x+h) - f(x)}{h} = 0$ that is, the derivative of a constant is zero.

For a linear function $y = f(x) = cx + b$, we find that

$$\lim_{t_1 \to 0} \frac{f(x+h) - f(x)}{h} = \lim_{h \to 0} \frac{ch}{h} = c$$

Moreover, we shall differentiate the function

$$y = f(x) = x^{\alpha},$$

at first assuming that α is a positive integer. Provided x_1 x, we have

$$\frac{f(x_1) - f(x)}{x_1 - x} = \frac{x_1^{\alpha} - x^{\alpha}}{x_1 - x};$$

the right-hand side of this equation is equal to $x_1^{\alpha-1} + x_1^{\alpha-2}x + \ldots + x^{\alpha-1}$, as we see either by direct division or by using the formula for the sum of a geometric progression. The new expression for the right-hand side of the equation is a continuous function, whence we can carry out the passage to the limit ($x_1 \rangle x$) by simply replacing x_1 everywhere by x. Each term is then $x^{\alpha-1}$ and since the number of terms is exactly α, we obtain

$$y' = f'(x) = \frac{\alpha\left(x^{\alpha}\right)}{dx} = \alpha x^{\alpha-1}$$

We arrive at the same result if α is a negative integer $-\beta$; however, we must assume that x is not zero. We then find that

$$\frac{f(x_1) - f(x)}{x_1 - x} = \frac{\frac{1}{x_1^{\beta}} - \frac{1}{x^{\beta}}}{x_1 - x} = -\frac{x^{\beta} - x_1^{\beta}}{x - x_1} \cdot \frac{1}{x^{\beta} x_1^{\beta}}$$

$$= -\frac{x^{\beta-1} + x^{3-2} x_1 + \ldots + x_1^{\beta-1}}{x_1^{\beta} x^{\beta}}$$

Once again, we can carry out the passage to the limit simply by replacing x_1] everywhere by x. Then, just as above, we obtain for the limit the expression

$$y' = -\beta \frac{x^{\beta-1}}{x^{2\beta}} = -\beta x^{-\beta-1}$$

Hence, for negative integral values of α, the derivative is again given by

$$y' = \alpha x^{\alpha-1}$$

Finally, we shall prove the same formula, where x is positive and α any rational number. We let $\alpha = p/q$, where both p and q are integers and, moreover, positive. (If one of them were negative, no essential changes in the proof would be required; for $\alpha = 0$, the result is already known, since x^{α} is then constant.) We now have

$$\frac{f(x_1) - f(x)}{x_1 - x} = \frac{x_1^{p/q} - x^{p/q}}{x_1 - x}$$

If we now let $x^{1/q} = \xi$ and $x_1^{1/q} = \xi_1$, we obtain

$$\frac{f(x_1)-f(x)}{x_1-x} = \frac{\xi_1^p-\xi^p}{\xi_1^q-\xi^q} = \frac{\xi_1^{p-1}+\xi_1^{p-2}\xi+\ldots+\xi^{p-1}}{\xi_1^{q-1}+\xi_1^{q-2}\xi+\ldots+\xi^{q-1}}$$

After this last transformation, we can immediately perform the passage to the limit ($x_1 \to x$ or, what amounts to the same thing. $\xi_1 \to \xi$), and thus obtain for the limiting value

$$y = \frac{p}{q}\frac{\xi_1^{p-1}}{\xi_1^{q-1}} = \frac{p}{q}\xi^{p-q} = \frac{p}{q}x^{(p-q)/q} = \frac{p}{q}x^{p/q-1}$$

which is formally the same result as before. We leave it to the reader to prove that the same differentiation formula holds also for negative rational superscripts. Once we have developed the theory in a more connected form, we shall return in 2.7.2 to the differentiation of powers.

Finally, as a further example, we consider the differentiation of the trigonometric functions: sin *x* and cos *x*. We use the elementary trigonometrical formula

$$\frac{\sin(x+h)-\sin x}{h} = \frac{\sin x \cos h + \cos x \sin h - \sin x}{h}$$

$$= \sin x\frac{\cos h-1}{h} + \cos x\frac{\sin h}{h}$$

Now, by 1.7, we know that

$$\lim_{h\to 0}\frac{\sin h}{h} = 1,\ \lim_{h\to 0}\frac{\cos h-1}{h} = 0$$

Thus, we find for the required derivative immediately

$$y' = \frac{d(\sin x)}{dx} = \cos x$$

The function $y = \cos x$ can be differentiated in exactly the same manner. Starting with

$$\frac{\cos(x+h)-\cos x}{h} = \cos x\frac{\cos h-1}{h} - \sin x\frac{\sin h}{h}$$

and taking the limit as $h \to 0$, we at once obtain the derivative

$$y' = \frac{d(\cos x)}{dx} = -\sin x$$

Some Fundamental Rules of Differentiation: Just as in the case of the integral, certain simple, but fundamental rules for forming the derivative follow immediately from the definition. If $\phi(x) = f(x) + g(x)$, then $\phi'(x) = f'(x) + g'(x)$; again, if $\psi(x) = cf(x)$ (where *c* is a constant), then $\psi'(x) = cf'(x)$. Because

$$\frac{\phi(x+h)-\phi(x)}{h} = \frac{f(x+h)-f(x)}{h} + \frac{g(x+h)-g(x)}{h}$$

and

$$\frac{\psi(x+h)-\psi(x)}{h} = c\frac{f(x+h)-f(x)}{h}$$

our statements follow directly by passage to the limit.

For example, according to these rules, the derivative of the function $\phi(x) = f(x) + ax + b$ (where a and b are constants) is given by the equation

$\phi'(x) = f'(x) + a.$

Differentiability and Continuity of Functions: It is useful to point out that, if we know that a function can be differentiated, we need not give any special proof of its continuity.

If a function is differentiable, then it is necessarily continuous.

In fact, if the difference quotient $\dfrac{f(x+h)-f(x)}{h}$ approaches a definite limit as h tends to zero, the numerator of the fraction, that is, $f(x+h) - f(x)$, must tend to zero with h, and this fact expresses the continuity of the function $f(x)$ at the point x.

However, the converse of this is definitely false; it is not true that every continuous function has a derivative at every point. The simplest example, which disproves this assumption, is the function $f(x) = |x|$, i.e., $f(x) = -x$ for $x > 0$ and $f(x) = x$ for $x > 0$; its graph. At the point $x = 0$, this function is continuous, but has no derivative. The limit of $[f(x+h) - f(x)]/h$ is equal to 1 if h tends to 0 through positive values and equal to - 1 if h tends to zero through negative values; if we do not restrict the sign of h, There does not exist a limit. We say that the function has different right hand and and left-hand derivatives at the point x, where by we mean the limiting values of $[f(x+h) - f(x)]/h$ as h approaches 0 through positive values only and negative values only, respectively. Thus, the differentiability of a function requires not merely that the right-hand and left-hand derivatives exist, but that they are equal. Geometrically, the inequality of the two derivatives means that the curve has a sharp corner.

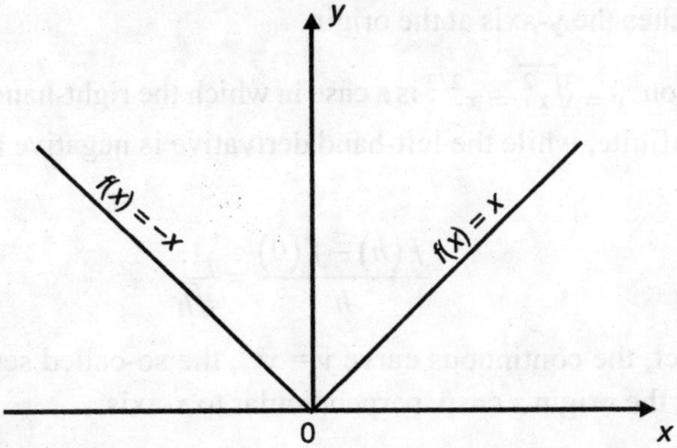

$f(x) = |x|$

As further examples of points where a continuous function is not differentiable, we consider the points where the derivative becomes infinite, i.e., the points at which there exists neither a right-hand nor a left-hand derivative, the difference quotient $[f(x+h) - f(x)]/h$ increasing beyond all bounds as $h \to 0$. For example, the function $y = f(x) = \sqrt[3]{x} = x^{1/3}$ is defined and continuous for all values of x. For all non-zero values of x, its derivative is given by the formula $y' = x^{-2/3}/3$. At the point $x = 0$, we have $[f(x+h) - f(x)]/h = h^{1/3}/h = h^{-2/3}$, and we see at once that, as $h \to 0$, the expression has no limiting value, but tends to ∞.

This state of affairs is often briefly described by saying that the function possesses an infinite derivative or the derivative ∞ at the point in question; however, we should remember that this merely means that, as h tends to 0, the difference quotient increases beyond all bounds and that the derivative in the sense, in which we have defined it, really does not exist. The geometrical meaning of an infinite derivative is that the tangent to the curve is vertical.

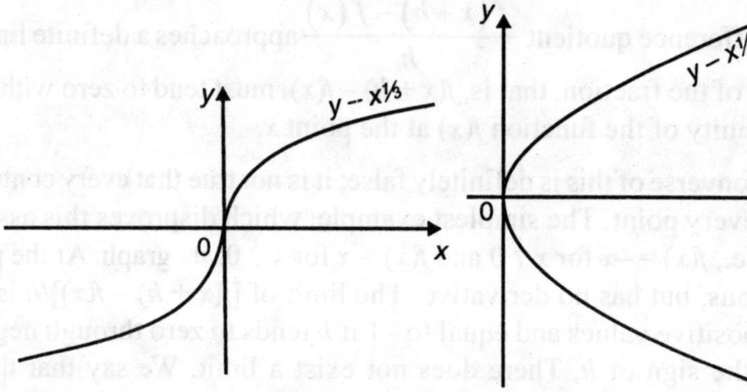

The function $y = f(x) = \sqrt{x}$ which is defined and continuous for $x \to 0$, is also non-differentiable at the point $x = 0$. Since y is undefined for negative values of x, we here consider only the right-hand derivative. The equation $\frac{f(h) - f(0)}{h} = \frac{1}{\sqrt{h}}$ shows us that this derivative is infinite; the curve touches the y-axis at the origin.

Finally, the function $y = \sqrt[3]{x^2} = x^{2/3}$ is a case in which the right-hand derivative at the point $x = 0$ is positive and infinite, while the left-hand derivative is negative and infinite, as follows from the relation

$$\frac{f(h) - f(0)}{h} = \frac{1}{\sqrt[3]{h}}$$

As a matter of fact, the continuous curve $y = x^{2/3}$, the so-called semi-cubical parabola or Neil's parabola, has at the origin a cusp, perpendicular to x-axis.

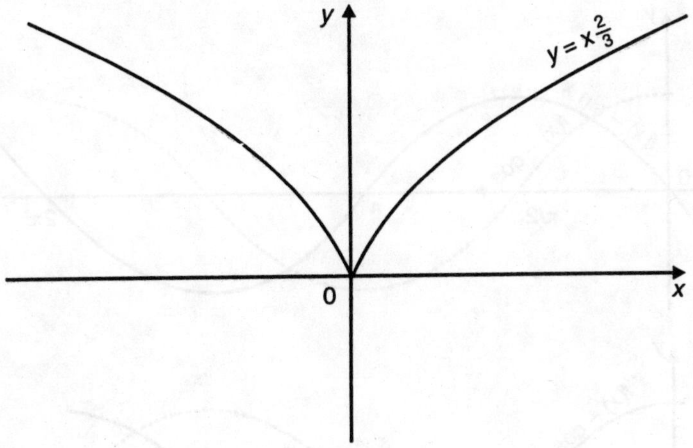

Fig. Cusp

Higher Derivatives and their Significance: The derivative $f'(x)$ of a function is itself a function of x, the graph of which we call the derived curve of the given curve. For example, the derived curve of the parabola $y = x$ is a straight line, represented by the function $y = 2x$. The derivede curve of $y = \sin x$ is $y = \cos x$; similarly, the derived curve of $y = \cos x$ is $y = -\sin x$. (Any of these latter curves can be obtained from the others by translation in the direction of the x-axis.

It is now quite a natural step to form the derived curves of the derived curves, i.e., to form the derivative of the function $f'(x) = \phi(x)$. This derivative

$$\phi'(x) = \lim_{h\to 0} \frac{f'(x+h) - f'(x)}{h}$$

provided that it really exists, we shall call the second derivative of the function $f(x)$ and denote it by $f''(x)$.

Similarly, we may attempt to form the derivative of $f''(x)$, the so-called third derivative of $f(x)$, which we then denote by $f'''(x)$. In the case of most functions of importance, there is nothing to stop us from imagining this process repeated as many times as we like and from thus defining the n-th derivative $f^{(n)}(x)$. At times, it will be convenient to call the function $f(x)$ its own zero-th derivative.

The terms second, third, n-th differential coefficient are also employed.

If the independent variable is interpreted as the time t and the motion of a point is represented by means of the function $f(t)$, the physical meaning of the second derivative is found to be the velocity with which the velocity $f'(t)$ changes, or, as it is usually called, the acceleration. Later on, we shall discuss in detail the geometrical interpretation of the second derivative . However, we may note here the following facts: At a point where $f''(x)$ is positive, $f'(x)$ increases with x; on the other hand, if $f''(x)$ is negative, $f'(x)$ decreases with x.

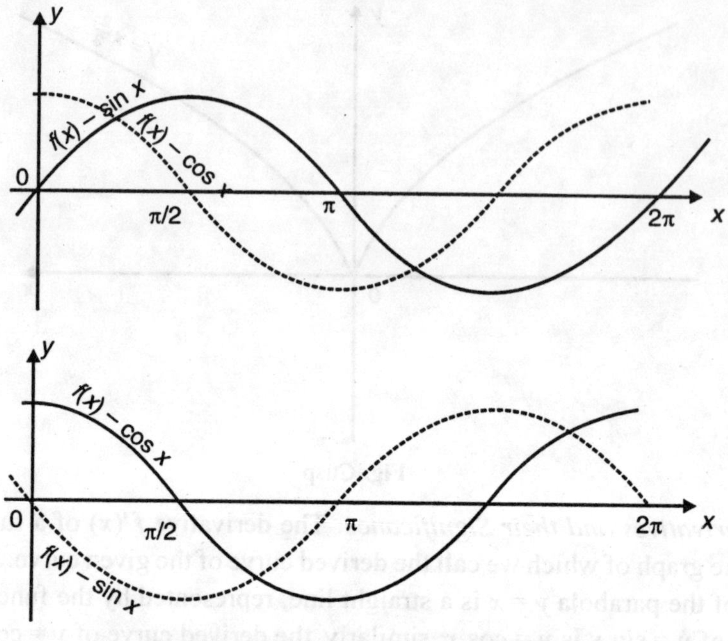

Fig. Derived curves of $\sin x$ and $\cos x$

The Derivative and the Difference Quotient: The fact that in the limiting process, which defines the derivative, the difference Δx tends to 0 is sometimes expressed by saying that the quantity Δx becomes infinitely small. This expression indicates that the passage to the limit is regarded as a process during which the quantity Δx is never zero, yet approaches zero as closely as we please.

In Leibnitz's notation, the passage to the limit in the process of differentiation is symbolically expressed by replacing the symbol Δ by the symbol *d*, so that we can define Leibnitz's symbol for the derivative by the equation

$$\frac{dy}{dx} = \lim_{\Delta x \to 0} \frac{\Delta y}{\Delta x}$$

However, if we wish to obtain a clear grasp of the meaning of the differential calculus, we must beware of regarding the derivative as the quotient of two quantities which are actually infinitely small. The difference quotient $\Delta y/\Delta x$ definitely must be formed with differences Δx which are not equal to 0. After forming this difference quotient, we must imagine the passage to the limit carried out by means of a transformation or some other device. We have no right to assume that first Δx goes through something like a limiting process and reaches a value which is infinitesimally small but still not 0, so that Δx and Δy are replaced by the infinitely small quantities or infinitesimals dx and dy, and the quotient of these quantities is then formed. Such a conception of the derivative is incompatible with the clarity of ideas demanded in mathematics; in fact, it is altogether meaningless. For a great many simple-minded people, it undoubtedly has a certain charm - the charm of mystery - which is always associated with the word infinite; and in the early days of the differential calculus, even Leibnitz himself was capable of combining these vague mystical ideas with a thoroughly clear understanding of the limiting process. It is true that this fog, which surrounded the foundations of the new science, did not prevent Leibnitz or his great successors from finding the right path. But this does not release us from the duty of avoiding every such hazy idea in our building-up of the differential and integral calculus.

However, the notation of Leibnitz is not merely attractive in itself, but is actually of great flexibility and the greatest use. The reason is that we can deal in many calculations and formal transformations with the symbols *dy* and *dx* in exactly the same way as if they were ordinary numbers. They enable us to give neater expression to many calculations which can be carried out without their use. In the following pages, we shall see this fact verified over and over again and shall find ourselves justified in making free and repeated use of it, provided we do not lose sight of the symbolical character of the signs *dy* and *dx*.

Leibnitz has also devised for the second and higher derivatives a notation of great suggestiveness and practical utility. He thinks of the second derivative as the limit of the second difference quotient in the following manner. In addition to the variable x, we consider $x_1 = x + h$ and $x_2 = x + 2h$. We then take the second difference quotient as meaning the first difference quotient of the first difference quotient, i.e., the expression

$$\frac{1}{h}\left(\frac{y_2 - y_1}{h} - \frac{y_1 - y}{h}\right) = \frac{1}{h^2}\left(y^2 - 2y_1 + y\right)$$

where $y = f(x)$, $y_1 = f(x_1)$ and $y_2 = f(x_2)$. If we also write $h = \Delta x$ and $y_2 - y_1 = \Delta y_1$, $y_1 - y = \Delta y$, we may appropriately call the expression in the last bracket the difference of the difference of y or the second difference of y and write symbolically *

$$y_1 - 2y_1 + y = \Delta y_1 - \Delta y_1 - \Delta y = \Delta(\Delta y) = \Delta^2 y$$

* Here $\Delta\Delta = \Delta$ is not a square, but merely a symbol for difference of difference or second difference.

In this symbolic notation, the second difference quotient is then $\Delta y/(\Delta x)$, where the denominator is really the square of Δx, while in the numerator the number 2 syrnbolically denotes the repetition of the difference process. This symbolism for the difference quotient led Leibnitz to introduce the notation

$$y'' = f''(x) = \frac{d^2 y}{dx^{2'}} y''' = f'''(x) = \frac{d^3 y}{dx^3}, \&c.$$

for the second and higher derivatives, and we shall find that this notation also stands the test of use.

We must emphasize that the statement that the second derivative may be represented as the limit of the second difference quotient requires proof. For we previously defined the second derivative not in this way, but as the limit of the first difference quotient of the first derivative. In actual fact, the two definitions are equivalent, provided the second derivative is continuous; however, the proof is not given, as we have no particular need for it here.

The Mean Value Theorem: There exists between the derivative $dy/dx = f'(x)$ and the difference quotient a simple relation which is important for many purposes is known as the mean value theorem and is obtained in the following manner. We consider the difference quotient

$$\frac{f(x_1) - f(x_2)}{x_1 - x_2} = \frac{\Delta f}{\Delta x}$$

of a function $f(x)$ and assume that the derivative exists everywhere in the interval $x_1 \succ x \succ x_2$ so that the graph of the curve has everywhere a tangent. The difference quotient will be represented by the direction of the secant in fact, it is the tangent of the angle α.

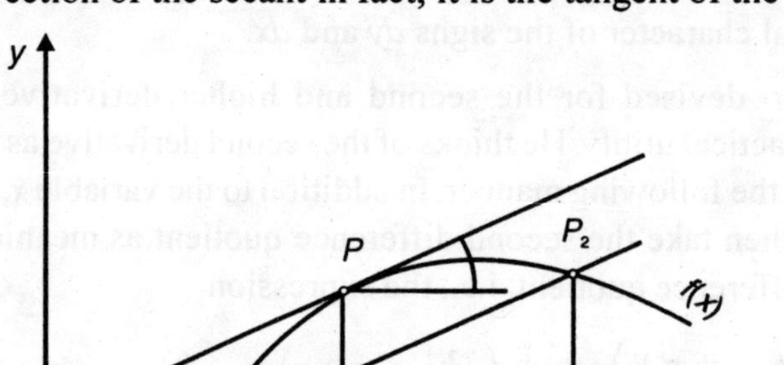

Fig. To illustrate the mean value theorem

Let us imagine that this secant is shifted parallel to itself. At least once, it will reach a position in which it is a tangent to the curve at a point between x_1 and x_2, namely at the point of the curve which is at the greatest distance from the secant. Hence there will be an intermediate value ξ such that

$$\frac{f(x_1) - f(x_2)}{x_1 - x_2} = f'(\xi)$$

This statement is called the mean value theorem of the differential calculus. We can also express it somewhat differently by noticing that the number ξ may be written in the form

$$\xi = x_1 + \theta(x_2 - x_1)$$

where θ is a certain number between 0 and 1. In applications of the mean value theorem, we shall often find that θ cannot be determined more accurately than this, but it will usually turn

out that a more accurate value is not required. When accurately formulated, the mean value theorem is:

If $f(x)$ is continuous in the closed interval $x_1 \leq x \leq x_2$ and differentiable at every point of the open interval $x_1 < s < x_2$, then there is at least one value θ, where $0 < \theta < 1$ such that

$$\frac{f(x_2)-f(x_1)}{x_2-x_1} = f'\{x_1+\theta(x_2-x_1)\}$$

If we replace x_1 by x and x_2 by $x + h$, we can express the mean value theorem by the formula

$$\frac{f(x+h)-f(x)}{h} = f'(\xi) = f'(x+\theta h), x<\xi<x+h$$

We wish to emphasize that, while it is essential that $f(x)$ should be continuous for all points of the interval, including the end-points, we need not assume that the derivative exists at the end-points. This apparently trivial remark is actually useful in many applications.

If at any point in the interior of an interval the derivative fails to exist, the mean value theorem is not necessarily true. This is shown by the example $f(x)=|x|$.

We can complete our intuitive argument by the following consideration. There is at least one point P on the curve which has the greatest possible distance from the chord joining the points on the curve with the abcissae x_1 and x_2 .

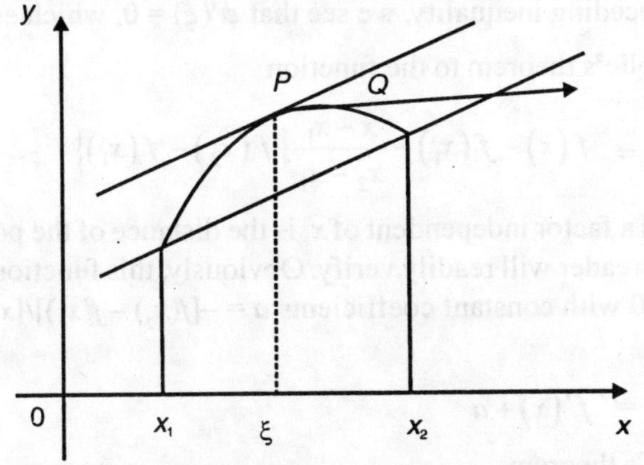

Fig. To illustrate the mean value theorem

At this point, the curve has by assumption a definite tangent. We shall now prove that this tangent must be parallel to the chord. By definition, the tangent is the limiting position of the secant and is obtained by joining P to a point Q on the curve and letting the point Q move towards P. Since, by assumption, Q is not further from the chord than P, the line PQ produced in the direction P to Q must either intersect the chord or run parallel to it; and this must be the case, no matter on which side of P lies the point Q. This, however, is only possible if the limiting

position is parallel to the chord. If we denote the abscissa of the point P by ξ, the slope $f'(\xi)$ of the tangent at P is then equal to the slope of the chord, $[f(x_1)-f(x_2)]/[x_1-x_2)]$, whence we may simply take for the number ξ in the theorem the abscissa of P.

The rigorous proof of the mean value theorem is usually developed as follows: We first establish Rolle's theorem - a special case of the mean value theorem: If a function $\phi(x)$ is continuous in the closed interval $x_1 \leq x \leq x_2$ and differentiable in the open interval $x_1 < x < x_2$, and, moreover, $\phi(x_1) = 0$ and $\phi(x_1) = 0$, then there exists at least one point ξ in the interior of the interval at which $f'(\xi)= 0$.

In fact, there must be at least one point ξ, interior to the interval, at which the function $\phi(x)$ takes on its greatest or its least value; to be specific, we assume that ξ is a point where $\phi(x)$ is a maximum so that for every x in the interval $\phi(x) \leq \phi(\xi)$. Then it is certainly true for every number, the absolute value $|h|$ of which is small enough, that $\phi(\xi)-\phi(\xi+h) \geq 0$ If h is positive,

$$\frac{\phi(\xi+h)-\phi(\xi)}{h} \leq 0$$

we now let h tend to zero through positive values and obtain $\phi'(\xi) \leq 0$. On the other hand, if h is negative,

$$\frac{\phi(\xi+h)-\phi(\xi)}{h} \geq 0$$

and thus, by letting h tend to zero through negative values, we obtain $\phi'(\xi) \geq 0$; comparing this result with the preceding inequality, we see that $\phi'(\xi) = 0$, which establishes our theorem.

We now apply Rolle's theorem to the function

$$\phi(x) \quad = \quad f(x)-f(x_1)-\frac{x-x_1}{x_2-x_1}\{f(x_2)-f(x_1)\}$$

which, apart from a factor independent of x, is the distance of the point $(x, f(x))$ of the curve from the secant, as the reader will readily verify. Obviously, this function satisfies the condition $\phi(x) = f(x) + ax + b = 0$ with constant coefficients $a = -[f(x_2)-f(x_1)]/[x_2-x_1]$ and b. We know already that

$$\phi'(x) \quad = \quad f'(x)+a$$

whence, by Rolle's theorem,

$$0 \quad = \quad \phi'(\xi) = f'(\xi)+a$$

for a suitably chosen intermediate value ξ, and therefore

$$f'(\xi) \quad = \quad -\alpha = \frac{f(x_2)-f(x_1)}{x_2-x_1}$$

and the mean value theorem has been proved.

As the first of many applications of the mean value theorem,we shall prove the following: Let the function $f(x)$ be continuous in the closed interval $a \leq x \leq b$ and have the derivative $f'(x)$ at every point of the open interval $a < x < b$. Then, if $f'(x)$ is positive everywhere in $a < x < b$, the function $f(x)$ is monotonic increasing in the interval $a \leq x \leq b$; and likewise, if $f(x)$ is negative in $a < x < b$, it is monotonic decreasing.

Let $f'(x) > 0$ and $x_1, x_2 > 0$ be any two values of x in the closed interval. Then, by the mean value theorem,

$$f(x_2)-f(x_1) = (x_2-x_1)f'(\xi)$$

where $x_1 < \xi < x_2$; Since both factors on the right hand side are positive, this proves that $f(x_2)>f(x_1)$, whence $f(x)$ is monotonic increasing.

The Approximate Representation of Arbitrary Functions by Linear Functions. Differentiation: The equation

$$\lim_{h\to 0}\frac{f(x+h)-f(x)}{h} = f'(x)$$

which defines the derivative is equivalent to the equations

$$f(x+h)-f(x) = hf'(x)+\in h$$

$$\text{or } y+\Delta y = f(x+\Delta x)=f(x)+f'(x)\Delta x+\in\Delta x$$

where ε is a quantity which tends to zero with $h = \Delta x$. If, for the moment, we think of the point x as being fixed and the increment Δx as being variable, then, by this formula, the increment of the function, that is, the quantity Δy, consists of two terms, namely a part $hf'(x)$, proportional to h and an error which can be made as small as we please relative to h by making h itself small enough. Thus, the smaller is the interval about the point x under consideration, the more accurately is the value of the function $f(x + h)$ (which is a function of h), represented by its linear part $f(x) + hf'(x)$. This approximate representation of the function $f(x + h)$ by a linear function of h is expressed geometrically by the substitution of its tangent for the curve at the point x.

Here we merely remark in passing that it is possible to use this approximate representation of the increment Δy by the linear expression $hf'(x)$ to construct a logically satisfactory definition of the notion of a differential as this was done, in particular, by Cauchy.

While the idea of the differential as an infinitely small quantity has no meaning and it is accordingly futile to define the derivative as the quotient of two such quantities, we may still try to assign a sense to the equation $f'(x) = dy/dx$ in such a way that the expression dy/dx need not be thought of as purely symbolic, but as the actual quotient of two quantities dy and dx. For this purpose, we first define the derivative $f'(x)$ by our limiting process, then think of x as fixed and consider the increment $h = \Delta x$ as the independent variable. We will call this quantity h the differential of x and write $h = dx$. We now define the expression $dy = y'dx$ as the differential of the function y; dy is therefore a number which has nothing to do with infinitely small quantities.

Thus, the derivative $y' = f'(x)$ is now really the quotient of the differentials dy and dx; however, there is nothing remarkable in this statement; in fact, it is merely a restatement of the verbal definition. The differential dy is accordingly the linear part of the increment Δy.

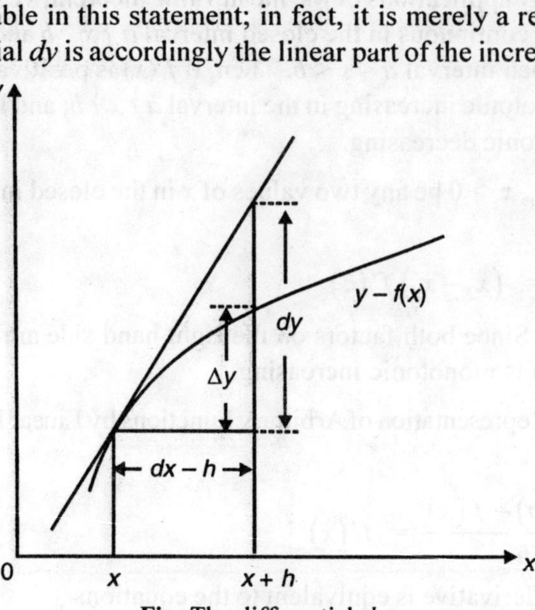

Fig. The differential *dy*

We shall not make any immediate use of these differentials. Nevertheless, it may be pointed out, for the sake of completeness, that we may also form second and higher differentials. In fact, if we think of h as chosen in any manner, but always the same for every value of x, then $dy = hf'(x)$ is a function of x, of which we can again form the differential. The result will be called the second differential of y and be denoted by the symbol $dy = df(x)$. The increment of $hf'(x)$ being $h\{f'(x+h) - f'(x)\}$, the second differential is obtained by replacing the quantity in braces by its linear part $hf''(x)$, so that $dy == hf''(x)$. We may naturally proceed further along the same lines, obtaining third, fourth, differentials of y, etc., which can be defined by the expressions $h^3f'''(x)$, $h^4f^{(4)}(x)$, etc.

Remarks on Applications to the Natural Sciences:. In applications of mathematics to natural phenomena, we never have to deal with sharply defined quantities. Whether a length is exactly a metre is a question which cannot be decided by any experiment and which consequently has no physical meaning. Again, there is no immediate physical meaning in saying that the length of a material rod is rational or irrational; we can always measure it with any desired degree of accuracy in rational numbers, and the real matter of interest is whether or not we can manage to perform such a measurement using rational numbers with relatively small denominators. Just as the question of rationality or irrationality in the rigorous sense of exact mathematics has no physical meaning, so the actual carrying out of limiting processes in applications will usually be nothing more than a mathematical idealization.

The practical significance of such idealizations lies chiefly in the fact that, if they are used, all analytical expressions become essentially simpler and more manageable. For example, it is vastly simpler and more convenient to work with the notion of instantaneous velocity, which is a function of only one definite instant of time, than with the notion of average velocity between two different instants. Without such an idealization, every rational investigation of nature would be condemned to hopeless complications and would break down at the very outset.

However, we do not intend to enter into a discussion of the relationship of mathematics to reality. We merely wish to emphasize, for the sake of our better understanding of the theory, that we have in applications the right to replace a derivative by a difference quotient and *vice versa*, provided only that the differences are small enough to guarantee a sufficiently close approximation. The physicist, the biologist, the engineer, or anyone else who has to deal with these ideas in practice, will therefore have the right to identify the difference quotient with the derivative within his limits of accuracy. The smaller is the increment $h = dx$ of the independent variable, the more accurately can it represent the increment $\Delta y = f(x + h) - f(x)$ by the differential $dy = hf'(x)$. As long as one keeps within the limits of accuracy required by a given problem, one is accustomed to speak of the quantities $dx = h$ and $dy = hf'(x)$ as infinitesimals. These physically infinitesimal quantities have a precise meaning. They are finite quantities, not equal to zero, which are chosen small enough for a given investigation, e.g., smaller than a fractional part of a wave-length or smaller than the distance between two electrons in an atom; in general, they are chosen smaller than the degree of accuracy required.

The Indefinite Integral, the Primitive Function and the Fundamental Theorems of Integral Calculus

As we have already mentioned above, the connection between the problems of integration and of differentiation is the corner-stone of the differential and integral calculus. We will now study this connection.

The Integral as a Function of the Upper Limit: The value of the definite integral of a function $f(x)$ depends on the choice of the two limits of integration a and b; it is a function of the lower limit a as well as of the upper limit b. In order to study this dependence more closely, we imagine the lower limit a to be a definite fixed number, denote the variable of integration no longer by x but by u (Integration of a Linear Function) and the upper limit by x instead of b, in order to suggest that we shall let the upper limit vary and wish to investigate the value of the integral as a function of the upper limit. Accordingly, we write

$$\int_a^x f(u)\,du = \Phi(x)$$

We call this function $\Phi(x)$ an indefinite integral of the function $f(x)$. When we speak of an and not of the indefinite integral, we suggest that instead of the lower limit a any other could be chosen, in which case we should ordinarily obtain a different value for the integral. Geometrically speaking, the indefinite integral for each value of x will be given by the area under the curve $y=f(u)$ and bounded by the ordinates $u = a$ and $u = x$, the sign being determined by the rules given earlier.

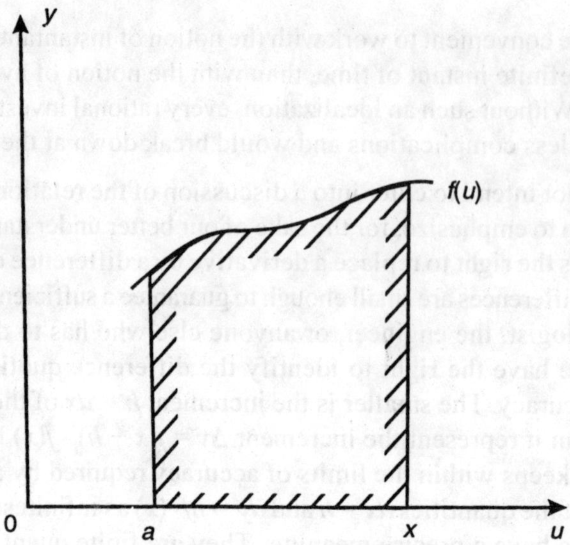

If we choose another lower limit α in place of the lower limit a, we obtain the indefinite integral

$$\psi(x) = \int_a^x f(u)\,du$$

The difference $\Psi(x) - \Phi(x)$ will obviously be

$$\int_\alpha^a f(u)\,du$$

which is a constant, since α and a are each fixed given numbers. Hence

$$\psi(x) = \Phi(x) + \text{const}$$

Different indefinite integrals of the same function differ only by an additive constant.

We may likewise regard the integral as a function of the lower limit and introduce the function

$$\Phi(x) = \int_x^b f(u)\,du$$

in which b is a fixed number. Here again two such integrals with different upper limits b and β differ only by an additive constant

$$\int_b^\beta f(u)\,du$$

The Derivative of the Indefinite Integral: We will now differentiate the indefinite integral $\Phi(x)$ with respect to the variable x. The result is the theorem:

The indefinite integral

$$\Phi(x) = \int_a^x f(u)\,du$$

of a continuous function $f(x)$ always possesses a derivative $\Phi'(x)$ and moreover

$$\Phi'(x) = f(x)$$

that is, differentiation of the indefinite integral of a given continuous function always gives us back that function.

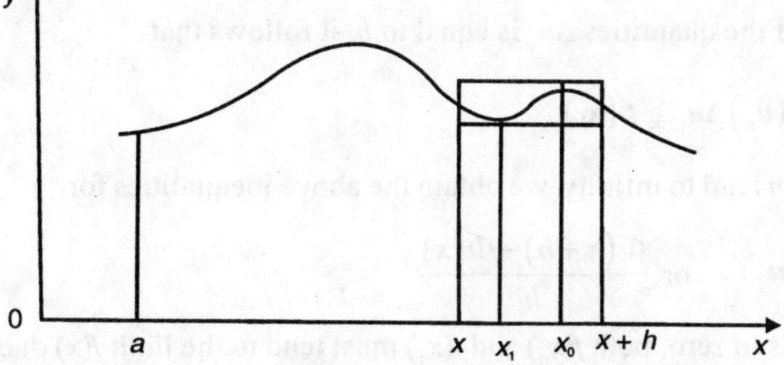

Fig. Differentiation of the indefinite integral

This is the basic idea of all of the differential and integral calculus. The proof follows extremely simply from the interpretation of the integral as an area. We form the difference quotient

$$\frac{\Phi(x+h)-\Phi(x)}{h}$$

and observe that the numerator

$$\Phi(x+h)-\Phi(x) = \int_a^{x+h} f(u)\,du - \int_a^x f(u)\,du = \int_x^{x+h} f(u)\,du$$

is the area between the ordinate corresponding to x and the ordinate corresponding to $x + h$.

Now let x_0 be a point in the interval between x and $x + h$, at which the function $f(x)$ takes its greatest value, and x_1 a point, at which it takes its least value in the interval. Then the area in question will lie between the values $hf(x_0)$ and $hf(x_1)$ which represent the areas of rectangles with the interval from x to $x+h$ as base and the altitudes $f(x_0)$ and $f(x_1)$, respectively. Expressed analytically,

$$f(x_0) \geq \frac{\Phi(x+h)-\Phi(x)}{h} \geq f(x_1)$$

This can also be proved directly from the definition of the integral without appealing to the geometrical interpretation (Compare also with the discussion in 2.7). For this purpose, we write

$$\int_{x}^{x+h} f(u)du = \lim_{n\to\infty} \sum_{n\to\infty}^{n} f(u_v)\Delta u_v$$

where $u_0 = x, u_1, u_2, \ldots, u_n = x + h$ are points of sub-division of the interval from x to $x + h$, and the largest of the absolute values of the differences $\Delta u_v = u_v - u_{v-1}$ tends to zero as n increases. Then $\Delta u_v/h$ is certainly positive, no matter whether h is positive or negative. Since we know that

$$F(x_0) \geqq f(u_v) \geqq f(x_1)$$

and the sum of the quantities Δu_v is equal to h, it follows that

$$F(x_0) \geqq \frac{1}{h}\Sigma f(u_v)\Delta u_v \geqq f(x_1)$$

thus, if we let n tend to infinity, we obtain the above inequalities for

$$\frac{1}{h}\int_{x}^{x+h} f(u)du \quad \text{or} \quad \frac{\Phi(x+h)-\Phi(x)}{h}$$

Now, if h tends to zero, both $f(x_0)$ and $f(x_1)$ must tend to the limit $f(x)$ due to the continuity of the function, whence we see at once that

$$\Phi'(x) = \lim_{h\to 0} \frac{\Phi(x+h)-\Phi(x)}{h} = f(x)$$

as stated by the theorem.

Owing to the differentiability of $\Phi(x)$ and 2.3.5, we have the theorem:

The integral of a continuous function $f(x)$ is itself a continuous function of the upper limit.

For the sake of completeness, we point out that, if we regard the definite integral not as a function of its upper limit but as a function of its lower limit, the derivative is not equal to $f(x)$, but instead equal to $-f(x)$; in symbols, if we set

$$\phi(x) = \int_{x}^{b} f(u)du$$

then

$$\phi'(x) = -f(x)$$

The proof follows immediately from the remark that

$$\int_{x}^{b} f(u)du = -\int_{b}^{x} f(u)du$$

The Primitive Function; General Definition of the Indefinite Integral: The theorem, which we have just proved, shows that the indefinite integral $\Phi(x)$ at once yields the solution of the problem: Given a function $f(x)$, find a function $F(x)$ such that

$$F'(x) = f(x)$$

This problem requires us to reverse the process of differentiation. It is a typical inverse problem such as occurs in many parts of mathematics and such as we have already found to be a fruitful mathematical tool for generating new functions. (For example, the first extension of the idea of natural numbers was made under the pressure of the necessity for reversing certain elementary computational processes. The formation of inverse functions has led and will lead us to new kinds of functions!)

A function $F(x)$ such that $F'(x) = f(x)$ is called a primitive function of $f(x)$, or simply a primitive of $f(x)$; this terminology suggests that the function $f(x)$ arises from $F(x)$ by differentiation.

This problem of the inversion of differentiation or of the determination of a primitive function is at first sight of quite a different character from the problem of integration. However, we know from the preceding section that:

Every indefinite integral $\Phi(x)$ of the function $f(x)$ is a primitive of $f(x)$.

However, this result does not completely solve the problem of finding primitive functions, because we do not yet know whether we have found all the solutions of the problem. The question concerning the group of all primitive functions is answered by the following theorem, sometimes referred to as the fundamental theorem of the differential and integral calculus:

The difference between two primitives $F_1(x)$ and $F_2(x)$ of the same function $f(x)$ is always a constant:

$$F_1(x) - F_2(x) = c$$

Thus, from any one primitive function $F(x)$, we can obtain all the others in the form

$$F(x) + c$$

by a suitable choice of the constant c. Conversely, for every value of the constant c the expression $F_1(x) = F(x) + c$ represents a primitive function of $f(x)$.

It is clear that for any value of the constant c the function $F(x) + c$ is a primitive, provided that $F(x)$ itself is one. From 2.3.4, we have

$$\frac{\{F(x+h)+c\} - \{F(x)+c\}}{h} = \frac{F(x+h) - F(x)}{h}$$

and since, by assumption, the right-hand side tends to $f(x)$ as $h \rightarrow 0$, so does the left-hand side, whence

$$\frac{d}{dx}\{F(x)+c\} = f(x) = F'(x)$$

Thus, in order to complete the proof of the theorem, there only remains to show that the difference of two primitive functions $F_1(x)$ and $F_2(x)$ is always a constant. For this purpose, we consider the difference

$$F_1(x) - F_2(x) = G(x)$$

and form the derivative

$$G'(x) = \lim_{h\to 0}\left\{\frac{F_1(x+h)-F_1(x)}{h} - \frac{F_2(x+h)-F_2(x)}{h}\right\}$$

By assumption, both the expressions on the right-hand side have the same limit $f(x)$ as h } 0; thus,for every value of x, we have $G'(n) = 0$. But a function, the derivative of which is everywhere zero, must have a graph the tangent of which is everywhere parallel to the x-axis, i.e., it must be a constant, whence $G(x) = c$, as has been stated above. We can prove this last fact by using the mean value theorem without relying on tuition. Applying the mean value theorem to $G(x)$, we find

$$G(x_2)-G(x_1)=(x_2-x_1)G'(\xi);x_1<\xi<x_2$$

However, we have seen that the derivative $G'(x)$ is equal to 0 for every value of x, whence, in particular, it is true for the value ξ, and it follows immediately that $G(x_1) = G(x_2)$. Since x_1 and x_2 are arbitrary values of x in the given interval, $G(x)$ must be a constant.

Combining the theorem which has just been proved with the result of 2.4.2, we can now state:

Every primitive function $F(x)$ of a given function $f(x)$ can be represented in the form

$$F(x) = c+\Phi(x)=c+\int_a^x f(u)\,du$$

where c and a are constants, and conversely, for any arbitrarily chosen constant values a and c, this expression always represents a primitive function.

It may readily be guessed that, as a rule, the constant c can be omitted, since, by changing the lower limit a, we change the primitive function by an additive constant. However, in many cases, we should not obtain all the primitive functions, if we were to omit c, as is shown by the example $f(x) = 0$. For this function, the indefinite integral of 2.4.1 is always 0, independently of the lower limit; yet any arbitrary constant is a primitive function of $f(x) = 0$. A second example is the function $f(x)=\sqrt{x}$, which is only defined for non-negative values of x. The indefinite integral is

$$\Phi(x) = \frac{2}{3}x^{3/2} - \frac{2}{3}\alpha^{3/2}$$

and we see that, independently of the choice of the lower limit a, the indefinite integral of $\Phi(x)$ is always obtained from $2x^{3/2}/3$ by addition of a constant which is less than or equal to zero, namely, by the constant $-2a^{3/2}$; yet such a function as $2x^{3/2} + 1$ is also a primitive of . Thus, in the general expression for the primitive function, we cannot dispense with the additive constant. The relationship which we have found suggests an extension of the idea of the indefinite integral. We shall henceforth call every expression of the form

$$c+\Phi(x) = c+\int_a^x f(u)\,du$$

an indefinite integral of $f(x)$. In other words, we shall no longer make any distinction between the primitive function and the indefinite integral. Nevertheless, if the reader is to have a proper understanding of the interrelationships of these concepts, it is absolutely necessary that he should clearly bear in mind that, in the first instance, integration and inversion of differentiation are two entirely different things, and also that it is only the knowledge of the relationship between them which allows us to apply the term indefinite integral also to the primitive function.

It is customary to represent the indefinite integral by a notation which in itself is perhaps not perfectly clear. We write

$$F(x) = c + \int_a^x f(u)du = \int f(x)dx$$

that is, we omit the upper limit x and the lower limit a and also the additive constant c and use the letter x for the integration variable. It would really be more consistent to avoid this last change, in order to prevent confusion with the upper limit x which is the independent variable in $F(x)$. When using the notation $\int f(x)dx$, we must never lose sight of the indeterminacy connected with it, i.e., the fact that that symbol always denotes only an indefinite integral.

The use of the Primitive Function in the Evaluation of Definite Integrals:. Suppose that we know any one primitive function $F(x) = \int f(x)dx$ for the function $f(x)$ and that we wish to evaluate the definite integral $\int_a^b f(u)du$. We know that the indefinite integral

$$\Phi(x) = \int_a^x f(u)du$$

being also a primitive of $f(x)$, can only differ from $F(x)$ by an additive constant. Consequently,

$$\Phi(x) = F(x) + c$$

and the additive constant c is at once determined, if we recollect that the indefinite integral $\Phi(x) = \int_a^x f(u)du$ must take the value 0 when $x = a$. We thus obtain

$$0 = \Phi(a) = F(a) + c$$

whence $c = -F(a)$ and $\Phi(x) = F(x) - F(a)$. In particular, we have for the value $x=b$

$$\int_a^b f(u)du = F(b) - F(a)$$

which yields the important rule:

If $F(x)$ is any primitive of the function $f(x)$, the definite integral of $f(x)$ between the limits a and b is equal to the difference $F(b) - F(a)$.

If we use the relation $F'(x) = f(x)$, this may be written in the form

$$F(b) - F(a) = \int_a^b F'(x)dx = \int_a^b \frac{dF(x)}{dx}dx$$

This formula can easily be proved and understood directly. We subdivide the interval $a \} x \} b$ into intervals $\Delta x_1, \Delta x_2, \dots, \Delta x_n$ and consider the sum $\Sigma(\Delta F/\Delta x_\nu)\Delta x_\nu$. On the one hand, this sum is simply $\Sigma\Delta F = F(b) - F(a)$, independently of the particular sub-division, whence its limit is $F(b) - F(a)$. On the other hand, its limit is also equal to $\int_a^b F'(x)\,dx$ as follows from the mean value theorem. In fact, $\Delta F/\Delta x_\nu = F'(\xi_\nu)$, where ξ_ν is a point between the ends $x_{\nu-1}$ and x_ν, of the interval Δx. The sum is therefore equal to $\Sigma\Delta x_\nu F'(\xi_\nu)$ and, by the definition of the integral, this tends to the limit $\int_a^b F'(x)\,dx$ as the subdivision is made finer, which establishes the theorem.

In applying our rule, we often employ the symbol | to denote the difference $F(b) - F(a)$, i.e., we write

$$\int_a^b f(x)\,dx = F(b) - F(a) = F(x)\Big|_a^b$$

indicating by the vertical line that in the preceding expression first the value b and then the value a is to be substituted for x and finally the difference of the resulting numbers found.

Examples:

We are now in a position to illustrate by a series of simple examples the relationship between the definite integral, the indefinite integral and the derivative, which we have just investigated. By virtue of the theorem of 2.4.2 , we can derive from each of the integration formulae, which have been proved directly in 2.1.2, a differentiation formula.

In 2.2.4, we have obtained the integration formula

$$\int_a^b x^\alpha\,dx = \frac{1}{\alpha+1}\left(b^{\alpha+1} - a^{\alpha+1}\right)$$

for every rational number $\alpha \neq 1$ and all positive values of a and b; if we replace the variable of integration by u and the upper limit by x, this may be written in the form

$$\int_a^x u^\alpha\,du = \frac{1}{\alpha+1}\left(x^{\alpha+1} - a^{\alpha+1}\right)$$

It follows from this by the fundamental theorem that the right hand side is a primitive function of the integrand, i.e., the differentiation formula

$$\frac{d}{dx}x^{\alpha+1} = (\alpha+1)x^a$$

is valid for every rational value of $\alpha \neq -1$ and all positive values of x. By direct substitution, we find that this last formula is also true for $\alpha = -1$, if $x > 0$. The result obtained exactly agrees with what we have already found in 2.3.3 by direct differentiation. Thus, by using the fundamental theorem after having carried out the integration, we could have saved ourselves the trouble of that differentiation.

Moreover, it follows from the integration formula

$$\int_a^x \cos u\,du = \sin x - \sin \alpha$$

given in 2.2.5 that $d/dx \sin x = \cos x$, in agreement with the result found at the end of 2.3.3.

However, conversely, we may regard every directly proved differentiation formula $F'(x) = f(x)$ as a link between a primitive function $F(x)$ and a derived function $f(x)$, that is, we may regard it as a formula for indefinite integration and then obtain from it the definite integral of $f(x)$ as in 2.4.4. This very method is frequently employed, as we shall see in 4.1. In particular, we may start from the results of 2.3.3 and obtain the integral formula of 2.1.3 by virtue of the fundamental theorem. For example, we know from 2.3.3 that $dx^{\alpha+1}/dx = (\alpha+1)x^{\alpha}$, whence $x^{\alpha+1}/(\alpha+1)$ is a primitive function or indefinite integral of x^{α}, provided that $\alpha \neq -1$, and thus, by 2.4.4 arrive at the above integration formula.

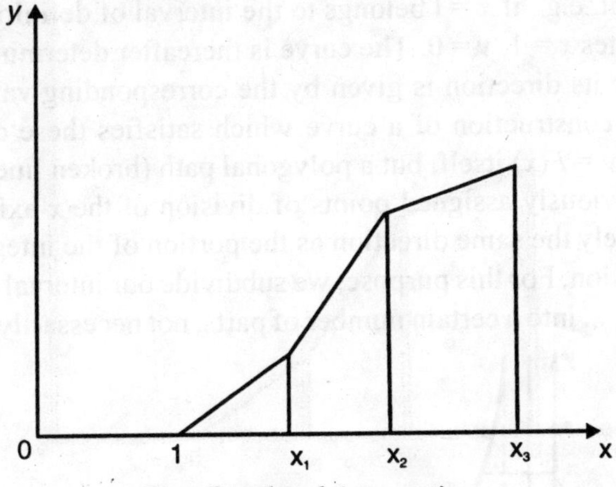

Fig. Graphical integration

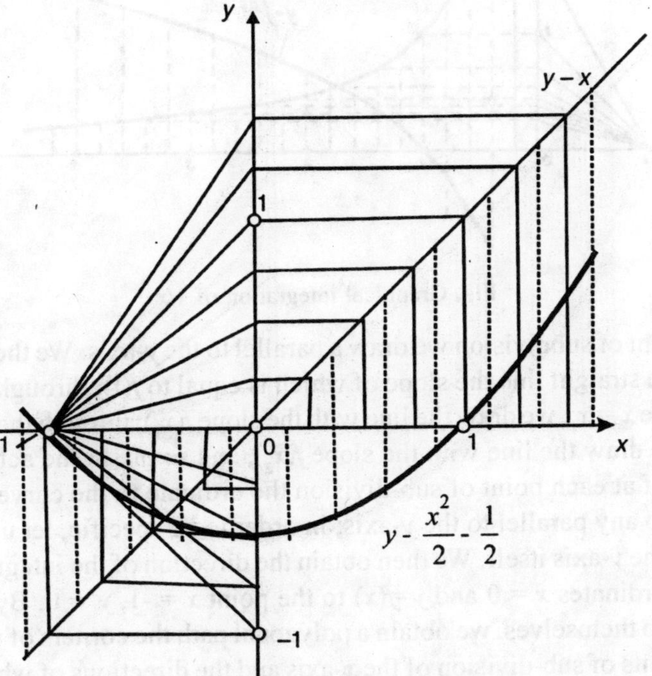

Fig. Simple Methods of Graphical Integration

Simple Methods of Graphical Integration

Since an indefinite integral or primitive function of $f(x)$ is a function $y = F(x)$, which not only can be visualized as an area, but, like any other function, can be represented graphically by a curve. Our definition immediately suggests the possibility of constructing this curve approximately and thus obtaining a graph of the integral function. To begin with, we must remember that this last curve is not unique, but on account of the additive constant can be shifted parallel to itself in the direction of the y-axis. We can therefore require that the integral curve shall pass through an arbitrarily selected point, e.g., if $x = 1$ belongs to the interval of definition of $f(x)$, through the point with the coordinates $x = 1, y = 0$. The curve is thereafter determined by the requirement that for each value of x its direction is given by the corresponding value of $f(x)$. In order to obtain an approximate construction of a curve which satisfies these conditions, we seek to construct not the curve $y = F(x)$ itself, but a polygonal path (broken line) the corners of which lie vertically above previously assigned points of division of the x-axis and the segments of which have approximately the same direction as the portion of the integral curve between the same points of sub-division. For this purpose, we subdivide our interval of the x-axis by means of the points $x = 1, x_1, x_2, \ldots$ into a certain number of parts, not necessarily all of the same length,

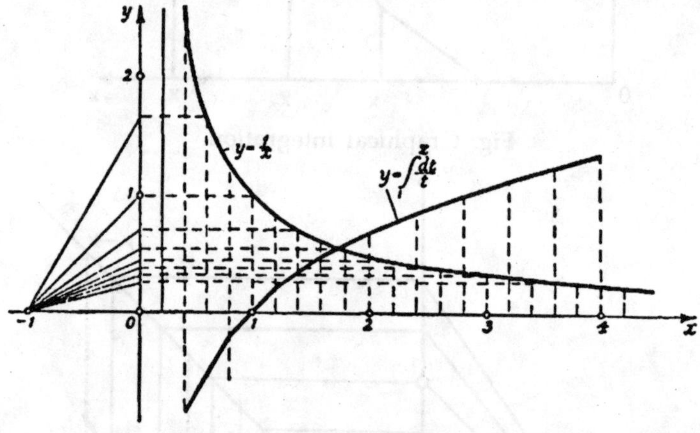

Fig. Graphical integration of $1/x$

and at each point of subdivision we draw a parallel to the y-axis. We then draw through the point $x = 1, y = 0$ the straight line the slope of which is equal to $f(1)$; through the intersection of this line with the line $x = x_1$, we draw the line with the slope $f(x_1)$; through the intersection of this line with $x = x_2$, we draw the line with the slope $f(x_2)$, and so on. In the actual construction of these lines, we erect at each point of sub-division the ordinate to the curve $y = f(x)$ and project these ordinates onto any parallel to the y-axis; in order to be specific, let us suppose that they are projected onto the y-axis itself. We then obtain the direction of the integral curve by joining the points with coordinates $x = 0$ and $y = f(x)$ to the point $x = -1, y = 0$. By transferring these directions parallel to themselves, we obtain a polygonal path the corners of which lie vertically above the given points of sub-division of the x-axis and the directions of which agrees with the direction of the integral curve at the starting point of each interval. This polygonal path can be made to represent the integral curve with any desired degree of accuracy by making the subdivisions of the interval fine enough. We can frequently improve the accuracy of the construction by choosing for the direction of each segment of the polygon that direction which does not belong to the starting but to the central point of the corresponding interval.

We mention here in passing that graphical integration (that is, finding the graph of a primitive $F\{x)$ of a function $f(x)$ which itself is given by a graph) can also be performed by means of a mechanical device, the so-called integraph. In this mechanism, a pointer is moved along the given curve and a pen automatically traces one of the curves $y = F(x)$ for which $F'(x) = f(x)$. The indeterminacy of the constant of integration is expressed by a certain arbitrariness in the initial position of the instrument.

In Fig., the construction described above is carried out for the function $f(x)=x$. By graphical integration, we obtain an approximation to the integral curve, which is the parabola $y = x$. In addition, an approximation to the integral function of the function $f(x) = 1/x$. We shall study this integral later in greater detail - it will turn out to be the logarithmic function. Finally, the reader would he well advised to work out some other examples on his own, e.g., the graphical integration of the functions $\sin x$ and $\cos x$.

Further Remarks on the Connection between the Integral and the Derivative

Before we pursue systematically the relationships found in 2.4, we shall illustrate them from another point of view, which is closely related to the intuitive idea of density and other physical concepts.

2.6.1 Mass Distribution and Density; Total Quantity and Specific Quantity: We assume that any mass is distributed along a straight line, the x-axis, the distribution being continuous, but not necessarily uniform. For example, we may think of a vertical column of air standing on a surface of area 1; we take as x-axis a line pointing vertically upwards and as origin the point on the Earth's surface. The total mass between two abscissae x_1 and x_2 is then determined in the following manner by means of a so-called sum-function $F(x)$. We measure the distance along the line from the initial point of the mass-distribution $x = 0$ and denote by $F(x)$ the total mass between the abscissa 0 and the abscissa x. The increment of mass from the abscissa x_1 to the abscissa x_2 is then given simply by

$$F(x_2) - F(x_1)$$

thus a sign is assigned to the increment and this sign changes if x_1 and x_2 are interchanged. The average mass per unit length in the interval x_1 to x_2 is

$$\frac{F(x_2) - F(x_1)}{x_2 - x_1}$$

If we assume that the function $F(x)$ is differentiable, then, as $x_2 \to x_1$, this value tends to the derivative $F'(x_1)$. This quantity is precisely what is usually called the specific mass or density of the distribution at the point x_1; as a rule, of course, its value depends on the particular point chosen. There exists accordingly between the density $f(x)$ and the sum-function $F(x)$ the relation

$$\int_0^x f(u)du; f(x) = F'(x)$$

The sum-function is a primitive function of the density, or, what amounts to the same thing, the mass is the integral of the density; conversely, the density is the derivative of the sum-function.

Exactly the same relation is very frequently encountered in physics. For example, if we denote by $Q(t)$ the total amount of heat needed to raise the unit mass of a substance from the temperature t_0 to the temperature t, then to raise the temperature from t_1 to t_2 requires the amount of heat

$$Q(t_2) - Q(t_1)$$

Between t_1 to t_2, the average amount of heat used per unit increase in temperature is then

$$\frac{Q(t_2) - Q(t_1)}{t_2 - t_1}$$

If we assume once again differentiability of the function $Q(t)$, we obtain in the limit the function

$$q(t) = \lim_{t_1 \to t} \frac{Q(t) - Q(t_1)}{t - t_1}$$

which we call the specific heat of the substance. In general, this specific heat is to be regarded as a function of the temperature. Here again, there exists between the specific heat and the total quantity of heat the characteristic relationship of integral and derivative

$$\int_a^b q(t)dt = Q(b) - Q(a)$$

We shall encounter the same relations in all cases where total and specific quantities are interrelated, e.g., electric charge with density of charge, or total force acting on a surface with force-density or pressure.

In Nature, usually what we know directly is not density or specific quantity, but total quantity, whence it is the integral which is primary (as the name primitive suggests) and the specific quantity is only arrived at after a limiting process, namely, differentiation.

Incidentally, it may be noted that if the masses considered are by their nature positive, the sum-function $F(x)$ must be a monotonically increasing function of x, and consequently the specific quantity, the density $f(x)$, must be non-negative. However, nothing stops us from considering also negative quantities (for example, negative electricity); then our sum-functions $F(x)$ need no longer be monotonic.

The Question of Applications: Perhaps, the relationship of the primitive sum-function to the density distribution becomes clearer when it is realized that, from the point of view of physical facts, the limiting processes of integration and differentiation represent idealizations and that they do not express anything exact in nature. On the contrary, in the realm of physical reality, we can form in place of the integral only a sum and in place of the derivative only a difference quotient of very small quantities. The quantities Δx remain different from 0; the passage to the limit $\Delta x \to 0$ is merely a mathematical simplification, in which the accuracy of the mathematical representation of the reality is not essentially impaired.

As an example, we return to the vertical column of air. According to the atomic theory, we find that we cannot think of the mass distribution as a continuous function of x. On the contrary, we will assume (and this, too, is a simplifying idealization) that the mass is distributed along the x-axis in the form of a large number of point-molecules lying very close to each other. Then the sum-function $F(x)$ will not be continuous, but it will have a constant value in the interval between two molecules and will take a sudden jump as the variable x passes the point occupied by a molecule. The amount of this jump will be equal to the mass of the molecule, while the average distance between molecules, according to results established in atomic theory, is of the order of 10^{-8} cm. Now, if we are performing upon this air column some measurement in which masses of the order 10^4 molecules are to be considered negligible, our function cannot be distinguished from a continuous function. In fact, if we choose two values x and as $x + \Delta x$, the difference Δx of which is less than 10^{-4} cm, then the difference between $F(x)$ and $F(x + \Delta x)$ will be the mass of the molecules in the interval; since the number of these molecules is of the order of 10^4, the values of $F(x)$ and $F(x + \Delta x)$ are equal as for as our experiment is concerned. We consider simply as density of distribution the difference quotient

$$\frac{\Delta F(x)}{\Delta x} = \frac{F(x+\Delta x)-F(x)}{\Delta x}$$

It is an important physical assumption that we do not obtain measurably different values for this quotient when Δx is allowed to vary between certain bounds, say between 10^{-4} and 10^{-5} cm. Now, imagine that $F(x)$ *is* measured and plotted for a large number of points about 10^{-4} cm. apart and that the points thus found are joined by straight lines; we obtain a polygon and, by rounding off the corners, obtain finally a curve with a continually turning tangent. This curve is the graph of some function, say $F_1(x)$. This new function cannot be distinguished within the limits of experimental accuracy from $F(x)$ and its derivative is within the same limits equal to $\Delta F/\Delta x$; we thus have found a continuous differentiable function which for the purposes of physics is the function $F(x)$.

It is perhaps appropriate to discuss yet another example of the concepts of sum-function and distribution density. In statistics, e.g., in the kinetic theory of matter or in statistical biology, these concepts frequently occur in a form in which the nature of the mathematical idealization is particularly clear. For example, let us consider the molecules of a gas confined in a vessel and observe their velocities at a given instant of time. Let the number of molecules be N and the number of those with velocities less than x be $N\Phi(x)$. Then $\Phi(x)$ denotes the ratio of the number of molecules moving with velocities between 0 and x to the total number of molecules. Of course, this sum-function is not continuous, but is sectionally constant and suddenly increases by $1/N$ when x, as it increases, passes a value which is equal to the velocity of some molecule.

The idealization which we shall make here is that we shall think of the number N as increasing beyond all bounds. We assume that, in this passage to the limit $N \rangle \rangle$, the sum-function $\Phi(x)$ tends to a definite continuous limit function $F(x)$. That this is really the case, i.e., that we can with sufficient accuracy replace $\Phi(x)$ by this continuous function $F(x)$, is obviously an important physical assumption; and it is another such assumption to assume that this sum-function $F(x)$ possesses a derivative $F'(x) = f(x)$, which we then call the density-distribution. The sum-function is connected with the density distribution by the equations

$$F(x) = \int_0^x f(u)\,du; F(b) - F(a) = \int_a^b f(x)\,dx$$

The density distribution is occasionally referred to as the specific probability that a molecule possesses the velocity x. The idealization we have just carried out has a great role in the kinetic theory of gases of Maxwell; it appears in exactly the same mathematical form in many problems of mathematical statistics.

THE ESTIMATION OF INTEGRALS AND THE MEAN VALUE THEOREM OF THE INTEGRAL CALCULUS

The full importance of which will not appear until somewhat later on. The point in question is the estimation of integrals.

The Mean Value Theorem of the Integral Calculus: The first and simplest of these estimation roles runs as follows: If in an interval $a \rangle x \rangle b$ the continuous function $f(x)$ is everywhere non-negative (is either positive or zero), then the definite integral

$$\int_a^b f(x)\,dx$$

is also non-negative. Similarly, the integral is not positive, if the function is nowhere positive in the interval. The proof of this theorem follows directly from the definition of the integral. This leads to the theorem: If

$$f(x) \geqq g(x)$$

everywhere in the interval $a \rangle x \rangle b$, then also

$$\int_a^b f(x)\,dx \geqq \int_a^b g(x)\,dx$$

In fact, by our first remark, the integral of the difference $f(x) - g(x)$ is non-negative and, by our addition rule,

$$\int_a^b (f(x) - g(x))dx = \int_a^b f(x)\,dx - \int_a^b g(x)\,dx$$

Let M be the greatest and m the smallest value of the function $f(x)$ in the interval ab. The function $M\text{-}f(x)$ is non-negative in the interval and the same is true for the function $f(x)\text{-}m$, whence we obtain immediately the double inequality

$$\int_a^b md \leq \int_a^b f(x)dx \leq \int_a^b Mdx$$

However,

$$\int_a^b mdx = m\int_a^b dx = m(b-a)$$

and likewise

$$\int_a^b Mdx = m(b-a)$$

whence

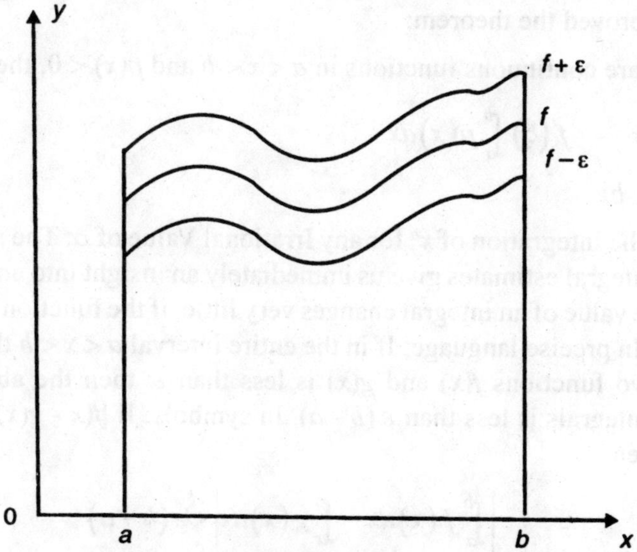

$$m(b-a) \leq \int_a^b f(x)dx \leq M(b-a)$$

Hence the integral under consideration can be represented as the product of (*b-a*) and some number μ between *m* and *M*:

$$\int_a^b f(x)dx = \mu(b-a), m \leq \mu \leq M$$

As a rule, there is no need to state the exact value of this mean value μ. However, we may say that it will be assumed by the function at least at one point ξ of the interval $a < \xi < b$, since in its interval of definition a continuous function assumes all values between its greatest and smallest values. As in the case of the mean value theorem of the differential calculus, the exact statement of the value ξ is in many cases unimportant. Hence we may set $\mu = f(\xi)$, where ξ is an intermediate value of *x*, and find then

$$\int_a^b f(x)dx = (b-a)f(\xi), a \leq \xi \leq b$$

This last formula is called the mean value theorem of the integral calculus.

We can generalize the theorem somewhat by considering instead of the integrand $f(x)$ an integrand of the form $f(x)p(x)$, where $p(x)$ is an arbitrary/ non-negative function which, like $f(x)$, is assumed to be continuous. Since $mp(x) \leq f(x)\,p(x) \leq Mp(x)$, we find immediately

$$m\int_a^b p(x)dx \leq \int_a^b f(x)p(x)dx \leq M\int_a^b p(x)dx$$

or, as a single equation,

$$\int_a^b f(x)p(x)dx = f(\xi)\int_a^b p(x)dx$$

where ξ is again a number between a and b.

Thus, we have proved the theorem:

If $f(x)$ and $p(x)$ are continuous functions in $a < x < b$ and $p(x) < 0$, then

$$\int_a^b f(x)p(x)dx = f(\xi)\int_a^b p(x)dx$$

where $a < \xi < b$.

Applications. The Integration of x^α for any Irrational Value of α: The mean value theorem and the equivalent integral estimates give us immediately an insight into an intuitive and easily understood fact: The value of an integral changes very little, if the function itself is everywhere changed very little. In precise language: If in the entire interval $a < x < b$ the absolute value of the difference of two functions $f(x)$ and $g(x)$ is less than ε, then the absolute value of the difference of their integrals is less than $\varepsilon(b - a)$. In symbols: If $|f(x - g(x) < \varepsilon$ throughout the interval $a < x < b$, then

$$\left|\int_a^b f(x)dx - \int_a^b g(x)dx\right| < \in(b-a)$$

or

$$-\in(b-a) + \int_a^b g(x)dx < \int_a^b f(x)dx < \int_a^b g(x)dx + \in(b-a)$$

Fig. illustrates very clearly this theorem. We draw for the curve $y = f(x)$ the parallel curves $y = f(x) + \varepsilon$ and $y = f(x) - \varepsilon$. By assumption, the function $g(x)$ keeps within the strip bounded by these parallel curves. It is clear from this that the areas, which are bounded by the curves $f(x)$ and $g(x)$, differ from each other by less than half the area of the strip, and the area of the strip is just

$$\int_a^b \{f(x) + \in\}dx - \int_a^b \{f(x) - \in\}dx = 2\in(b-a)$$

There is no need for intuition. Since

$$-\in + g(x) < f(x) < \in + g(x)$$

it follows, by arguments analogous to those in 2.7.1,

$$\int_a^b\{-\in+g(x)\}dx<\int_a^b f(x)dx<\int_a^b\{g(x)+\in\}dx$$

which, as the result of the fundamental rules of integration, takes the form

$$-\in(b-a)+\int_a^b g(x)dx<\int_a^b f(x)dx<\int_a^b g(x)dx+\in(b-a)$$

here we have merely replaced the integral of a sum by the corresponding sum of integrals and have taken into consideration that

$$\int_a^b \in dx \;=\; \in(b-a)$$

As an indication of the importance of this theorem, we shall show that with its help we can integrate the function x^α for any irrational value of α, or more precisely, calculate the definite integral $\int_a^b x^\alpha dx$ Here we assume that $0<\alpha<b$.

We represent the index α as the limit of a sequence of rational numbers $\alpha_1, \alpha_2, \ldots, \alpha_n, \ldots$ so that $\alpha=\lim_{n\to\infty}\alpha_n$ we can here assume that none of the values α_n is equal to -1, since α_n itself is different from -1. Now, we use for the power x^α the definition

$$x^\alpha \;=\; \lim_{n\to\infty} x^\alpha n$$

and note that, no matter how small a positive number ε we choose, we can always find an n so large that in the entire interval $\tau<x<b$ * we have $\left|x^\alpha-x^{\alpha_n}\right|<\varepsilon$

This can be proved quite simply as follows. Remembering that x^α is monotonic and setting $\delta_n=\alpha_n-\alpha$, we have

$$\left|x^\alpha-x^{\alpha_n}\right| \;=\; x^\alpha\left|1-x^{\delta_n}\right|\le\left(\alpha^\alpha+b^\alpha\right)\left(\left|1-\alpha^{\delta_n}\right|+\left|1-b^{\delta_n}\right|\right)$$

in fact, x^α lies between a^α and b^α, so that $x^\alpha \} a^\alpha+b^\alpha$, and likewise $1-x^{\alpha_n}$ lies between $1-a^{\delta_n}$ and $1-b^{\delta_n}$, so that $\left|1-x^{\delta_n}\right|\le\left(\left|1-a^{\delta_n}\right|+\left|1-b^{\delta_n}\right|\right)$ From $\lim_{n\to\infty}a^{\delta_n}=\lim_{n\to\infty}b^{\delta_n=1}$ it follows that

$$\lim_{n\to\infty}\left|1-a^{\delta_n}\right| \;=\; \lim_{n\to\infty}\left|1-b^{\delta_n}\right|=0$$

hence, if n is chosen large enough, the right-hand side of the inequality is less than ε. This yields $\left|x^{\alpha_n}-x^\alpha\right|<0$ simultaneously for all values of x in the interval $a<x<b$.

Now we need only apply the relationship, to the functions $f(x)=x^\alpha$ and $g(x)=x^{\alpha_n}$, yielding

$$-\varepsilon(b-a)+\int_a^b x^{\alpha_n}dx<\int_a^b x^n dx<\int_a^b x^{\alpha_n}dx+\varepsilon(b-a)$$

However, the integrals on the right-hand and left-hand sides may be evaluated in accordance with. earlier results, which yields

$$-\varepsilon(b-a)+\frac{1}{\alpha_n+1}\left(b^{a_n+1}-a^{a_n+1}\right)$$

$$<\int_a^b x^{\alpha}\,dx<\frac{1}{\alpha_n+1}\left(b^{a_n+1}-a^{a_n+1}\right)+c(b-a)$$

If we now let the number ε decrease steadily and tend to 0, the corresponding values of n increase beyond all bounds; the number s a_n, a^{a_n}, and b^{α_n} must then converge to α, a^{α} and b^{α}, respectively, and we immediately obtain the result

$$\int_a^b x^a\,dx = \frac{1}{\alpha+1}\left(b^{\alpha+1}-a^{\alpha+1}\right)$$

In other words, the integration formula, which holds for rational values of α, also holds for irrational values of α.

It follows from this, by virtue of the fundamental theorem, that, for positive values of x, the differentiation formula for rational values

$$\frac{d}{dx}x^{\alpha+1} = (\alpha+1)x^a$$

is also valid for irrational values of α.

Chapter 4

Integration of the Elementary Functions

THE CORRESPONDING INTEGRAL FORMULAE

General Rules for Integration

The fundamental theorem of and the definition of the indefinite integral reveal the possibility of writing down an integration formula corresponding to each differentiation formula. The following rules of integration are completely equivalent to the first three rules of differentiation.

Multiplication by a constant: If c is a constant, then

$$\int cf(x)dx = c\int f(x)dx$$

Integration of a sum: It is always true that

$$\int\{f(x)+g(x)\}dx = \int f(x)dx+\int g(x)dx$$

To the third rule of differentiation corresponds the rule for the integration of a product or, as it is usually called, the rule for integration by parts. On integration, the product rule yields

$$\int\{f(x)g(x)\}'dx = \int f(x)g'(x)dx+\int g(x)f'(x)dx$$

The indefinite integral on the left hand side is obviously $f'(x)g(x)$ (except possibly for an additive constant), whence we can write the rule for integration by parts in the form:

$$\int f(x)g'(x)dx = f(x)g(x)-\int g(x)f'(x)dx$$

This last integration formula, the counterpart of the rule for the differentiation of a product, has been given here only for the sake of completeness.

Integration of the Simplest Functions

Corresponding to the differentiation formulae for special functions which we have found above, we now set down the equivalent integration formulae. The formula

$$\frac{d}{dx}x^n = nx^{n-1}$$

expressed as an integration formula, becomes

$$\int x^{n-1}dx = \frac{x^n}{n}, n \neq 0$$

In fact, this formula merely means that the derivative of the right hand side is equal to the integrand on the left hand side. If we replace n by $n+1$, we obtain the formula

$$\int x^{n-1}dx = \frac{x^n}{n}, n \neq 0$$

This formula holds for every integral index n (when $n < 0$, it holds of course only if $x \succ 0$) except when $n = -1$, when the denominator $n + 1$ would vanish. In 3.6, this exceptional case will be studied in detail. The fundamental theorem of the integral calculus allows at once to use our integral formulae for the determination of areas, that is, the values of definite integrals. By 2.4.5, we immediately obtain

$$\int_a^b x^n dx = \frac{1}{n+1}\left(b^{n+1} - a^{n+1}\right), n \neq -1$$

where, if n is negative, we assume that a and b have the same sign, since otherwise the integrand would be discontinuous in the interval of integration.

There correspond to the differentiation formulae for $\sin x$, $\cos x$ the integration formulae:

$$\int \cos x dx = \sin x, \int \sin x dx = -\cos x,$$

$$\int \frac{1}{\cos^2 x} dx = \tan x, \int \frac{1}{\sin^2 x} dx = -\cot x$$

They yield by way of the fundamental rule of 2.4 the value of the definite integral between any limits, an only restriction being that when the last two formulae are used, the interval of integration must not contain any point of discontinuity of the integrand. For example,

$$\int_a^b \cos x dx = \sin x\Big|_a^b = \sin b - \sin a$$

It hardly need be emphasized that, with the help of the first two rules of integration, we are now in a position to integrate any polynomial in x and, in fact, any linear combination with arbitrary constant coefficients of the functions integrated here. However, note that, according to the fundamental theorem, rules of integration and differentiation must be equivalent to each other; it is therefore possible to prove the general integration rules of this section and then to read off the differentiation rules of the preceding section. The reader is well advised to carry out this suggestion.

The Corresponding Integral Formulae

Expressed in the language of the indefinite integral, the formulas which we have just derived become

$$\int \frac{1}{\sqrt{(1-x^2)}}dx = ar\sin x, \int \frac{1}{\sqrt{(1-x^2)}}dx = -ar\cos x$$

$$\int \frac{1}{1+x^2}dx = ar\tan x, \qquad \int \frac{1}{1+x^2}dx = -ar\cot x$$

There arises no contradiction between the pair of formulae on the left and on the right hand sides, which express each indefinite integral in the form of two functions, which appear to be entirely different. We must remember that, in the case of the indefinite integral, an arbitrary additive constant remains at at our disposal. If we choose these constants so that they differ by $\pi/2$ and recall that $\pi/2$ – arcos x = arsin$x$ and likewise $\pi/2$ – arcot x = artan x, this formal disagreement is immediately removed. The indefiniteness simply depends on the fact that the indefinite integral is not a single definite function, but an entire family of functions which differ from each other by arbitrary additive constants. The equation for an indefinite integral does not specify the value, but only one value of it. As we have already remarked, it would be more correct to express this fact by always including the undetermined constant, thus not writing

$$\int f(x)dx = F(x)$$

but

$$\int f(x)dx = F(x)+c$$

However, for the sake of convenience, this more detailed form is usually avoided; the reader should therefore be all the more careful to bear in mind the indefiniteness which is always associated with the shorter form.

There follow immediately from the formulae for indefinite integration formulae for definite integration. In particular,

$$\int_a^b \frac{dx}{1+x^2} = ar\tan x\Big|_a^b = ar\tan b - ar\tan a$$

If we set $a = 0$, $b = 1$ and recall that tan 0 = 0 and tan $\pi/4 = 1$, we obtain the remarkable formula

$$\frac{\pi}{4} = \int_0^1 \frac{1}{1+x^2}dx$$

The number π, which originally arose from the study of the circle, is by this formula brought into a very simple relationship to the rational function $1/(1+x)$ and is expressed and defined by the area.

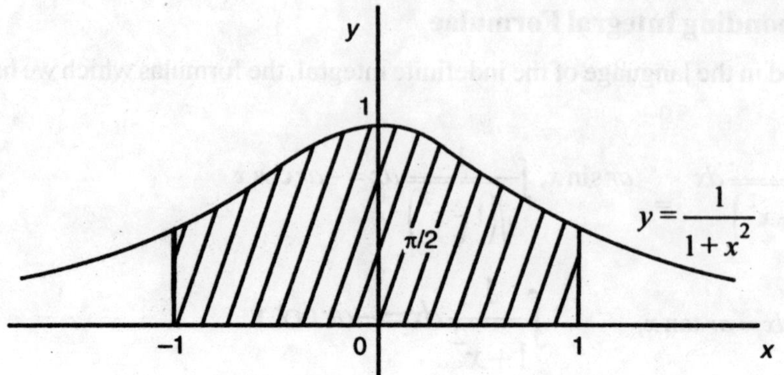

Fig. π/2 illustrated by an area

As a very simple example, we consider the function $y = x^\alpha$, where we $\alpha = p/q$, q being a positive integer and p a positive or negative integer, so that α is an arbitrary, positive or negative rational number. Let x be positive. By the chain rule with

$y = \varphi^p, \varphi = x^{1/q}$

we have the formula

$$y' = p\varphi^{p-q} \cdot \frac{1}{q} x^{(1-q)/q} = \frac{p}{q} x^{p/q-1}$$

so that for arbitrary, rational values of α we obtain the differentiation formula

$$\frac{d}{dx} x^\alpha = \alpha x^{\alpha-1}$$

in agreement with the result already found in another way in 2.3.3. *As* a second example, we consider

$$y = \sqrt{(1-x^2)} \text{ or } y = \sqrt{\varphi}$$

where $\varphi = I - x$ and $-1 < x < 1$. The chain rule yields

$$y' = \frac{1}{2\sqrt{\varphi}} \cdot (-2x) = \frac{x}{\sqrt{(1-x^2)}}$$

Further examples are given in brief calculations:

1. $\quad y = \arcsin\sqrt{(1-x^2)}$

$$\frac{dy}{dx} = \frac{1}{\sqrt{\left\{1-\left(1-x^2\right)\right\}}} \cdot \frac{d\sqrt{\left(1-x^2\right)}}{dx}$$

$$= \frac{1}{|x|} \cdot \frac{-x}{\sqrt{\left(1-x^2\right)}} = \mp \frac{1}{\sqrt{\left(1-x^2\right)}}$$

2. $y = \sqrt{\left(\frac{1+x}{1-x}\right)}$

$$\frac{dy}{dx} = \frac{1}{\sqrt[2]{\left(\frac{1+x}{1-x}\right)}} \cdot \frac{a\left(\frac{1+x}{1-x}\right)}{dx}$$

$$= \frac{\sqrt{(1-x)}}{\sqrt[2]{(1+x)}} \cdot \frac{2}{(1-x)^2} = \frac{1}{(1+x)^{1/2}(1-x)^{3/2}}$$

The chain rule for differentiation can also be expressed in the form of an integration formula in agreement with the fact that there corresponds to each differentiation formula a completely equivalent integration formula. Nevertheless, we will pass over this formula for the present, since we have no immediate need of it here and, moreover, it is discussed in detail lin in 4 .2.

Further Remarks on the Integration and Differentiation of x^α, when α is Irrational: In view of the elementary definition of the power x^α by the equation

$$x^\alpha = \lim x^{r_n}$$

where the numbers r_n form a sequence of rational numbers with the limit α, we might be tempted to effect the differentiation of x^α by direct passage to the limit in the differentiation formula

$$\frac{d}{dx} x^{r_n} = r_n x^{r_n - 1}$$

We are not entitled to do so unless we have the might conclude that there follows from the relation $x^{r_n} \to x^\alpha$ the relation $\frac{d}{dx} x^{r_n} \to \frac{d}{dx} x^\alpha$. However, there is a very serious objection to such a passage to the limit. In fact, in any arbitrarily small neighbourhood of a given curve one may draw other curves the direction of which at arbitrarily selected points differs from the direction of the original curve by any desired amount; for example, we may approximate to a straight line by a wave lying arbitrarily near to it, the angle between the wave and the line reaching a value as large as 45. In other words, the above example shows that from the fact that two functions differ only very little from each other, we cannot immediately conclude that their derivatives also are everywhere nearly equal to one another. This objection forbids us to perform the apparently obvious passage to the limit in the absence of further justification .

However, in this respect, the integral behaves quite differently from the derivative. We have already observed that, if two functions differ by less than ε throughout the interval from a to b, their integrals must differ by less than $\varepsilon(b-a)$. We used there this result to establish the validity of the differentiation formula

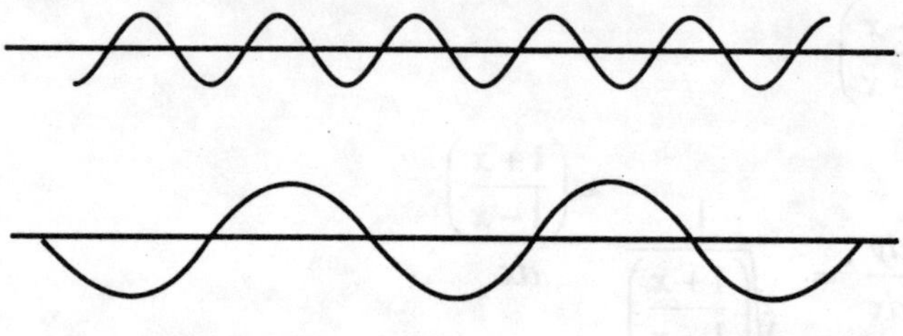

Fig. Approximation to a straight line by wavy curves

$$\frac{1}{\alpha+1}\frac{d}{dx}x^{\alpha+1} = x^{\alpha}$$

or, replacing $\alpha+1$ by α,

$$\frac{d}{dx}x^{\alpha} = ax^{\alpha-1}$$

Hence, in this indirect way, there is verified the relation $\frac{d}{dx}x^{r_n} \to \frac{d}{dx}x^{a}$ given above. This discussion is a characteristic example of the interrelation of the differential and integral calculus. Yet, in principle, it is preferable to replace (as we shall do in 3.6.5) the elementary definition of x^{α} by another, essentially simpler definition which will lead us once more to the same result and this time directly,

THE LOGARITHM AND THE EXPONENTIAL FUNCTION

The systematic relations between the differential and the integral calculus lead naturally to a convenient method of approach to the exponential function and the logarithm. Although we have already investigated these functions, we now define them afresh and redevelop their theory without making any use of our previous definition and the results based on it. We shall begin with the logarithm and then obtain the exponential function as its inverse.

Definition of the Logarithm. The Differentiation Formula:

We have seen that, in general, indefinite integration of the power x^n for integral indices n leads to a power of x. The only exception is the function $1/x$, which does not appear as the derivative of any of the functions which we have dealt with so far. It is natural to assume that the indefinite integral of the function $1/x$ represents a new sort of function; thus, pursuing this idea, we will investigate the function

$$y = \int_1^x \frac{d\xi}{\xi} = f(x)$$

for $x > 0$. We call it the logarithm of x or, more accurately, the natural logarithm of x, and write it $y = \log_e x$ or $y = \ln x$. We have denoted the variable of integration by ξ, in order to avoid confusion with the upper limit x.

The choice of the number 1 as lower limit is arbitrary; however, it will soon prove its convenience.

During the following argument, it will appear that the logarithm, thus defined, is the same as the logarithm which we defined previously in an elementary way. But, as we once more emphasize, the results of the following study are independent of those obtained earlier.

Geometrically, our logarithmic function means the area shown shaded Fig.; it is bounded above by the rectangular hyperbola $y = 1/\xi$, below *by* the ξ-axis, and on the sides by the lines $\xi = 1$ and $\xi = x$. This area is to be reckoned positive, if $x > 1$, negative if $x < 1$. For $x = 1$, the area vanishes, whence we have $\log 1 = 0$.

According to the above definition, the derivative of the logarithm is given by the formula

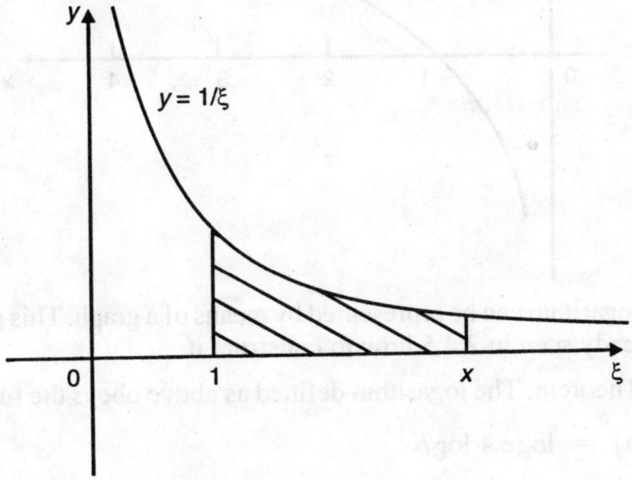

Fig. Log x illustrated by an area

$$\frac{d(\log x)}{dx} = \frac{1}{x}$$

We emphasize specially that we assume throughout that the argument x is positive; the logarithm of 0 or of any negative value cannot be formed in accordance with the formula above, because the integrand $1/\xi$ becomes infinite when $\xi = 0$. On the other hand, if we choose some negative number, say –1, as the lower limit, we can form the integral with a negative upper limit x, i.e. we can consider the expression

$$\int_{-1}^{x}\frac{d\xi}{\xi}\ (x<0)$$

Owing to the significance of the integral as the limit of a sum or as an area, we see that for $x<0$

$$\int_{-1}^{x}\frac{d\xi}{\xi} = \int_{1}^{-x}\frac{d\xi}{\xi} = \int_{1}^{|x|}\frac{d\xi}{\xi} = \log|x|$$

Hence we can, in general, write the formula for indefinite integration as

$$\int\frac{dx}{x} = \log|x|$$

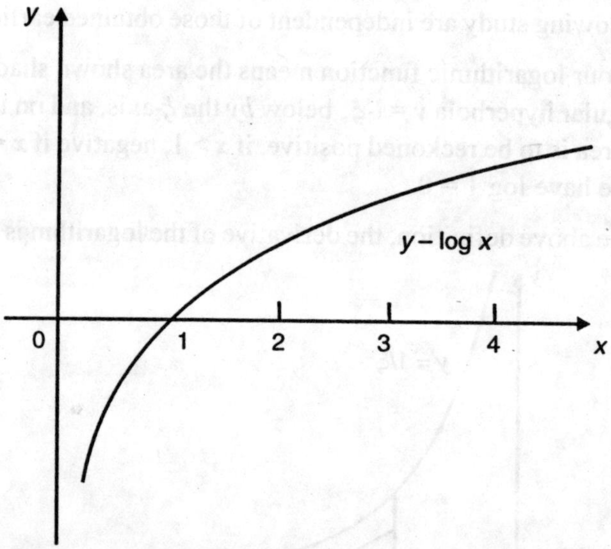

Naturally, the logarithm can be represented by means of a graph. This graph, the logarithmic curve. We have already seen in 2.4.5 how to construct it.

The Addition Theorem: The logarithm defined as above obeys the fundamental law:

$$\log(ab) = \log a + \log b$$

The proof of this addition theorem follows directly from the differentiation formula. In fact, writing $z=\log(ax)$ and applying the chain rule, we obtain

$$\frac{dz}{dx} = \frac{1}{ax}.a = \frac{1}{x}$$

However,

$$\frac{d}{dx}\log x = \frac{1}{x}$$

and, since the functions z and $\log x$ have the same derivative, they differ only by a constant, so that $z = \log x + c$ or

$$\log ax = \log x + c$$

This being true for all positive values of x, we first put $x = 1$ to find c; since $\log 1 = 0$, this yields

$$\log a = c$$

Substituting this value for c, we have

$$\log ax = \log x + \log a$$

whence, for $x = b$,

$$\log ab = \log a + \log b$$

which was to be proved.

For arbitrary positive numbers $a_1, a_2, \ldots, a_n$, the equation

$$\log(a_1 a_2 \ldots a_n) = \log a_1 + \log a_2 + \ldots + \log a_n$$

follows from the addition theorem for the logarithm.

In particular, if all the numbers $a_1, a_2, \ldots, a_n$ are equal to one and the same number a, we have

$$\log a^n = n \log a$$

Similarly, it follows that

$$\log a + \log \frac{1}{a} = \log 1 = 0$$

so that

$$\log a = -\log \frac{1}{a}$$

Moreover, if we set $\sqrt[n]{a} = a$, it follows that $\log a = n \log \alpha$, or

$$\log \sqrt[n]{a} = \log a^{1/n} = \frac{1}{n} \log a$$

Hence, by repeated use of the addition theorem, we find that, when m is a positive integer,

$$\frac{m}{n} \log a = \log \sqrt[n]{a^m} = \log a^{m/n}$$

The equation

$$\log a^r = r \log a$$

is thus proved for all positive rational values of r, and it is obviously correct for $r = 0$. It is also valid for negative rational values of r, because then

$$\log a^{r} = \log\frac{1}{a^{-r}} = -\log a^{-r} = r\log a$$

Monotonic Character and Values of the Logarithm: Obviously, the value of the logarithm increases and decreases with x, whence the logarithm is a monotonic function.

Since the derivative $1/x$ becomes smaller and smaller as x increases, the function increases more and more slowly as x increases. Nevertheless, as x increases beyond all bounds, the function $\log x$ does not tend to a positive limit, but becomes infinite, that is, for every positive number A, no matter how large, there are values of x for which $\log x > A$. This fact follows very readily from the addition theorem. In fact, $\log 2^n = n \log 2$, and since $\log 2$ is a positive number, we can, by taking $x = 2^n$ with sufficiently large values of n. *make* $\log x$ as large as we please.

Since $\log (1/2^n) = -n \log 2$, we see that, as x tends to zero through positive values, $\log x$ is negative and increases numerically beyond all bounds.

In summary, we have: The function $\log x$ is a monotonic function which assumes all values between – and + as the independent variable x ranges over the continuum of positive numbers.

The Inverse Function of the Logarithm (*the Exponential Function*): Since the function $y = \log x\ (x > 0)$ is a monotonic function of x which assumes all real values, its inverse function, which we shall at first denote by $x = E(y)$, must be a single-valued monotonic function defined for every real value of y; it is differentiable, since $\log x$ itself is differentiable. We interchange the notation for the dependent and independent variables and study the function $E(x)$ in detail. In the first place, obviously, it must be positive for every value of x. Moreover, we must have

$$E(0) = 1$$

In fact, this equation is equivalent to the statement that $\log 1 = 0$.

From the addition theorem for the logarithm follows immediately the multiplication theorem

$$E(\alpha)E(\beta) = E(\alpha+\beta)$$

In order to prove this, we merely need note that the equations

$$E(\alpha)=a, E(\beta)=b, E(\alpha+\beta)=c$$

are equivalent to

$$E(\alpha)=a, E(\beta)=b, E(\alpha+\beta)=c$$

Since, by the addition theorem for the logarithm, $\alpha+\beta = \log ab$, it must be true that $c = ab$, which proves the multiplication theorem.

We derive from this theorem a fundamental property of the function $y = E(x)$, which gives us the right to call our function the exponential function and to write it symbolically in the form

$$y = e^x$$

In order to obtain this property, we observe that there must be a number - which we shall call e - for which

$$\log e = 1$$

Its identity with the number e considered in in 1.6.5 will be proved in 3.6.6.

This is equivalent to the definition

$$E(1) = e$$

Using the multiplication theorem for the function $E(x)$, we have

$$E(n) = e^n$$

and, in the same way, for positive integers m and n,

$$E\left(\frac{m}{n}\right) = e^{m/n}$$

which we could also have found directly from the addition theorem for the logarithm.

The equation $E(r) = e^r$ thus proved for positive rational numbers r holds also for negative rational numbers by virtue of the equation

$$E(r)E(-r) = E(0) = 1$$

The function $E(x)$ is therefore a function which is continuous for all values of x and for rational values of x coincides with e^x. These facts give us the right to call our function e^x also for arbitrary irrational values of x. (It should be noted that here the continuity of e^x is an immediate consequence of its definition as the inverse function of a continuous, monotonic function, while, if the elementary definition is adopted, the continuity must be proved.)

The exponential function is differentiated according to the formula

$$\frac{d}{dx}e^x = e \text{ or } y' = y$$

This formula expresses the important fact that the derivative of the exponential function is the function itself.

The proof is very simple. In fact, we have $x = \log y$, whence, by the formula for the differentiation of the logarithm, we have $dx/dy = 1/y$, and then, by the rule for inverse functions,

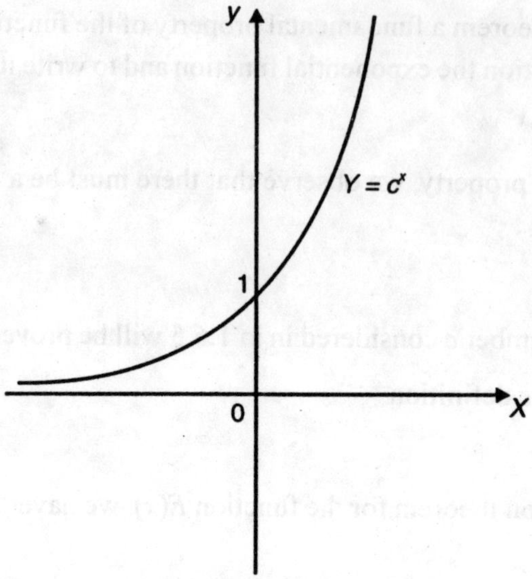

Fig. The exponential function

$$\frac{dy}{dx} = y = e^x$$

as stated.

The graph of the exponential function e^x - the so-called exponential curve - is obtained by reflection of the logarithmic curve in the line which bisects the first quadrant.

The General Exponential Function a^x and the General Power x^α:

The exponential Function a^x for an arbitrary positive base a *is* now simply defined by the equation

$$y = a^x = e^{x\log a}$$

which agrees with the earlier definition by virtue of the relation

$$e^{\log a} = a$$

By the chain rule, we immediately obtain

$$\frac{d}{dx}a^x = \frac{d}{dx}e^{x\log} = e^{x\log a}\cdot\log a$$

$$= a^x \log\alpha$$

The inverse function of the exponential function $y = a^x$ is called the logarithm to the base a and is written

$$x = \log_\alpha y$$

while the logarithmic function previously introduced, when it is required to distinguish it, is spoken of as the natural logarithm, or the logarithm to the base e.

It follows immediately from the definition that

$$\log y = x \log a = \log_a yK.\log \alpha$$

which shows that the logarithm of y for an arbitrary positive base $a < 1$ is obtained by multiplying the natural logarithm of y by the reciprocal of the natural logarithm of a, the modulus of the system of logarithms to the base a.

If we take $a = 10$, we obtain the ordinary Briggian logarithms, which have already been encountered in elementary mathematics and which are advantageous for use in numerical computations.

Instead of our previous definition of the general power $x^{a'}$ $(x > 0)$, we shall now define this power by the equation

$$x^{\alpha} = e^{\alpha \log x}$$

The rule for differentiating the power x^{α} follows immediately from the definition, using the chain rule; for

$$\frac{d}{dx}x^{\alpha} = e^{\alpha \log x}, \frac{\alpha}{x} = \alpha x^{\alpha-1}$$

in agreement with our previous result .

The Exponential Function and the Logarithm represented as Limits: We are now in a position to state important limiting relations referring to the quantities introduced above. We begin with the formula for differentiating the function $f(x) = \log x$:

$$\frac{1}{x} = f'(x) = \lim_{h\to 0}\frac{f(x+h)-f(x)}{h} = \lim_{h\to 0}\frac{\log(x+h)-\log x}{h}$$

$$= \lim_{h\to 0}\frac{1}{h}\log\left(1+\frac{h}{x}\right)$$

If we set $1/x = z$, this becomes

$$\lim_{h\to 0}\frac{1}{h}\log(1+zh) = x$$

Since the function e^x is continuous for all values of x, this implies that

$$e^x = \lim_{h\to 0}{}^{e\{\log(1+zh)/h\}} = \lim_{h\to 0}{}^{(1+zh)^{1/h}}\ldots \qquad (a)$$

In particular, if we give h the sequence of values 1, 1/2, 1/3,... , $1/n$, ... , we have

$$\lim_{h\to\infty}\left(1+\frac{z}{n}\right)^n = e^x\ldots. \qquad (b)$$

If we assign to z the value 1, Equation states the important fact:

As h tends to zero, the expression $(1 + h)^{1/h}$ tends to the number e:

$$\lim_{h\to 0}(1+h)^{1/h} = e$$

Equation yields

$$\lim_{h\to\infty}\left(1+\frac{1}{n}\right)^{n} = e$$

which proves that the number e is the same as the number denoted earlier by the symbol e.

It follows from the differentiation formula for a^x

$$a^{\alpha}\log\alpha = \lim_{h\to 0}\frac{\alpha^{x+h}-\alpha^{n}}{h}$$

that for $x = 0$

$$\log\alpha = \lim_{h\to 0}\frac{\alpha^{n}-1}{h}$$

which expresses the logarithm of a directly as a limit.

We append to this equation the remark that we can complete by it the earlier obtained relation

$$\int_a^b x^4 dx = \frac{1}{\alpha+1}\left(b^{\alpha+1}-\alpha^{\alpha+1}\right)$$

We have always been obliged to exclude the case $a = -1$. However, we can now discover what happens when the number a tends to the limit -1. If we put $a=1$, the left-hand side will, by our definition of the logarithm, have the limit

$$\int_1^b \frac{dx}{x} = \log b$$

whence the right-hand side has the same limit when $a < -1$.

We have here carried out the passage to the limit -1 under the integral sign without further investigation; cf. the earlier discussion.

Moreover, this fact agrees with the formula

$$\log b = \lim_{h\to 0}\frac{b^{n}-1}{h}$$

we only need write $a + 1 = h$.

We have thus cleared up the exceptional case $a = -1$ in the integration formula which we have so often used. The formula above is still meaninglesS when $a = -1$, but it retains as a limit formula its significance as $a < -1$.

Final Remarks: We review here briefly the train of thought followed in this section, We have first defined the natural logarithm $y = \log x$ as for $x > 0$ by means of an integral, whence we immediately deduced the differentiation formula, the addition theorem, and the existence of an inverse. We then investigated the inverse function $y = e^x$, where the number e was seen to be the number the logarithm of which is 1, and derived its differentiation formula as well as limit expressions for it and for the logarithm. The introduction of the functions $y = x^\alpha = e^{\alpha \log x}$ and $y = a^x = e^{x \log a}$ followed naturally.

In the discussion given here, in contrast to the elementary treatment, the question of continuity causes no difficulty, since the logarithm is defined as an integral and therefore as a continuous and differentiable function, the inverse function of which is also continuous.

Some Applications of the Exponential Function

We shall consider now some different problems involving the exponential function and thus gain insight into the fundamental importance of this function for all kinds of applications.

Definition of the Exponential Function by Means of a Differential Equation: We can define the exponential function by a simple theorem the use of which will save us many detailed investigations of particular cases:

If a function $y = f(x)$ satisfies an equation of the form

$$y' = \alpha y$$

where a is a constant other than zero, then y has the form

$$y = f(x) = ce^{ax}$$

where c if also a constant; conversely, every function of the form $ce^{\alpha x}$ satisfies the equation $y' = ay$.

The last equation is usually briefly referred to as a differential equation, since it expresses a relation between the function and its derivative. In order to make the theorem clear, we note first of all that in the simplest case $a = 1$ the above equation becomes $y' = y$. We know that $y = e^x$ satisfies this equation and it is clear that the same is also true of $y = ce^x$, where c is an arbitrary constant. Conversely, we can easily see that no other function satisfies the differential equation. For if y is such a function, we consider the function $u = ye^{-x}$. We must then have

$$u' = y'e^{-x} - ye^{-x} = e^{-x}\left(y' - y\right)$$

But the right-hand side vanishes, since we have assumed that $y' = y$, whence $u' = 0$, so that, by 2.4.3, u is a constant c and $y = ce^x$, as we wished to prove.

The case of any non-zero value of α can be treated in exactly the same manner as the special case $\alpha = 1$. If we introduce the function $u = ye^{-ax}$, we obtain the equation $u' = y'e^{-ax} — ae^{-ax}$. Hence we find from the assumed differential equation that $u' = 0$, so that $u = c$ and $y = ce^{ax}$. The converse is clear.

We will now apply this theorem to a number of examples and thereby make it more intelligible.

Interest, Compounded Continuously. Radio-active Disintegration: A capital sum, or principal, which has its interest added to it at regular periods of time, increases by jumps at these interest periods in the following manner. If 100α is the rate of interest per cent and moreover the interest accrued is added to the principal at the end of each year, then after x years the accumulated amount of an original principal of 1 will be

$$(1+\alpha)^x$$

However, if the principal had the interest added to it not at the end of each year, but at the end of each n-th part of a year, then after x years the principal would amount to

$$\left(1+\frac{\alpha}{n}\right)^{nx}$$

Taking $x = 1$, for the sake of simplicity, i.e, reckoning the interest at 100α per cent for one year, we find that, if the interest is computed in this latter way, the principal 1 amounts after one year to

$$\left(1+\frac{\alpha}{n}\right)^{n}$$

If we now let n increase beyond all bounds, i.e., if we let the interest be calculated at shorter and shorter intervals, the limiting case will signify in a sense that the interest is compounded continuously, at each instant; and we see that the total amount after one year will be e^{α} times the original principal. Similarly, if the interest is calculated in this manner, an original principal of 1 will have grown after x years to an amount $e^{\alpha x}$, where x may be any number, integral or otherwise.

The discussion in 3.7.1 forms a framework within which examples of this type are readily understood. We consider a quantity y, which increases (or decreases) with time. Let the rate at which this quantity changes be proportional to the total quantity. Then, if we take time as the independent variable x, we obtain a law of the form $y' = \alpha y$ for the rate of increase, where α, the factor of proportionality, is positive or negative, according to as the quantity is increasing or decreasing, Then, in accordance with 3.7.1, the quantity y itself will be given by a formula

$$y = ce^{qx}$$

where the meaning of the constant c is immediately obvious, if we consider the instant $x =$ 0. At that instant, $e^{\alpha x} = 1$ and we find that $c = y_0$ is the. quantity at the beginning of the time under consideration, so that we may write

$$y = y_0^{ex}$$

A characteristic example of the use of these ideas is the case of radioactive disintegration. The rate at which the total quantity y of a radioactive substance is diminishing at any instant is proportional to the total quantity present at that instant; this is *a priori* plausible, as each portion of the substance decreases as rapidly as every other portion. Hence the quantity y of the substance, expressed as a function of the time, satisfies a relation of the form $y' = -ky$, where k is to be taken as positive, since we are dealing with a diminishing quantity. The quantity of substance is thus expressed as a function of the time by $y = 3.8y_0e^{-kx}$, where y_0 is the amount of the substance at the beginning of the time (time $x = 0$).

After a certain time τ, the radioactive substance will be diminished to half its original quantity. This so-called half-period is given by the equation

$$\frac{1}{2}y_0 = y_0e^{-kr}$$

whence we immediately obtain $\tau = \dfrac{\log 2}{k}$

Cooling or Heating of a Body by a Surrounding Medium: Another typical example of the occurrence of the exponential function is offered by the cooling of a body, e.g., a metal plate, which is immersed in a very large bath of a given temperature. In considering this cooling, we assume that the surrounding bath is so large that its temperature is unaffected by the cooling process. Moreover, we assume that at each instant all parts of the immersed body are at the same temperature and that the rate at which the temperature changes is proportional to the difference between the temperature of the body and that of the surrounding medium (Newton's law of cooling).

If we denote the time by x and the temperature difference by $y = y(x)$, this law of cooling is expressed by the equation

$$y' = -ky$$

where k is a positive constant the value of which depends on the body itself. From this instantaneous relationship, which expresses the effect of the cooling process at a given instant, we now wish to derive an integral law which will allow us to find the temperature at an arbitrary time x from the temperature at an initial time $x = 0$. The theorem of 3.7.1 immediately gives us this integral law in the form

$$y = ce^{-kx}$$

where k is the above-mentioned constant depending on the body. This shows that the temperature decreases exponentially and tends to become equal to the external temperature. The rapidity with which this happens is expressed by the number k. As before, we find the meaning of the constant c by considering the instant $x = 0$; this yields $y_0 = c$, so that our law of cooling can finally be written is the form

$$y = y_0e^{-kx}$$

Obviously, the same discussion will also apply to the heating of a body. The only difference is that the initial difference of the temperature y_0 is in this case negative instead of positive.

Variation of the Atmospheric Pressure with the Height above the Surface of the Earth: As another example of the occurrence of the exponential formula we shall derive the law according to which the atmospheric pressure varies with the height. We use here: (1) the physical fact that the atmospheric pressure is equal to the weight of the column of air vertically above a surface of area 1 and (2) Boyle's law, according to which the pressure of the air (p) at a given constant temperature is proportional to the density of the air (σ). Boyle's law, expressed in symbols, is $p = \alpha\sigma$, where α is a constant which depends on a specific physical property of the air and, in addition, it is proportional to the absolute temperature—here, we are not concerned with this, as we shall assume that the temperature is constant. Our problem is to determine $p = f(h)$ as a function of the height (h) above the surface of the Earth.

If we denote by p_0 the atmospheric pressure at the surface of the Earth, i.e., the total weight of the air column supported by unit area, and by $\sigma(\lambda)$ the density of the air at the height λ above the Earth, the weight of the column up to the height h will he given by the integral $\int_0^h \sigma(\lambda)\,d\lambda$. Hence the pressure at height h will be

$$p = f(h) \;=\; p_0 - \int_0^h \sigma(\lambda)\,d\lambda$$

By differentiation, this yields the relationship between the pressure $p = f(h)$ and the density $\sigma(h)$:

$$\sigma(h) \;=\; -f'(h) = -p'$$

We now use Boyle's law to eliminate the quantity σ from this equation and obtain an equation

$$p' \;=\; -\frac{1}{a}p$$

which only involves the unknown pressure function.

$$p = f(h) = p_0 e^{-h/a}$$

If we denote, as above, the pressure at the Earth's surface, i.e., $f(0)$, by p_0, it follows immediately that $c = p_0$, and consequently

$$h \;=\; a\log\frac{p_0}{p}$$

These two formulae find frequent application. For example, if the constant a is known, they enable us to find the height of a place from the barometric pressure or to find the difference of the heights of two locations by measuring the local atmospheric pressures. Again, if the atmospheric pressure and the height h are known, we can determine the constant a, which is of great importance in gas theory.

Progress of a Chemical Solution: We will now consider an example from chemistry, namely the so-called molecular reaction. Let a substance be dissolved in a relatively large amount of solvent, say a quantity of cane sugar in water. If a chemical reaction takes place, the chemical law of mass action in this simple case states that the rate of reaction is proportional to the quantity of reacting substance present. If we assume that the cane-sugar is being transformed by catalytic action into invert sugar and if we denote by $u(x)$ the quantity of cane sugar which at time x is still unchanged, the velocity of reaction will be $-du/dx$, and, in accordance with the law of mass action, there holds an equation of the form

$$\frac{du}{dx} = -ku$$

where k is a constant which depends on the reacting substance. From this instantaneous law, we immediately obtain, as in 3.7.1, an integral law, which gives us the amount of cane sugar as a function of the time

$$u(x) = ae^{-kx}$$

This formula show us clearly how the chemical reaction tends asymptotically to its final state $u = 0$, that is, the complete transformation of the reacting substance. The constant a is obviously the quantity present at $x = 0$.

Making and Breaking of an Electric Circuit: As a final example, consider the growth of a (direct) electric current when a circuit is completed (or its decay, when it is broken). If R is the resistance of the circuit and I the impressed electro-motive force (Voltage), the current I will gradually increase from its original value 0 to the steady final value E/R. Hence we must consider I as a function of the time. The growth of the current depends on the self-induction of the circuit ; the circuit has a characteristic constant L - the coefficient of self-induction - of such a nature that as the current increases, an electro-motive force of magnitude LdI/dx, opposed to the external electromotive force E, is developed. We obtain from Ohm's law, according to which the product of the resistance and the current is at each instant equal to the actual effective voltage, the relation

$$IR = E - L\frac{dI}{dx}$$

Hence, we put

$$f(x) = I(x) - \frac{E}{R}$$

we find immediately that $f'(x) = -R/Lf(x)$, so that, by the theorem in 3.7.1, $f(x) = f(0)e^{-Rx/L}$. Recalling that $I(0) = 0$, we see that $f(0) = -E/R$ and thus obtain the expression for the current as a function of the time

$$I = f(x) + \frac{E}{R} = \frac{E}{R}\left(1 - e^{-Rx/L}\right)$$

We see from this expression that as the circuit is closed the current tends asymptotically to its steady value E/R.

The Hyperbolic Functions

Analytical Definition: In many applications, the exponential function does not enter by itself, but in combinations of the form

$$\frac{1}{2}\left(e^x+e^{-x}\right) \quad \text{or} \quad \frac{1}{2}\left(e^x-e^{-x}\right)$$

It is convenient to introduce these and similar combinations as special functions with the notation:

$$\sin x=\frac{e^x-e^{-x}}{2}, \cos x=\frac{e^x+e^{-x}}{2}$$

$$\tanh x=\frac{e^x-e^{-x}}{e^x+e^{-x}}, \coth x=\frac{e^x+e^{-x}}{e^x-e^{-x}}$$

and call them the hyperbolic sine, hyperbolic cosine, hyperbolic tangent, and hyperbolic cotangent, respectively. The functions sinh x, cosh x and tanh x are defined for all values of x, while in the case of coth x the point $x = 0$ must be excluded. This notation is designed to express a certain analogy with the trigonometric functions; it is this analogy, which we are about to study in detail. as it justifies special consideration of these new functions. The graphs of the hyperbolic functions; the dotted lines in figure are the graphs of $y = e^x/2$ and $y = e^{-x}/2$, from which the graphs of sinh x and cosh x may readily be constructed. We see that cosh x is an even function, i.e., a function which remains unchanged when x is replaced by $-x$, while sinh x is an odd function, i.e., a function which changes sign when x is replaced by -ii.

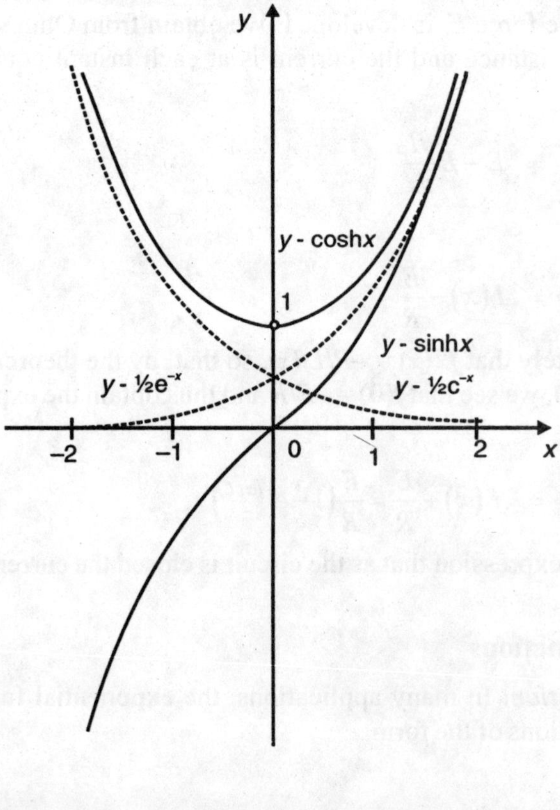

The function

$$\cosh x = \frac{e^x + e^{-x}}{2}$$

is (by its definition) positive for all values of x. It has its minimum when $x = 0$, while $\cosh 0 = 1$.

There exists between $\cosh x$ and $\sinh x$ the fundamental relation

$$\cosh^2 x - \sinh^2 x = 1$$

which follows immediately from the definitions of these functions.If we now denote the independent variable by t instead of x and write

$$x = \cosh t, \qquad y = \sinh t$$

we have

$$x^2 - y^2 = 1$$

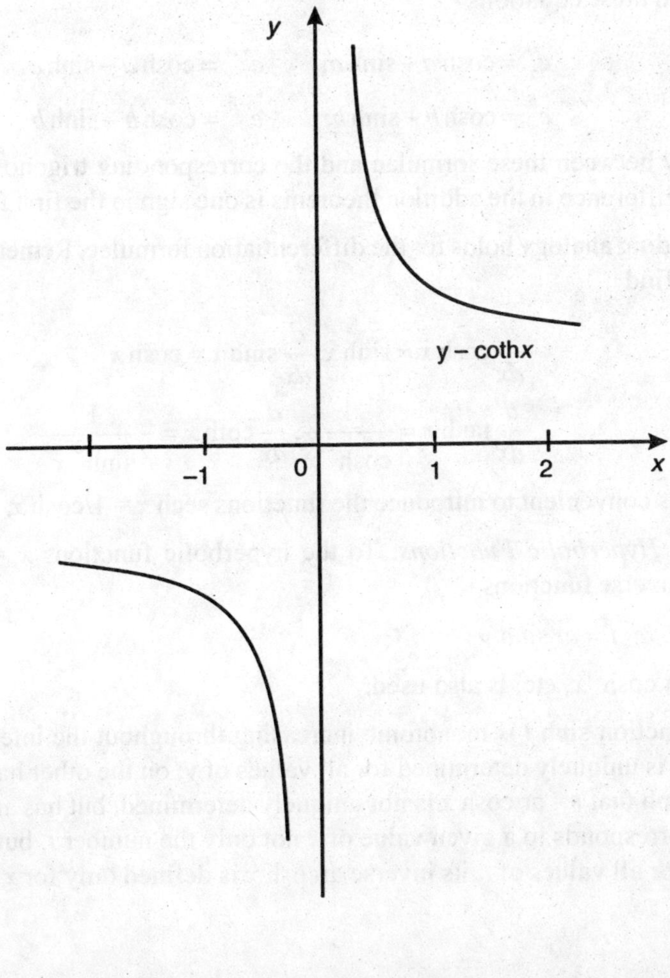

thus, the point with the coordinate $x = \cosh t, y = \sinh t$ moves along the rectangular hyperbola $x - y = 1$ as t runs through the whole scale of values from $-\infty$ to $+\infty$. According to the defining equation, $x \geq 1$ and we may easily convince ourselves that y runs through the whole scale of values $-\infty$ to $+\infty$ with t; in fact, if t tends to infinity so does e^t, while e^{-t} tends to zero. We may therefore state more exactly that as t runs from $-\infty$ to $+\infty$, the equations $x = \cosh t, y = \sinh t$ give us one branch, namely, the right-hand one, of the rectangular hyperbola.

Addition Theorems and Formulae for Differentiation: From the definitions of our functions follow the formulae, known as addition theorems:

$$\cosh(a+b) = \cosh a \cosh b + \sinh a \sinh b,$$
$$\sinh(a+b) = \sinh a \cosh b + \cosh a \sinh b$$

The proofs are obtained at once by writing

$$\cosh(a+b) = \frac{e^a e^b + e^{-a} e^{-b}}{2}, \sinh(a+b) = \frac{e^a e^b - e^{-a} e^{-b}}{2}$$

and setting in these equations

$$e^a = \cosh a + \sinh a, \qquad e^{-a} = \cosh a - \sinh a,$$
$$e^b = \cosh b + \sinh b, \qquad e^{-b} = \cosh b - \sinh b$$

The analogy between these formulae and the corresponding trigonometrical formulae is clear. The only difference in the addition theorems is one sign in the first formula.

A corresponding analogy holds for the differentiation formulae. Remembering that $d(e^x)/dx = e^x$, we readily find

$$\frac{d}{dx}\cosh x = \sinh x, \frac{d}{dx}\sinh x = \cosh x$$
$$\frac{d}{dx}\tanh x = \frac{1}{\cosh^2 x}, \frac{d}{dx}\coth x = -\frac{1}{\sinh^2 x}$$

At times, it is convenient to introduce the functions $\operatorname{sech} x = 1/\cosh x$, $\operatorname{cosech} x = 1/\sinh x$.

The Inverse Hyperbolic Functions: To the hyperbolic functions $x = \cosh t$, $y = \sinh t$ correspond the inverse functions

$t = ar \cosh x, \qquad t = ar \sinh y$

The notation $\cosh^{-1} x$, etc. is also used.

Since the function $\sinh t$ is monotonic increasing throughout the interval $-1 < t < +1$, its inverse function is uniquely determined for all values of y; on the other hand, we learn from a glance at the graph that $t =$ ar cosh x is not uniquely determined, but has an ambiguity of sign, because there corresponds to a given value of x not only the number t, but also the number $-t$. Since $\cosh t < 1$ for all values of t, its inverse arcosh x is defined only for $x < 1$.

We can express these inverse functions very easily in terms of the logarithm by regarding the quantity $e^t = u$ in the definitions

$$x = \frac{e^t + e^{-t}}{2}, y = \frac{e^t - e^{-t}}{2}$$

as unknowns and solving these (quadratic) equations for u. Then

$$u = x \pm \sqrt{(x^2 - 1)}, u = y + \sqrt{(y^2 + 1)};$$

since $u = e^t$ can have only positive values, the square root in the second equation must be taken with the positive sign, while in the first equation either sign is possible. In the logarithmic form

$$t = \log\left(x \pm \sqrt{(x^2 - 1)}\right) = ar\cosh x$$

$$t = \log\left(y + \sqrt{(y^2 + 1)}\right) = ar\sinh y$$

In the case of arcosh x, the variable x is restricted to the interval $x \geq 1$, while arsinh y is defined for all values of y.

The formula gives us two values, $\log\left\{x + \sqrt{(x^2 - 1)}\right\}$ and $\log\left\{x - \sqrt{(x^2 - 1)}\right\}$ for arcosh x, corresponding to the two branches of arcosh x. Since

$$\left\{x + \sqrt{(x^2 - 1)}\right\}\left\{x - \sqrt{(x^2 - 1)}\right\} = 1$$

the sum of these two values of arcosh x is zero, which agrees with an earlier remark.

The inverses of the hyperbolic tangent and hyperbolic cotangent can be defined analogously and can also be expressed in terms of logarithms. We denote these functions by artanh x and arcotanh x; expressing the independent variable everywhere by x, we readily obtain

$$ar\tanh x = \frac{1}{2}\log\frac{1+x}{1-x} \text{ in the interval } -1 < x < 1$$

$$\text{arcoth}x = \frac{1}{2}\log\frac{x+1}{x-1} \text{ in the intervals } x < -1, x > 1$$

The differentiation of these inverse functions should be carried out by the reader himself; he may use either the rule for differentiating an inverse function or the chain rule in conjunction with the above expressions for the inverse functions in terms of logarithms. If x is the independent variable, the results are

$$\frac{d}{dx} ar\cosh x = \pm\frac{1}{\sqrt{(x^2-1)}}, \quad \frac{d}{dx} ar\sinh x = \frac{1}{\sqrt{(x^2+1)}}$$

$$\frac{d}{dx} ar\tanh x = \frac{1}{1-x^2}, \frac{d}{dx} ar\coth x = \frac{1}{1-x^2}$$

The last two formulae do not contradict each other, since the first holds only for $-1 < x < 1$ and the second only for $x < -1$ and $1 < x$. The two values of

$d$ar cosh x/dx, expressed by the sign in the first formula, correspond to the two different branches of the curve

$$y = ar\cosh x = \log\left\{x \pm \sqrt{(x^2-1)}\right\}$$

Further Analogies: In the above representation of the rectangular hyperbola by the quantity t, we did not attempt to bring out any geometrical meaning of the parameter t itself. We shall now return to this matter and thus gain still more insight into the analogy between the trigonometric and hyperbolic functions. If we represent the circle by the equation $x + y = 1$ by means of a parameter t in the form $x = \cos t$, $y = \sin t$, we can interpret the quantity t as an angle or a length of arc measured along the circumference; however, we may also regard f as twice the area of the circular sector, corresponding to that angle, the area being reckoned positive or negative a cording *to as* the angle is positive or negative.

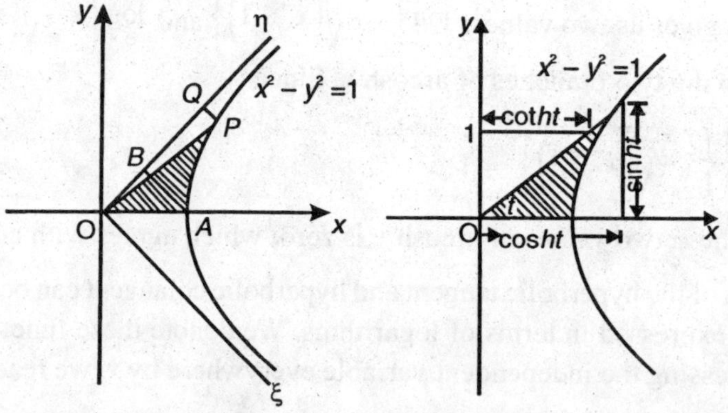

Fig. Parametric representation of the hyperbola and hyperbolic functions

We now make the analogous statement that for the hyperbolic functions the quantity t is twice the area of the hyperbolic sector.

Just as the notation t = arcos x recalls that t is an arc of a circle of reference, so t = ar cosh x recalls that t is a certain area connected with a rectangular hyperbola.

The proof is obtained without difficulty, if we refer the hyperbola to its asymptotes as axes by means of the coordinate transformation

$$x - y = \sqrt{2}\xi, x + y = \sqrt{2}\eta$$

or

$$x = \frac{1}{\sqrt{2}}(\xi+\eta), y = \frac{1}{\sqrt{2}}(r_1-\xi)$$

with these new coordinates, the equation of the hyperbola becomes $\xi\eta = 0$. We thus see immediately that the area in question is equal to the area of the figure *ABQP;* in fact, the two right-angled triangles *OPQ* and *OAB* have the same area, according to the equation of the hyperbola. ' The two points *A* and *P* obviously have the coordinates

$$\xi = \frac{1}{\sqrt{2}}, \eta = \frac{1}{\sqrt{2}} \text{ and } \xi = \frac{x-y}{\sqrt{2}}, \eta = \frac{x+y}{\sqrt{2}}$$

respectively, and we thus obtain for double the area of our figure

$$2\int_{1/\sqrt{2}}^{(x+y)/\sqrt{2}} (1/2\eta)\,d\eta = \log(x+y) = \log\left\{x \pm \sqrt{\left(x^2-1\right)}\right\}$$

A comparison of this expression with the formula in 3.8.3 for the inverse function t = ar cosh x shows us that our statement about the quantity t is true.

In conclusion, it may be pointed out that above, the hyperbolic functions can be diagrammatically represented on the circle.

The Order of Magnitude of Functions

The various functions which exhibit very important differences as regards their behaviour for large values of their arguments or, as we also say, of the order of magnitude of their increase. In view of the great importance of this, we shall discuss here briefly this matter, even though it is not directly linked to the idea of the integral or of the derivative.

The Concept of Order of Magnitude: The simplest Cases: If the variable x increases beyond all bounds, then, for $a > 0$, the functions a, $\log x$, e^x, e^{ax} will also increase beyond all bounds. However, as regards the manner of this increase, we can immediately point out an essential difference between them. For example, the function x will become infinite to a higher order than x; this means that, as x increases, the quotient x/x itself increases beyond all bounds. Similarly, we shall say that the function x^α becomes infinite to a higher order than x^β, if $\alpha > \beta >$ 0, etc.

Quite generally, we shall say of two functions $f(x)$ and $g(x)$, the absolute values of which increase with x beyond all bounds, that $f(x)$ becomes infinite of a higher order than $g(x)$, if, as x increases, the quotient $|f(x)/g(x)|$ increases beyond all bounds; we shall say that $f(x)$ becomes infinite of a lower order than $g(x)$, if the quotient $|f(x)/g(x)|$ tends to zero as x increases; and we shall say that the two functions become infinite of the same order of magnitude if as x increases the quotient $|f(x)/g(x)|$ has a limit other than 0 or at least remains between two fixed positive bounds. For example, the function $ax + bx + c = f(x)$, where $a \not= 0$, will be of the same order of magnitude as the function $x = g(x)$; in fact, the quotient $\left|\frac{f(x)}{g(x)}\right| = \left|\frac{ax^3+bx^2+c}{x^3}\right|$ has the limit $|a|$.

On the other hand, the function $x + x + 1$ becomes infinite of a higher order of magnitude than the function $x + x + 1$.

A sum of two functions $f(x)$ and $\phi(x)$, where $f(x)$ is of higher order of magnitude than $\phi(x)$, has the same order of magnitude as $f(x)$. In fact,

$$\left|\frac{f(x)+\phi(x)}{f(x)}\right| = \left|1+\frac{\phi(x)}{f(x)}\right|$$

and, by hypothesis, this expression tends to 1as x increases.

We might be tempted to measure the order of magnitude of functions by a scale, assigning to the quantity x the order of magnitude 1 and to the power x^{α} ($\alpha > 0$) the order of magnitude α. Obviously, then a polynomial of the n-th degree has the order of magnitude n; a rational function, the degree of the numerator of which is greater than that of the denominator by h, has then the order of magnitude h.

The Order of Magnitude of the Exponential Function and the Logarithm: However, it turns out that any attempt to fix the order of magnitude of an arbitrary function by the above scale must end in failure. In fact, there are functions which become infinite of higher order than the power x^{α} of x, no matter however large is the chosen α; again, there are functions which become infinite of lower order than the power x^{α}, no matter how small the positive number α is chosen. Hence, these functions will not fit in anywhere into our scale. Without entering into a detailed theory of order of magnitude we shall prove the theorem: If a is an arbitrary number greater than 1, then the quotient a^x tends to infinity as x increases. In order to prove this statement, we construct the function

$$\phi(x) = \log\frac{a^x}{x} = x\log a - \log x$$

obviously, it is sufficient to show that this increases beyond all bounds if x tends to +1. For this purpose, we consider the derivative

$$\phi'(x) = \log a - \frac{1}{x}$$

and note that for $x \rangle 0$ this is not less than the positive number 1/2 log a, whence, for $x < c$

$$\phi(x)-\phi(c) = \int_c^x \phi'(t)\,dt \geq \int_c^x \frac{1}{2}\log a\,dt \geq \frac{1}{2}(x-c)\log a$$

$$\phi(x) \geq \phi(c) + \frac{1}{2}(x-c)\log a$$

and the right-hand side becomes infinite as x increases.

We shall give a second proof of this important theorem. If we write $\sqrt{a} = b = 1 + h$, we have $b > 1$ and $h > 0$. Let n be an integer such that $n < \xi < n+1$; we may take $x > 1$, so that $n < 1$. Applying the Lemma, we have

$$\sqrt{\left(\frac{a^x}{x}\right)} = \frac{b_x}{\sqrt{x}} = \frac{(1+h)^x}{\sqrt{x}} > \frac{(1+h)^n}{\sqrt{(n+1)}} > \frac{1+nh}{\sqrt{(n+1)}} > \frac{nh}{\sqrt{2n}} = \frac{h}{\sqrt{2}}\sqrt{n}$$

so that

$$\frac{a^x}{x} > \frac{h^2}{2} \cdot n$$

whence it tends to infinity with x.

There follow many results from the fact just proved. For example, for every positive index α and every number $a > 1$, the quotient a^x/x^α tends to infinity as x increases, that is: The exponential function becomes infinite of a higher order of magnitude than any power of x.

In order to see this, we need only show that the α-th root of the expression, that is,

$$\frac{a^{x/\alpha}}{x} = \frac{1}{\alpha} \frac{a^{x/a}}{x/a} = \frac{1}{a} \cdot \frac{a^x}{y} \left(y = \frac{x}{a}\right)$$

tends to infinity. However, this follows immediately from the preceding theorem, when x is replaced by $y = x/\alpha$.

We can prove in a similar fashion the theorem: For every positive value of α, the quotient $(\log x)/x^\alpha$ tends to zero when x tends to infinity, that is, the logarithm becomes infinite at a lower order of magnitude than any arbitrarily small positive power of x.

The proof follows immediately, if we set $\log x = y$, so that our quotient is transformed into $y/e^{\alpha y}$. We then set $e^\alpha = a$; then a is a number > 1 and our quotient y/a^y approaches 0 as y increases. Since y approaches infinity with x, our theorem is proved.

Another very simple proof maybe suggested: For $x > 1$ and $\varepsilon > 0$,

$$\log x = \int_1^x \frac{d\xi}{\xi} < \int_1^x \xi^2 - 1 d\xi = \frac{1}{\in}\left(x^2 - 1\right)$$

if we choose ε smaller than α and divide both members of the inequality by x^α, then as x it follow that $(\log x)/x^\alpha < 0$.

Based on these results, we can construct functions of an order of magnitude by far higher than that of the exponential function and other functions e^{ex} of an order of magnitude far lower than that of the logarithm. For example, the function is of higher order than the exponential function, and the function $\log \log x$ of lower order than the logarithm; obviously, we can repeat these iteration processes as often as we like, piling up the symbols e or log to any extent we please.

General Remarks: These considerations show that it is not possible to assign by systematic reasoning to all functions definite numbers as orders of magnitude in such a way that, when two functions are compared, the function of the higher order of magnitude has the higher number.

For example, if the function x is of the order of magnitude 1 and the function $x^{1+\varepsilon}$ of the order of magnitude $1+\varepsilon$, then the function $x \log x$ must be of an order of magnitude which is greater than I and less than $1+\varepsilon$ no matter how small ε is chosen. But there exists no such number. Moreover, it is readily seen that functions need not possess a clearly defined order of magnitude.

For example, the function $\dfrac{x^2(\sin x)^2 + x + 1}{x^2(\cos x)^2 + x}$

approaches no definite limits as x increase; on the contrary, for $x = n\pi$ (where n is an integer) its value is $1/n\pi$, while, for $x = (n+1/2)\pi$, it is $(n+1/2)\pi + 1 + 1/(n+)\pi$. Although both the numerator and denominator become infinite, the quotient neither remains between positive bounds, nor tends to zero, nor tends to infinity. Hence, the numerator is neither of the same order as the denominator, nor of lower order, nor of higher order. This apparently startling situation merely means that our definitions are not designed in such a way that we can compare any pair of functions. This is not a defect; we have no desire to compare the orders of such functions like the numerator and denominator above, since knowledge of the value of one of them gives us no useful information about the other one.

The Order of Magnitude of a Function in the Neighbourhood of an Arbitrary Point: Just as we can investigate the behaviour of a function when x increases without limit, we may also ask ourselves whether and how functions which become infinite at the point $x = \xi$ may be distinguished with regard to their behaviour at that point. Moreover, we state that the function $f(x) = 1/|x - \xi|$ becomes infinite at the first order at the point $x = \xi$, and correspondingly that the function $1/|x - \xi|^{\alpha}$ becomes infinite at the order α, provided that a is positive.

We then recognize that the function $e^{1/|x-\xi|}$ becomes infinite at higher order, and the function $\log|x - \xi|$ infinite at lower order than all these powers, i.e., that there hold the limiting relations

$$\lim_{x \to \xi}\left(|x-\xi|^{\alpha} \cdot e^{1/|x-\xi|}\right) = \infty$$

and

$$\lim_{x \to \xi}\left(|x-\xi|^{\alpha} \cdot \log|x-\xi|\right) = 0$$

In order to see this, we merely set $1/|x - \xi| = y$; our statements then reduce to the known theorem, since

$$|x-\xi|^{a} \cdot e^{1/|x-\xi|} = e^{v}/y^{\alpha} \text{ and}$$

$$|x-\xi|^{\alpha} \cdot \log|x-\xi| = -(\log y)/y^{\alpha}$$

and y increases beyond all bounds as x tends to ξ The method of reduction of the behaviour at a finite point to the behaviour at infinity by the substitution $1/|x - \xi| = y$ frequently proves useful.

The Order of Magnitude of a Function tending to Zero: Just as we seek to describe the approach of a function to infinity more definitely by means of the concept of order of magnitude, we may also specify the way in which a function approaches zero.

$$\lim_{x \to 0}\left(x^{-a} \cdot \log x\right) = 0$$

applies for every positive α.

In the same way, we say that, for $x = \xi$, the quantity $x - \xi$ vanishes to the first order, the quantity $|x-\xi|^{\alpha}$ to the order α. With the above results, it is easy to prove the relations

$$\lim_{x \to 0}\left(|x|^{\alpha} \cdot \log|x|\right) = 0$$

$$\lim_{x \to 0}\left(|x|^{-\alpha} \cdot e^{-1/|x|}\right) = 0$$

which are usually expressed as follows:

The function $1/\log|x|$ vanishes as $x < 0$ to a lower order than any power of x; the exponential function $e^{-1/|x|}$ vanishes to a higher order than any power x.

Chapter 5

Further Development of the Integral Calculus

The rules for differentiation, have given us great power over the problem of differentiating given functions. However, almost always, its inverse problem of integration exceeds it greatly in importance, whence we must now study the art of integrating given functions. The results obtained by means of our differentiation formulae may be summed up as follows:

Every function which is formed from the elementary functions by means of a closed expression*, can be differentiated, and its derivative, if it is also a closed expression, formed from the elementary functions.

* By this we mean a function which can be built up from the elementary functions by repeated application of the rational operations and the processes of compounding and inversion. However. in this context, it should be emphasized that the distinction between elementary functions and other functions is itself quite arbitrary.

On the other hand, we have encountered any exactly corresponding fact relating to the integration of elementary functions. We do know that every elementary function and, in fact, every continuous function can be integrated, and we have integrated a large number of elementary functions either directly or by inversion of differentiation formulae and have found their integrals to be expressions involving only elementary functions. But we are still far from being able to find a general solution of the problem: Given a function $f(x)$, which is expressed in terms of the elementary functions by any closed expression, find an expression for its indefinite integral $F(x) = \int f(x)dx$ which is itself a closed expression in terms of the elementary functions.

It is a fact that, in general, this problem is insoluble; it is by no means true that every elementary function has an integral which itself is an elementary function. Nevertheless, it is extremely important that we should be able to actually carry out such integrations whenever they are possible, and that we should acquire a certain amount of technical skill in the integration of given functions.

In this connection, we expressly warn the beginner against merely memorizing the many formulae obtained by using these technical devices. Instead, students should direct their efforts towards gaining a clear understanding of the methods of integration and learning how to apply them. Moreover, they should remember that even when integration by these devices is impossible, the integral does exist (at least for all continuous functions) and can actually be computed to as high a degree of accuracy as is desired by means of the numerical methods which will be developed.

We shall endeavour to deepen and extend our ideas of integration and the integral, quite apart from the problem of integration techniques.

Elementary Integrals

First of all, we repeat that an equivalent integration formula corresponds to everyone of the earlier proved differentiation formulae. Since these elementary integrals are used over and over again as material in the art of integration, we have collected them in the table below. The right-hand column contains elementary functions, the left-hand column the corresponding derivatives. If we read the table from the left hand side to the right hand side, we find in the right hand column an indefinite integral of the function in the left hand column.

We also remind the reader of the fundamental theorems of the differential and integral calculus, proved in 2.4, and, in particular, of the fact that the definite integral is obtained from the indefinite integral $F(x)$ by the formula

$$\int_a^b f(x)dx = F(x)\Big|_a^b = F(b) - F(a)$$

Finally, for the technique of integration, the reader should have the elementary rules of integration 2.1.3 at his finger-tips.

In the following sections, we shall attempt to reduce the calculation of integrals of given functions in some way or other to the elementary integrals collected in the table above. Apart from devices which the beginner certainly could not acquire systematically, but which, on the contrary, occur only to those with long experience, this reduction is based essentially on two useful methods. Each of these methods enables us to transform a given integral in many ways; the objective of such transformations is to reduce the given integral, in one step or in a sequence of steps, to one or more of the above elementary integration formulae.

The Method of Substitution

The first of these useful methods for attacking integration problems is the introduction of a new variable (i.e. the method of substitution or transformation). The corresponding integral formula is just the chain rule of the differential calculus, expressed in the integral form.

The Substitution Formula: We assume that a new variable u is introduced into a function $F(x)$ by means of the equation $x = \phi(u)$, so that $F(x)$ becomes a function of u:

$$F(x) = F(\phi(u)) = G(u)$$

By the chain rule of the differential calculus,

$$\frac{dG}{du} = \frac{dF}{dx}\phi'(u)$$

If we now write

$$F'(x) = f(x) \text{ and } G'(u) = g(u)$$

or the equivalent expressions

$$F(x) = \int f(x)dx \text{ and } G(u) = \int g(u)du$$

then, on the one hand, the chain rule takes the form

$$g(u) = \int(x)\phi'(u)$$

and, on the other hand, $G(u) = F(x)$ by definition, that is,

$$\int g(u)du = \int f(x)dx$$

and we obtain the integral formula, equivalent to the chain rule,

$$\int f(\phi(u))\phi'(u)du = \int f(x)dx, \{x = \phi(u)\}$$

This is the basic formula for the substitution of a new variable in an integral. It means that, if we wish to find an indefinite integral of a function of u, which is given in the special form $f(\phi(u))\phi'(u)$, we can instead find the indefinite integral of the function $f(x)$ as a function of x and after integration return to the variable u by setting $x = \phi(u)$.

For example, if we apply the formula to the integrand $\phi'(u)/\phi(u)$, we obtain

$$\int\frac{\phi'(u)}{\phi(u)}du = \int\frac{dx}{x} = \log|x| = \log|\phi(u)|$$

or, replacing u by x,

$$\int\frac{\phi'(x)}{\phi(x)}dx = \log|\phi(x)|$$

If we substitute in this important formula particular functions such as $\phi(x) = \log x$ or $\phi(x) = \sin x$, we obtain*

$$\int\frac{dx}{x\log x} = \log|\log x|$$

$$\int\cot x dx = \log|\sin x|, \int\tan x dx = -\log|\cos x|$$

A further example is

$$\int\varphi(u)\varphi'(u)du = \int x dx = \frac{1}{2}x^2 = \frac{1}{2}[\varphi(u)]^2$$

where $f(x) = x$. This yields for $\phi(u) = \log u$

$$\int \frac{\log u}{u} du = \frac{1}{2}(\log u)^2$$

These and the following formulae are verified by showing that differentiation of the result returns the integrand. Moreover, the formulae are, of course, only asserted as true in as far as the expressions occurring in them have a meaning.

We finally consider the example

$$\int \sin^n u \cos u du$$

Here $x = \sin u = \phi(u)$, whence

$$\int \sin^n u \cos u du = \int x^n dx = \frac{x^{n+1}}{n+1} = \frac{\sin^{n+1} u}{n+1}$$

However, in many cases, we shall use the above formula in the reverse direction, starting with the right hand side - the integral $\int f(x)dx$. We now have to evaluate or simplify a prescribed indefinite integral $F(x) = \int f(x)dx$ by introduction of the new integration variable u by means of the transformation formula $x = \phi(u)$, then working out the indefinite integral

$$G(u) = \int f(\phi(u))\phi'(u)du$$

and finally replacing the variable u in this integral by x. In order to carry out this last step, we must be certain that a definite value u actually corresponds to the value x, i.e., that the function $x = \phi(u)$ has an inverse. Accordingly, we now make the following assumption, in which we regard x as the primary variable: In the interval under consideration, $u = \psi(x)$ is a monotonic, differentiable function, the derivative $f'(x)$ of which does not vanish anywhere in the interval. We denote the inverse function, which under these conditions is definite and single-valued, by $x = \phi(u)$; its derivative is then given by $\phi'(u) == 1/\psi'(x)$. As the basic formula for the substitution of a new variable u in an integral, we obtain

$$\int f(x)dx = \int f(\phi(u))\phi'(u)du \quad (u = \psi(x))$$

The indefinite integral $\int f(x)dx$ can be obtained by calculation of the indefinite integral $\int f(\phi(u))\phi'(u)du$ and finally introducing x instead of u for the independent variable by means of the equation $u = \psi(x)$.

Hence it is not sufficient merely to express the old variable x in terms of the new variable u and then to integrate with respect to this new variable; before integrating, we must multiply by this derivative of the original variable x with respect to the new variable u.

The corresponding formula for definite integration between two limits is

$$\int_a^b f(x)dx = \int_{\psi(a)}^{\psi(b)} f(\phi(u))\phi'(u)du$$

In the new integral, we have to choose those limits of integration which are obtained by subjecting the old integration limits to the transformation $x = \phi(u)$, $u = \psi(x)$.

In most applications, the integrand $f(x)$ will appear at the outset as a function of a function, say, $f(x) = h(u)$, where $u = \psi(x)$. It is then more convenient to write our integral formula in a slightly different form by identifying the expression $f\{\phi(u)\}$ with the expression $h(h)$. If we make for u the substitution $u = \psi(x)$, $x = \phi(u)$, then our transformation formula is simply

$$\int h\{\psi(x)\}dx = \int h(u)\frac{dx}{du}du$$

As a first example, consider the integration of the function $f(x) = \sin 2x$ and introduce $u = \psi(x) = 2x$ and $h(u) = \sin u$. We have

$$\frac{du}{dx} = \psi'(x) = 2$$

If we now substitute $u = 2x$ into the integral as the new variable, then it is not transformed into $\int \sin u du$, but into

$$\frac{1}{2}\int \sin u du = -\frac{1}{2}\cos u = -\frac{1}{2}\cos 2\pi$$

of course, this may of be verified at once by differentiating the right hand side.

If we integrate for x between the limits 0 and $\pi/4$, the corresponding limits for u are 0 and $\pi/2$, and we obtain

$$\int_0^{\pi/4} \sin 2x dx = \frac{1}{2}\int_0^{\pi/2} \sin u du = -\frac{1}{2}\cos u\Big|_0^{\pi/2} = \frac{1}{2}$$

Another simple example is the integral $\int_1^4 \frac{dx}{\sqrt{x}}$. Here we take $u = \psi(x) = \sqrt{x}$ whence $x = \phi(u) = u$. Since $\phi'(u) = 2u$, we find

$$\int_1^4 \frac{dx}{\sqrt{x}} = 2\int_1^2 \frac{u du}{u} = 2\int_1^2 du = 2$$

Another Proof of the Substitution Formula: Our integration formula can also be explained in another and more direct manner by aiming at the formula for definite integration and basing the proof on the meaning of the definite integral as a limit of a sum. In order to calculate the integral

$$\int_a^b h(\psi(x))dx$$

(for the case $a < b$), we begin with an arbitrary subdivision of the interval $a < x < b$, and then make the sub-division finer and finer. We choose these sub-divisions in the following way. If the function $u = \psi(x)$ is assumed to be monotonic increasing, there is a (1,1) correspondence between the interval $a < x < b$ on the x-axis and an interval $\alpha < u < \beta$ of the values of $u = \psi(z)$, where $\alpha = \psi(a)$, $\beta = \psi(b)$. We subdivide this interval into n parts of length Δu; there is a corresponding subdivision of the x-interval into subintervals which, in general, are not all of the same length. (The assumption that these subintervals are all equal it by no means essential for the proof!) We denote the points of division of the x-interval by $x_0 = a, x_1, x_2, \ldots, x_n = b$ and the lengths of the corresponding sub-intervals by

$$\Delta x_1, \Delta x_2, \ldots, \Delta x_n$$

The integral under consideration is then the limit of the sum

$$\sum_{\nu=1}^{n} h\{\psi(\xi_\nu)\}\Delta x_\nu$$

where the value ξ_ν is arbitrarily selected from the ν-th subinterval of the x-subdivision. This limit exists (for $\Delta u \to 0$) and is the integral, since on account of the uniform continuity of $x = \phi(u)$ the greatest of the lengths Δx tends to 0 with Δu.

We now write this sum in the form $\sum_{\nu=1}^{n} h(u_\nu)\frac{\Delta x_\nu}{\Delta u}\Delta u$ where $u_\nu = \psi(\xi_\nu)$. By the mean value theorem of the differential calculus, $\frac{\Delta x_\nu}{\Delta u} = \phi'(\eta_\nu)$ where η_ν is a suitably chosen intermediate value of the variable u in the ν-th sub-interval of the u-sub-division and $x = \phi(u)$ denotes the inverse function of $u = \psi(x)$. If we now select the value ξ_ν in such a way that ξ and η coincide, i.e., $\xi_\nu = \phi(\eta_\nu)$, $\eta_\nu = \psi(\xi_\nu)$, then our sum takes the form

$$\sum_{\nu=1}^{n} h(\eta_\nu)\phi'(\eta_\nu)\Delta u$$

If we make here the passage to the limit, we immediately obtain the expression

$$\int_\alpha^\beta h(u)\frac{dx}{du}du$$

as the limiting value, that is, as the value of the integral under consideration in agreement with the formula given above.

Hence we have proved the theorem:

Let $h(u)$ be a continuous function of u in the interval $\alpha < u < \beta$. If the function $u = \psi(x)$ is continuous and monotonic, has a continuous, non-vanishing derivative du/dx in $a < x < b$, and $\psi(a) = \alpha, \psi(b) = \beta$, then

$$\int_a^b h(\psi(x))dx = \int_a^b h(u)dx = \int_\alpha^\beta h(u)\frac{dx}{du}du$$

This formula exhibits the advantage of Leibnitz's notation. In order to carry out the substitution $u=\psi(x)$, we only need write $(dx/du)du$ in place of dx, changing the limits from the original values of x to the corresponding values of u.

Examples. Integration formulae: With the help of the substitution rule, we canevaluate a given integral $\int f(x)dx$ in many cases by reducing it by means of a suitable substitution $x = \phi(u)$ to one of the elementary integrations in our table. Whether such substitutions exist and how to find them are questions which have no general answer; this is rather a matter in which practice and ingenuity, in contrast to a systematic method, come into their own.

As an example, we shall work out the integral $\int\frac{dx}{\sqrt{(a^2-x^2)}}$ using the substitution $x = \phi(u)$ $= au$, $u=\psi(x)=x/a$, $dx=adu$, by which, with No. 13 in the table above, we obtain

$$\int\frac{dx}{\sqrt{(a^2-x^2)}} = \int\frac{adu}{a\sqrt{(1-u^2)}} = \text{arc sinu} = \text{arc sin}\frac{x}{a},\ \text{for } |x|<|a|$$

For the sake of brevity, we have written the symbols dx and du separately, i.e., $dx = \phi$ '$(u)du$ instead of $dx/du = \phi$ '(u).

The same substitution yields also

$$\int\frac{dx}{a^2+x^2} = \int\frac{adu}{a^2(1+u^2)} = \frac{1}{a}\arctan u = \frac{1}{a}\arctan\frac{x}{a}$$

$$\int\frac{dx}{\sqrt{(a^2+x^2)}} = ar\sinh\frac{x}{a}$$

$$\int\frac{dx}{\sqrt{(x^2-a^2)}} = ar\cosh\frac{x}{a},\ \text{for } |x|>|a|$$

$$\int\frac{dx}{a^2-x^2} = \begin{cases}\frac{1}{a}ar\tanh\frac{x}{a}\ \text{for}|x|<|a| \\ \frac{1}{a}ar\coth\frac{x}{a}\ \text{for}|x|>|a|\end{cases}$$

formulae which occur very frequently and which are easily verified by differentiating the right hand side.

In conclusion, we again emphasize the point: In our substitution process, we have made the assumption that the substitution has a unique inverse $x = \phi(u)$, and indeed that $\psi'(x)$ is nowhere equal to zero in the interval under consideration. If our assumption is not fulfilled, application of the substitution formula may easily lead to wrong conclusions. If $\psi'(x) = 0$ at isolated points of the interval of integration only, we can avoid these difficulties by subdividing this interval in such a way that $\psi'(x)$ vanishes only at the ends of a sub-interval; we can then apply the substitution to each sub-interval separately.

An application of this method leads at once to the following result, which applies to many special cases: If the derivative $\psi'(x)$ vanishes at a finite number of points, but the function $\psi(x)$remains monotonic, then the substitution formula remains valid.

Further Examples of the Substitution Method

We collect here several examples which the reader should study carefully by way of practice!

By the substitution $u = 1 \} x$, $du = \} 2xdx$, we find

$$\int \frac{xdx}{\sqrt{\left(1 \pm x^2\right)}} = \pm\sqrt{\left(1 \pm x^2\right)}$$

$$\int \frac{xdx}{1 \pm x^2} = \pm\frac{1}{2}\log\left|1 \pm x^2\right|$$

In these formulae, we must take either the sign + or the sign - in all three places.

By the substitution $u = ax + b$, $du = adx$ $(a \} 0)$, we find

$$\int \frac{dx}{ax+b} = \frac{1}{a}\log|ax+b|$$

$$\int (ax+b)^{\alpha}\, dx = \frac{1}{a(\alpha+1)}(ax+b)^{\alpha+1} \qquad (\alpha \neq -1)$$

$$\int \sin(ax+b)\,dx = -\frac{1}{a}\cos(ax+b)$$

in a similar manner, using the substitution $u = \cos x$, $du = -\sin x\, dx$, we obtain

$$\int \tan x dx = -\log|\cos x|$$

and by means of the substitution $u = \sin x$, $du = \cos x\, dx$

$$\int \cot x dx = \log|\sin x|$$

Using the analogous substitutions $u = \cosh x$, $du = \sinh x\, dx$ and $u = \sinh x$, $du = \cosh x dx$, we obtain

$$\int \tanh x dx = \log|\cosh x|$$

$$\log|\sinh x| = \int \coth x dx$$

By the substitution $u = (a/b) \tan x$, $du = (a/b) \sec x\, dx$, we find the two formulae

$$\int \frac{dx}{a^2 \sin^2 x + b^2 \cos^2 x} = \frac{1}{b^2} \int \frac{1}{a^2/b^2 \tan^2 x + 1} \cdot \frac{dx}{\cos^2 x} = \frac{1}{ab} \arctan\left(\frac{a}{b} \tan x\right)$$

and

$$S_m(\alpha) = \frac{1}{m+1} \left[\frac{\sin \frac{(m+1)\alpha}{2}}{\sin \frac{\alpha}{2}} \right]^2$$

We evaluate the integral

$$\int \frac{dx}{\sin xa}$$

by writing $\sin x = 2 \sin x/2 \cos x/2 = 2 \tan x/2 \cos x/2$ and setting $u = \tan x/2$ so that du $\sec x/2 dx$; the integral then becomes

$$\int \frac{dx}{\sin x} = \int \frac{du}{u} = \log\left|\tan \frac{x}{2}\right|$$

If we replace x by $x + \pi/2$, this formula becomes

$$\int \frac{dx}{\cos x} = \log\left|\tan\left(\frac{x}{2} + \frac{\pi}{4}\right)\right|$$

If we also apply the known trigonometrical formulae $2 \cos x = 1 + \cos 2x$ and $2 \sin x = 1 - \cos 2x$, the substitution $u = 2x$ yields

$$\int \cos^2 dx = \frac{1}{2}(x + \sin x \cos x) \text{ and } \int \sin^2 x dx = \frac{1}{2}(x - \sin x \cos x)$$

By the substitution $x = \cos u$, equivalent to $u = \text{arcos}\, x$ or, more generally,

$x = a \cos u$ $(a \succ 0)$, we can reduce

$$\int \sqrt{(1 - x^2)} dx \text{ and } \int \sqrt{(a^2 - x^2)} dx$$

respectively, to these formulae. We thus obtain

$$\int \sqrt{(a^2 - x^2)} dx = -\frac{a^2}{2} \arccos \frac{x}{a} + \frac{x}{2} \sqrt{(a^2 - x^2)}$$

Similarly, the substitution $x = a\cosh u$ yields

$$\int\sqrt{\left(x^2-a^2\right)}dx = -\frac{a^2}{2}\operatorname{arcos}h\frac{x}{a}+\frac{x}{2}\sqrt{\left(x^2-a^2\right)}$$

and the substitution $x = a\sinh u$

$$\int\sqrt{\left(a^2+x^2\right)}dx = \frac{a^2}{2}\operatorname{ar}\sin h\frac{x}{a}+\frac{x}{2}\sqrt{\left(a^2+x^2\right)}$$

The substitution $u = a/x$, $dx = -(a/u)\,du$ leads to

$$\int\frac{dx}{x\sqrt{\left(x^2-a^2\right)}} = -\frac{1}{a}\arcsin\frac{a}{x}$$

$$\int\frac{dx}{x\sqrt{\left(x^2+a^2\right)}} = -\frac{1}{a}\operatorname{ar}\sinh\frac{a}{x}$$

$$\int\frac{dx}{x\sqrt{\left(a^2-x^2\right)}} = -\frac{1}{a}\operatorname{ar}\cosh\frac{a}{x}$$

Finally, we consider the three integrals

$$\int\sin mx\sin nx dx,\ \int\sin mx\cos nx dx,\ \int\cos mx\cos nx dx$$

where m and n are positive integers. By the well-known trigonometrical formulae

$$\sin mx\sin nx = \frac{1}{2}\left\{\cos(m-n)x-\cos(m+n)x\right\}$$

$$\sin mx\cos nx = \frac{1}{2}\left\{\sin(m+n)x+\sin(m-n)x\right\}$$

$$\cos mx\cos nx = \frac{1}{2}\left\{\cos(m+n)x+\cos(m-n)x\right\}$$

we can divide each of these integrals into two parts. If we now use the substitutions $u = (m+n)x$ and $u = (m-n)x$, respectively, we obtain directly the system of formulae:

$$\int\sin mx\sin nx dx = \begin{cases} -\frac{1}{2}\left\{\frac{\cos(m+n)x}{m+n}+\frac{\cos(m-n)x}{m-n}\right\} \text{if } m\neq n \\ -\frac{1}{2}\left(x-\frac{\sin 2mx}{2m}\right) \text{if } m=n \end{cases}$$

$$\int\sin mx\cos nx dx = \begin{cases} -\frac{1}{2}\left\{\frac{\cos(m+n)x}{m+n}+\frac{\cos(m-n)x}{m-n}\right\} \text{if } m\neq n \\ -\frac{1}{2}\left(\frac{\cos 2mx}{2m}\right) \text{if } m=n \end{cases}$$

$$\int \cos mx \cos nx dx = \begin{cases} \frac{1}{2}\left\{\frac{\sin(m+n)x}{m+n}+\frac{\sin(m-n)x}{m-n}\right\} \text{if } m \neq n \\ \frac{1}{2}\left(\frac{\cos 2mx}{2m}+x\right) \text{if } m=n \end{cases}$$

In particular, if we now integrate from $-\pi$ to $+\pi$, we obtain from these formulae the extremely important relations

$$\int_{-\pi}^{+\pi} \sin mx \sin nx dx = \begin{cases} 0 \text{ if } m \neq n \\ \pi \text{ if } m=n \end{cases}$$

$$\int_{-\pi}^{+\pi} \sin mx \cos nx dx = 0$$

$$\int_{-\pi}^{+\pi} \cos mx \cos nx dx = \begin{cases} 0 \text{ if } m \neq n \\ \pi \text{ if } m=n \end{cases}$$

These are the orthogonality relations of the trigonometric functions which we shall meet again in Chap ix .

Integration by Parts

The second useful method for dealing with integration problems is given by the differentiation formula for a product:

$$(fg)' = f'g + fg'$$

General Remarks:

If we write this formula as an integral formula, we obtain

$$f(x)g(x) = \int g(x)f'(x)dx + \int f(x)g'(x)dx$$

or

$$\int f(x)g'(x)dx = f(x)g(x) - \int g(x)f'(x)dx$$

This formula is referred to as the formula for integration by parts. The calculation of one integral is thereby reduced to the calculation of another integral. In fact, if we split up the integrand of an integral $\int w(x)dx$ into a product $\omega(x) = f(x)\,\phi(x)$ and find the indefinite integral

$$g(x) = \int \phi(x)dx$$

of the one factor $\phi(x)$, so that $\phi(x) = g'(x)$, then our formula reduces the integral

$$\int \omega(x)dx = \int f(x)\phi(x)dx = \int f(x)g'(x)dx$$

to the integral $\int g(x)f'(x)dx$, which in some cases is more readily evaluated than the original form. Since a given function $\omega(x)$, which occurs as an integrand, can be regarded as a product $f(x)\phi(x) = f(x)g'(x)$ in a great many different ways, this formula provides us with a very effective tool for the transformation of integrals.

Written as a formula for definite integration, the formula for integration by parts is

$$\int_a^b f(x)g'(x)dx = f(x)g(x)\Big|_a^b - \int_a^b g(x)f'(x)dx$$

$$= f(b)g(b) - f(a)g(a) - \int_a^b g(x)f'(x)dx$$

In fact, in order to obtain the formula for definite integration from the formula for indefinite integration (2.4.4), we only need replace, first of all, the variable appearing on both sides in the formula for the indefinite integral by the value $x=b$, then by the value $x = a$ and write down the difference of these two expressions.

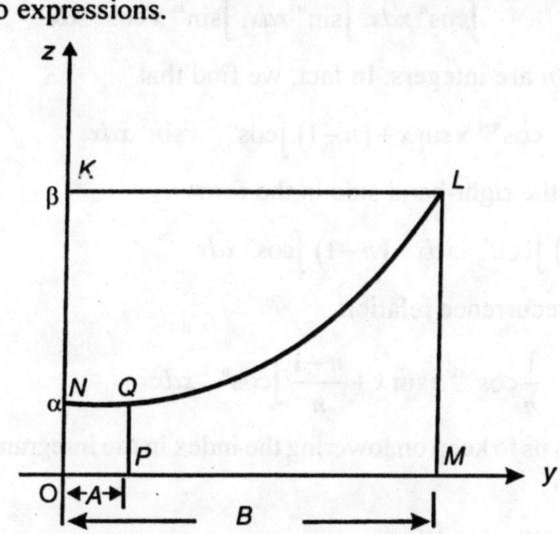

A simple interpretation of this formula, at least with suitable restrictions on the functions involved, can be given. Suppose that $y = f(x)$ and $z = g(x)$ are monotonic, and that $f(x)=A$, $f(b)=B$, $g(a)=\alpha$, $g(b)=\beta$; we can then form the inverse of the first function and substitute it into the equation, thus obtaining z as a function of y. Let this function be monotonic increasing. Since $dy = f'(x)dx$ and $dz = g'(x)dx$, the formula for integration by parts can be rewritten

$$\int_A^B z dy + \int_\alpha^\beta y dz = B\beta - A\alpha$$

in agreement with the relation made clear by Fig.,

$$\text{area } NQLK + \text{area } PMLQ = \text{area } OMLK - \text{area } OPQN$$

The following example may serve as a first illustration:

$$\int \log x dx = \int \log x.1.dx$$

We write the integrand in this way, in order to indicate that we intend to set $f(x) = \log x$ and $g'(x)=1$, so that we have $f'(x) = 1/x$ and $g(x) = x$. Our formula then becomes

$$\int \log x dx = x\log x - \int \frac{x}{x}dx = x\log x - x$$

This last expression is therefore the integral of the logarithm, as may be verified at once by differentiation.

Recurrence Formulae: In many cases, the integrand is not only a function of the independent variable, but also of an integral index n; on integrating by parts, instead of the value of the integral, we obtain another similar expression in which the index n has a smaller value. We thus arrive, after a number of steps, at an integral, which we can process by means of our table of integrals. Such a procedure is called a recurrence process. The following examples illustrate this method: By repeated integration by parts, we can calculate the trigonometrical integrals

$$\int \cos^n x dx, \int \sin^n x dx, \int \sin^m x \cos^n x dx$$

provided that m and n are integers. In fact, we find that

$$\int \cos^n x dx = \cos^{n-1} x \sin x + (n-1) \int \cos^{n-2} x \sin^2 x dx$$

here we can rewrite the right-hand side in the form

$$\cos^{n-1} x \sin x + (n-1) \int \cos^{n-2} x dx - (n-1) \int \cos^n x dx$$

and thus obtain the recurrence relation

$$\int \cos^n x dx = \frac{1}{n} \cos^{n-1} x \sin x + \frac{n-1}{n} \int \cos^{n-2} x dx$$

This formula enables us to keep on lowering the index in the integrand until we finally arrive at the integral

$$\int \cos x dx = \sin x \text{ or } \int dx = x$$

according to whether n is odd or even. In a similar way, we obtain the analogous recurrence formulae

$$\int \sin^n x dx = -\frac{1}{n} \sin^{n-1} x \cos x + \frac{n-1}{n} \int \sin^{n-2} x dx$$

and

$$\int \sin^m x \cos^n x dx = \frac{\sin^{m+1} x \cos^{n-1} x}{m+n} + \frac{n-1}{m+n} \int \sin^m x cos^{n-2} x dx$$

In particular, these formulae enable us to compute the integrals

$$\int \sin^2 x dx = \frac{1}{2}(x - \sin x \cos)$$

and

$$\int \cos^2 x dx = \frac{1}{2}(x + \sin x \cos x)$$

which we have already obtained by the method of substitution.

It need hardly be mentioned that the corresponding integrals for the hyperbolic functions can be calculated in exactly the same manner.

Further recurrence relations are obtained by the transformations:

$$\int (\log x)^m\,dx = x(\log x)^m - m\int (\log x)^{m-1}\,dx$$

$$\int x^m e^x dx = x^m e^x - m\int x^{m-1}e^x dx$$

$$\int x^m \sin x dx = -x^m \cos x + m\int x^{m-1}\cos x dx$$

$$\int x^m \cos x dx = x^m \sin x - m\int x^{m-1}\sin x dx$$

$$\int x^\alpha (\log x)^m\,dx = \frac{x^{\alpha+1}(\log x)^m}{a+1} - \frac{m}{a+1}\int x^a (\log x)^{m-1}\,dx \quad (a \neq -1)$$

Wallis' Product: The recurrence formula for the integral $\int \sin^n x dx$ leads on an elementary way to a most remarkable expression for the number π as an infinite product. We suppose that $n > 1$ and insertthe limits 0 and $\pi/2$ in the formula

$$\int \sin^n x dx = -\frac{1}{n}\sin^{n-1} x\cos x + \frac{n-1}{n}\int \sin^{n-2} x dx$$

obtaining

$$\int_0^{\pi/2} \sin^n x dx = \frac{n-1}{n}\int_0^{\pi/2} \sin^{n-2} x dx \text{ for n} > 1$$

Repeated application of the recurrence formula to the right-hand side yields for the cases $n = 2m$ and $n = 2m + 1$

$$\int_0^{\pi/2} \sin^{2m} x dx = \frac{2m-1}{2m}\cdot\frac{2m-3}{2m-2}\cdots\frac{1}{2}\cdot\int_0^{\pi/2} dx$$

$$\int_0^{\pi/2} \sin^{2m+1} x dx = \frac{2m}{2m+1}\cdot\frac{2m-2}{2m-1}\cdots\frac{2}{3}\cdot\int_0^{\pi/2} \sin x dx$$

whence

$$\int_0^{\pi/2} \sin^{2m} x dx = \frac{2m-1}{2m}\cdot\frac{2m-3}{2m-2}\cdots\frac{1}{2}\cdot\frac{\pi}{2}$$

$$\int_0^{\pi/2} \sin^{2m+1} x dx = \frac{2m}{2m+1}\cdot\frac{2m-2}{2m-1}\cdots\frac{2}{3}.$$

Division now yields

$$\frac{\pi}{2} = \frac{2.2}{1.3}\cdot\frac{4.4}{3.5}\cdot\frac{6.6}{5.7}\cdots\frac{2m.2m}{(2m-1)\cdot(2m+1)}\frac{\int_0^{\pi/2} \sin^{2m} x dx}{\int_0^{\pi/2} \sin^{2m+1} x dx}$$

The quotient of the two integrals on the right hand side converges to 1 as m increases, as we recognize from the following considerations. In the interval $0 < x < \pi/2$, we find

$$0 < \sin^{2m+1} x \leq \sin^{2m} x \leq \sin^{2m-1} x$$

whence

$$0 < \int_0^{\pi/2} \sin^{2m+1} xdx \leq \int_0^{\pi/2} \sin^{2m} xdx \leq \sin^{2m-1} xdx$$

If we here divide each term by $\int_0^{\pi/2} \sin^{2m+1} xdx$ and note that by the first formula, proved above,

$$\frac{\int_0^{\pi/2} \sin^{2m-1} xdx}{\int_0^{\pi/2} \sin^{2m+1} xdx} = \frac{2m+1}{2m} = 1 + \frac{1}{2m}$$

we find that

$$1 \leq \frac{\int_0^{\pi/2} \sin^{2m} xdx}{\int_0^{\pi/2} \sin^{2m+1} xdx} \leq 1 + \frac{1}{2m}$$

which yields the statement above, whence there applies the relation

$$\frac{\pi}{2} = \lim_{m\to\infty} \frac{2}{1}\frac{2}{3}\frac{4}{3}\frac{4}{5}\frac{6}{5}\frac{6}{7}\cdots\frac{2m}{2m-1}\frac{2m}{2m+1}$$

This product formula (due to Wallis), with its simple law of formation, presents a remarkable relation between the number π and the integers. If we observe that

$$\lim_{m\to\infty} \frac{2m}{2m+1} = 1$$

we can write

and, if we take the square root and then multiply numerator and denominator by 2.4 ... ,... (2m — 2), we find

$$\sqrt{\frac{\pi}{2}} = \lim_{m\to\infty} \frac{2.4...(2m-2)}{3.5...(2m-1)}\sqrt{2m} = \lim_{m\to\infty} \frac{2^2.4^2...(2m-2)^2}{(2m-1)!}\sqrt{2m}$$

$$= \lim_{m\to\infty} \frac{2^2.4^2...(2m)^2}{(2m)!}\frac{\sqrt{2m}}{2m}$$

Hence, we finally obtain

$$\lim_{m\to\infty} \frac{(m!)^2 2^{2m}}{(2m)!\sqrt{m}} = \sqrt{\pi}$$

Integration of Rational Functions

The most important general class of functions, which are integrable in terms of elementary functions, are the rational functions

$$R(x) = \frac{f(x)}{g(x)}$$

where $f(x)$ and $g(x)$ are the polynomials:

$$f(x) = a_m x^m + a_{m-1}x^{m-1} + ... + a_0$$

$$g(x) = b_n x^n + a_{n-1}x^{n-1} + ... + b_0 \left(b_n \neq 0\right)$$

We recall that every polynomial can be integrated at once and that the integral is itself a polynomial, whence we need only consider those rational functions for which the denominator is not a constant. Moreover, we can always assume that the degree of the numerator is less than the degree (n) of the denominator, because otherwise we can divide the polynomial $f(x)$ by the polynomial $g(x)$ and obtain a remainder of degree less than n; in other words, we can write $f(x)=q(x)g(x)+r(x)$, where $q(x)$ and $r(x)$ are also polynomials and $r(x)$ is of degree less than n. The integration of $f(x)/g(x)$ is then reduced to the integration of the polynomial $q(x)$ and of the proper fraction $r(x)/g(x)$. Moreover, we note that the $f(x)/g(x)$ can be represented as the sum of the functions $a_v x^v/g(x)$, so that we need only consider integrands of the form $x^v/g(x)$.

4.5.1 The Fundamental Types: We shall not at once proceed to the integration of the most general rational function of the above type, but instead shall study only those functions in which the denominator $g(x)$ is of a particularly simple type, namely, $g(x) = (\alpha x + \beta)^n$ – a power of a linear $\alpha x+\beta\,(\alpha \neq 0)$ – or $g(x) = (ax + 2bx + c)^n$ - a power of a definite * quadratic expression. In the first case, we introduce a new variable $\xi = \alpha x + \beta$. Then $d\xi/dx = \alpha$ and $x = (\xi - \beta)/\alpha$ is also a linear function of ξ. Each numerator $f(x)$ becomes a polynomial $\phi(\xi)$ of the same degree, whence

$$\int \frac{f(x)}{(ax+\beta)^n}dx = \frac{1}{\alpha}\int \frac{\phi(\xi)}{\xi^n}d\xi$$

A quadratic expression $Q(x) = \alpha x + 2bx + c$ is said to be definite, if it takes for for all real values of x values with one the same sign, i.e., if the equation $Q(x) = 0$ has no real roots. For this, it is necessary and sufficient that $ac - b$ is positive.

In the second case, we write

$$ax^2 + 2bx + c = \frac{1}{a}\left(ax + b^2\right) + \frac{d^2}{a}\left(d^2 = ac - b^2, d > 0\right)$$

observing that, since we have assumed our expression to be definite, $ac - b$ must be positive and $a \neq 0$. After introducing the new variable

$$\xi = \frac{ax+b}{d}$$

we obtain an integral with the denominator

$$\left[\frac{d^2}{a}\left(1+\xi^2\right)\right]^n$$

Hence, in order to integrate rational functions the denominators of which are powers of a linear expression or of a definite quadratic expression, it is sufficient to be able to integrate the functional types

$$\frac{1}{x^n}, \frac{x^{2\nu}}{\left(x^2+1\right)^n}, \frac{x^{2\nu+1}}{\left(x^2+1\right)^n}$$

In fact, we shall see that, in general, even these types need not be treated, for we can reduce the integration of every rational function to the integration of the very special forms of these three functions, obtained by taking $\nu = 0$. Hence, we will now consider the integration of the three expressions

$$\frac{1}{x^n}, \frac{1}{\left(x^2+1\right)^n}, \frac{x}{\left(x^2+1\right)^n}$$

Integration of the Fundamental Types: The integration of the first type of function $1/x^n$ yields immediately $\log|x|$, if $n = 1$ and $-1/\{(n-1)x^{n-1}\}$, if $n > 1$, so that in both cases the integral is again an elementary function. Functions of the third type can be integrated immediately after introduction of the new variable $\xi = x + 1$, whence we obtain $2xdx = d\xi$ and

$$\int \frac{x}{\left(x^2+1\right)^n} dx = \frac{1}{2}\int \frac{d\xi}{\xi^n} = \begin{cases} \frac{1}{2}\log\left(x^2+1\right) \text{ if } n = 1, \\ -\frac{1}{2(n-1)\left(x^2+1\right)^{n-1}} \text{ if } n > 1 \end{cases}$$

Finally, in order to calculate the integral

$$I_n = \int \frac{dx}{\left(x^2+1\right)^n}$$

where n has any value exceeding 1, we use the recurrence method. In fact, if we set

$$\frac{1}{\left(x^2+1\right)^n} = \frac{1}{\left(x^2+1\right)^{n-1}} - \frac{x^2}{\left(x^2+1\right)^n}$$

we can transform the right-hand side by integration by parts, using formulae in 4.3 with

$$f(x) = x, \; g'(x) = \frac{x}{\left(x^2+1\right)^n}$$

Then, as we have just discovered above,

$$g(x) = -\frac{1}{2}\frac{1}{(n-1)(x^2+1)^{n-1}}$$

whence we obtain

$$I_n = \int\frac{dx}{(x^2+1)^n} = \frac{x}{2(n-1)(x^2+1)^{n-1}} + \frac{2n-3}{2(n-1)}\int\frac{dx}{(x^2+1)^{n-1}}$$

The calculation of the integral I_n is thus reduced to that of the integral I_{n-1}. If $n-1>1$, we apply the same process to the latter integral and continue the process until we finally arrive at the expression

$$\int\frac{dx}{(x^2+1)} = \text{arc tan}x$$

We thus see that the integral * I_n can be explicitly expressed in terms of rational functions and the function artan x.

The integral of the function $1/(x-1)^n$ can be calculated in the same way; we reduce it by the corresponding recurrence method to the integral

$$\int\frac{dx}{1-x^2} = \text{ar tanh}\,x\ \{or\ \text{ar coth}\,x\}$$

Incidentally, we could also have integrated the function $1/(x+1)^n$ directly using the substitution $x = \tanh t$; we should then have obtained $dx = \sec t\ dt$ and $1/(1+x) = \cos t$, so that

$$\int\frac{dx}{(x^2+1)^n} = \int\cos^{2n-2} t dt$$

and we have already learnt in.4.4.3 how to evaluate this integral.

Partial fractions: We are now in a position to integrate the most general rational functions by virtue of the fact that every such function can be represented as the sum of so-called partial fractions, i.e., as the sum of a polynomial and a finite number of rational functions, each one of which has either a power of a linear expression as its denominator and a constant for its numerator or a power of a definite quadratic expression as its denominator and a linear function as its numerator. If the degree of the numerator $f(x)$ is less than that of the denominator $g(x)$, the polynomial does not occur. We can now integrate each partial fraction, since, by 4.5, the denominator can be reduced to one of the special form x^n and $(x+1)^n$, and the fraction is then a combination of the fundamental types integrated in 4.5.2.

We shall not give the general proof of the possibility of this resolution into partial fractions. On the contrary, we shall confine ourselves to stating the theorem in an form intelligible for the reader and to showing by examples how the resolution into partial fractions can be carried out in typical cases. In practice, one deals only with comparatively simple functions, since otherwise the computations become far too complicated.

As we know from elementary algebra, every polynomial $g(x)$ can be written in the form

$$g(x) = a(x-a_1)^{l_1}(x-a_2)^{l_2}\ldots(x^2+2b_1x+c_1)^{r_1}(x^2+2b_2x+c_2)^{r_2}\ldots+$$

Here, the numbers $a_1, a_2, \ldots$ are the real and distinct roots of the equation $g(x)=0$ and the positive integers $l_1, l_2,$ indicate how often they are repeated; the factors $x+2b_\nu x+c_\nu$ indicate definite quadratic expressions, of which no two are the same, with conjugate complex roots and the positive integers $r_1, r_2, \ldots$ give the numbers of times that these roots are repeated.

We assume that the denominator is either given to us in this form or else that we have brought it to this form by calculating the real and imaginary roots. Moreover, let us assume that the numerator $f(x)$ has a lower degree than the denominator. Then the theorem on the resolution into partial fractions is:

For each factor $(x - \alpha)^l$, where α is any one of the real roots and l is the number of times it is repeated, we can determine an expression of the form

$$\frac{A_1}{(x-a)}+\frac{A_2}{(x-a)^2}+\ldots+\frac{A_l}{(x-a)^l}$$

and for each quadratic factor $Q(x)=x+2bx+c$ in our product, which is raised to the power r, we can determine an expression of the form

$$\frac{B_1+C_1^x}{Q}+\frac{B_2+C_2^x}{Q^2}+\ldots+\frac{B_r+C_r^x}{Q^r}$$

in such a way that the function $f(x)/g(x)$ is the sum of all these expressions. In other words, the quotient $f(x)/g(x)$ can be represented by a sum of fractions, each of which belongs to one or another of the types integrated in 4.5.2.

We present here a brief outline of the method by which the possibility of this decomposition into partial fractions is proved. If $g(x)=(x-a)^k h(x)$ and $h(a)\neq 0$, then obviously there vanishes on the right hand side of the equation

the numerator for $x = -a$, whence it has the form $h(a)(x-a)^m f_1(x)$, where $f_1(x)$ is also a polynomial, the integer $m \geq 1$ and $f_1(a) \neq 0$. Writing $f(a)/h(a) = \beta$, we obtain

$$\frac{f(x)}{g(x)} = \frac{\beta}{(x-a)^k}=\frac{f_1(x)}{(x-a)^{k-m}h(x)}$$

Continuing the process, we can keep on reducing the degree of the power of $(x-a)$ in the denominator until finally no such factor is left. We repeat for the remaining fraction the process for some other roots of $g(x)$ and do so as many times as g has distinct factors. This is done not only for the real, but also for the complex roots until we arrive at the complete decomposition into partial fractions.

In particular cases, the splitting into partial fractions can be done easily by inspection. If, for example, $g(x) = x - 1$, we see at once that

$$\frac{1}{x^2-1} = \frac{1}{2}\frac{1}{x-1} - \frac{1}{2}\frac{1}{x+1}$$

whence

$$\int\frac{dx}{x^2-1} = \frac{1}{2}\log\left|\frac{x-1}{x+1}\right|$$

More generally, if $g(x) = (x-\alpha)(x-\beta)$, that is, if $g(x)$ is a non-definite quadratic expression with two real series α and β, we have

$$\frac{1}{(x-\alpha)(x-\beta)} = \frac{1}{\alpha-\beta}\frac{1}{x-\beta}\frac{1}{x-\alpha} - \frac{1}{\alpha-\beta}\frac{1}{x-\beta}$$

whence

$$\int\frac{dx}{(x-\alpha)(x-\beta)} = \frac{1}{\alpha-\beta}\log\left|\frac{x-\alpha}{x-\beta}\right|$$

Example. *The Bi-molecular Reaction*: A simple example of the application of this easy reduction to partial fractions is given by the so-called bimolecular reaction. Let there be given two reagents the original concentrations of which in mols per until volume are *a* and *b*, where we assume that $a < b$; we will assume that there is formed at time *t* in the unit volume a quantity *x* (mols) of the reaction product. Then, according to the law of mass action, in the simplest case - reaction between one molecule of each of the reagents - the rate of increase of the quantity *x* is given by the equation $dx/dt = k(a-x)\{b-x)$. The problem is to determine the function $x(t)$. Conversely, if we think of the time *t* as a function of we have

integration now yields

$$\frac{dt}{dx} = \frac{1}{k(a-x)(b-x)} = \frac{1}{k(b-a)}\left(\frac{1}{a-x} - \frac{1}{b-x}\right)$$

$$kt = \frac{1}{a-b}\log\frac{a-x}{b-x} + c, \text{for } x < a < b$$

We determine the constant of integration *c* by the condition that at time $t = 0$ there has not yet formed a product of reaction, so that

$$\frac{1}{a-b}\log\frac{a}{b} + c = 0$$

Thus, we finally obtain

$$kt = \frac{1}{a-b}\log\frac{1-\frac{x}{a}}{1-\frac{x}{b}}$$

and, by solving for x, the required function

$$x = \frac{ab\left(1-e^{(a-b)kt}\right)}{b-ae^{(a-b)kt}}$$

4.5.5 Further Examples of Resolution into Partial Fractions. The Method of Undetermined Coefficients: If $g(x) = (x-\alpha_1)(x-\alpha_2)\ldots\ldots(x-\alpha_n)$, where $\alpha_i \rangle \alpha_k$, $i \rangle k$, i.e., if the equation $g(x) = 0$ has only single real roots, the expression in terms of partial fractions has the simple form

$$\frac{1}{g(x)} = \frac{a_1}{x-a_1}+\frac{a_2}{x-a_2}+\ldots+\frac{a_n}{x-a_n}$$

We obtain explicit expressions for the coefficients $a_1, a_2, \ldots$ if we multiply both sides of this equation by $(x-a_1)$, cancel the common factor $(x-a_1)$ in the numerator and denominator on the left hand side and in the first term on the right hand side, and then put $x = a_1$. This yields

$$\alpha_1 = \frac{1}{(\alpha_1-\alpha_2)(\alpha_1-\alpha_3)\ldots(\alpha_1-\alpha_n)}$$

The reader will observe that the denominator on the right hand side is $g'(a_1)$, i.e., the derivative of the function $p(x)$ at the point $x = \alpha_1$.

As a typical example of a denominator $g(x)$ with multiple roots, consider the function $1/\{x(x-1)\}$. The preliminary statement

$$\frac{1}{x^2(x-1)} = \frac{a}{x-1}+\frac{b}{x}+\frac{c}{x^3}$$

as shown in 4.5.3 leads us to the required result. If we multiply both sides of this equation by $x(x-1)$, we obtain

$$1 = (a+b)x^2-(b-c)x-c$$

which is true for all values of x, from which we have to determine the coefficients a, b, c. This condition cannot hold unless all the coefficients of the polynomial $(a+b)x - (b—c)x - c - 1$ are zero, i.e., we must have

$a + b = b - c = c + 1 = 0$ or $c = -1$, $b = -1$, $a = 1$.

Thus, we obtain the resolution

$$\frac{1}{x^2(x-1)} = \frac{1}{x-1}-\frac{1}{x}-\frac{1}{x^2}$$

and consequently

$$\int \frac{dx}{x^2(x-1)} = \log|x-1| - \log|x| + \frac{1}{x}$$

We shall now split up the function $1/[x(x + 1)]$ – an example of the case when the zeroes of the denominator are complex - in accordance with the equation

$$\frac{1}{x(x^2+1)} = \frac{a}{x} + \frac{bx+c}{x^2+1}$$

We obtain for the coefficients $a + b = c = a - 1$, so that

$$\frac{1}{x(x^2+1)} = \frac{1}{x} - \frac{x}{x^2+1}$$

and consequently

$$\int \frac{dx}{x(x^2+1)} = \log|x| - \frac{1}{2}\log(x^2+1)$$

We consider as a third example the function $1/(x^4 + 1)$. Even Leibnitz found this to be a troublesome integration. We can represent the denominator as the product of two quadratic factors:

$$x^4+1 = (x^2+1)^2 - 2x^2 = \{x^2+1+\sqrt{2}x\}(x^2+1-\sqrt{2}x)$$

We now know that the resolution into partial fractions will have the form

$$\frac{1}{x^4+1} = \frac{ax+b}{x^2+\sqrt{2}x+1} + \frac{cx+d}{x^2-\sqrt{2}x+1}.$$

In order to determine the coefficients *a, b, c, d,* we have the equation

$$(a+c)x^3 + (b+d-a\sqrt{2}+c\sqrt{2})x^2 + (a+c-b\sqrt{2}+d\sqrt{2})x + (b+d-1) = 0$$

which is satisfied by the values

$$a = \frac{1}{2\sqrt{2}}, b = \frac{1}{2}, c = -\frac{1}{2\sqrt{2}}, d = \frac{1}{2}$$

whence

$$\frac{1}{x^4+1} = \frac{1}{2\sqrt{2}} \cdot \frac{x+\sqrt{2}}{x^2+\sqrt{2}x+1} - \frac{1}{2\sqrt{2}} \cdot \frac{x-\sqrt{2}}{x^2-\sqrt{2}x+1}$$

and, applying the method in 4.5.1, we obtain

$$\int \frac{dx}{x^4+1} = \frac{1}{4\sqrt{2}}\log|x^2+\sqrt{2}x+1| - \frac{1}{4\sqrt{2}}\log|x^3-\sqrt{2}x+1|$$

$$+\frac{1}{2\sqrt{2}}\arctan\left(\sqrt{2}x+1\right)+\frac{1}{2\sqrt{2}}\arctan\left(\sqrt{2}x-1\right)$$

which is readily verified by differentiation.

Integration of certain other Classes of Functions

Preliminary Remarks on the Rational Representation of the Trigonometric and Hyperbolic Functions: The integration of certain other general classes of functions can be reduced to the integratioa of rational functions. We shall better understand this reduction, if we begin by stating certain elementary facts about the trigonometric and hyperbolic functions. If we set $t = \tan x/2$, elementary trigonometry yields the simple formulae

$$\sin x = \frac{2t}{1+t^2}, \cos x = \frac{1-t^2}{1+t^2}$$

in fact,

$$\frac{1}{1+t^2} = \cos^2\frac{x}{2} \text{ and } \frac{t^2}{1+t^2} = \sin^2\frac{x}{2}$$

whence we obtain the above equations from the elementary formulae

$$\sin x = 2\cos^2\frac{x}{2}\tan\frac{x}{2} \text{ and } \cos x = \cos^2\frac{x}{2} - \sin^2\frac{x}{2}$$

These equations show that sin x and cos x can be expressed rationally in terms of the quamtity t=tan$x$/2. Differentiation now yields

$$\frac{dt}{dx} = \frac{1}{2\cos^2 x/2} = \frac{1+t^2}{2}$$

whence

$$\frac{dx}{dt} = \frac{2}{1+t^2}$$

whence the derivative dx/dt is also a rational expression in t.

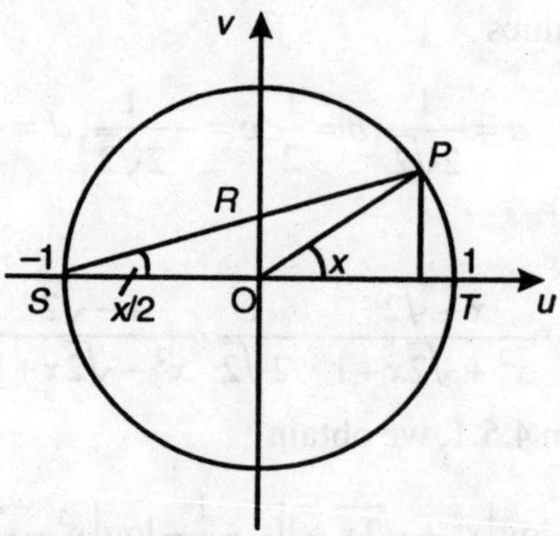

Fig. Parametric representation of the trigonometric functions

Fig. shows the geometrical representation and the geometrical meaning of our formulae. It shows the circle $u + v = 1$ in a *uv*-plane. If x denotes the angle *POT*, then $u = \cos x$ and $v = \sin x$. By a theorem in elementaly geometry, the angle *OSP* with its vertex at the point u=-1, v=0 is equal to $x/2$, and we can deduce from the figure the geometrical meaning of the parameter:

$$t = \tan x/2 = OR.$$

If the point P starts from S and moves once around the circle in the positive direction, i.e., if x moves through the interval from $-\pi$ to $+\pi$, the quantity t will move exactly once through the entire range of the values from $-\infty$ to $+\infty$.

We may correspondingly express the hyperbolic functions $\cosh x = (e^x+e^{-x})/2$ and $\sinh x=(e^x-e^{-x})/2$ as rational functions of a third quantity. The most obvious way is to set $e^x = \tau$, so that we have for these hyperbolic functions the rational expressions

$$\cosh x = \frac{1}{2}\left(\tau+\frac{1}{\tau}\right), \sin x = \frac{1}{2}\left(\tau-\frac{1}{\tau}\right)$$

Once again, $dx/dt = 1/\tau$. However, we obtain a closer analogy with the trigonometric functions by introduction of the quantity $t = \tanh x/2$; we then arrive at the formulae

$$\sinh x = \frac{2t}{1-t^2}, \cosh x = \frac{1+t^2}{1-t^2}$$

Differentiation of $t = \tanh x/2$ yields this time the rational expression

$$\frac{dx}{dt} = \frac{2}{1-t^2}$$

for the derivative dx/dt. Once again, the quantity t has a geometrical meaning similar to that which it has in the case of the trigonometric functions, as we see at once from Fig..

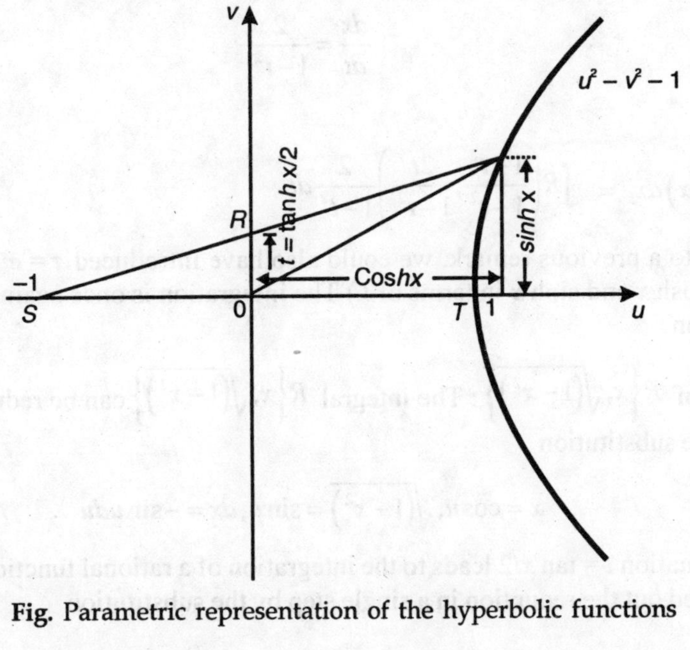

Fig. Parametric representation of the hyperbolic functions

However, while in the case of the trigonometrio functions, t must run through the entire range of values from $-\infty$ to $+\infty$, in order to yield all pairs of values of $\cos x$ and $\sin x$, in the case of the hyperbolic functions, t is limited to the interval $-1 < t < 1$.

After these preliminary remarks, we turn to our integration problem.

Integration of $R(\cos x, \sin x)$: Let $R(\cos x, \sin x)$ denote an expression which ia rational in the two functions $\sin x$ and $\cos x$, i.e., an expression which is formed rationally from these two functions and given constants such as

$$\frac{2\sin^2 x + \cos x}{3\cos^2 x + \sin x}$$

If we apply the substitution $t = \tan x/2$, the integral

$$\int R(\cos x, \sin x)\,dx$$

becomes the integral

$$\int R\left(\frac{1-t^2}{1+t^2}, \frac{2t}{1+t^2}\right)\frac{2}{1+t^2}\,dt$$

and we have now under the integral sign a rational function of t. Thus, we have obtained theoretically the integral of our expression, since we can now perform the integration by the methods of the preceding section.

Integration of $R(\cosh x, \sinh x)$: In the same way, if $R(\cosh x, \sinh x)$ is an expression which is rational in terms of the hyperbolic functions $\cosh x$ and $\sinh x$, we can integrate it by means of the substitution $t = \tanh x/2$. Recalling that

$$\frac{dx}{dt} = \frac{2}{1-t^2}$$

we find

$$\int R(\cosh x, \sinh x)\,dx = \int R\left(\frac{1+t^2}{1-t^2}, \frac{2t}{1-t^2}\right)\frac{2}{1-t^2}\,dt$$

(According to a previous remark, we could also have introduced $\tau = e^x$ as a new variable and expressed $\cosh x$ and $\sinh x$ in terms of t.) The integration is once again reduced to that of a rational function.

Integration of $R\left\{x, \sqrt{(1-x^2)}\right\}$: The integral $R\left\{x, \sqrt{(1-x^2)}\right\}$ can be reduced to the type in 4.6.2 by using the substitution

$$x = \cos u, \sqrt{(1-x^2)} = \sin u, dx = -\sin u\,du$$

the transformation $t = \tan x/2$ leads to the integration of a rational function. By the way, we could have carried out the reduction in a single step by the substitution

$$t=\sqrt{\left(\frac{1-x}{1+x}\right)};x=\frac{1-t^2}{1+t^2};\sqrt{\left(1-x^2\right)}=\frac{2t}{1+t^2};\frac{dx}{dt}=\frac{-4t}{\left(1+t^2\right)^2}$$

that is, we could have obtained directly a rational function by introduction of a new variable t=tan$u$/2.

Integration of $R\left\{x,\sqrt{\left(x^2-1\right)}\right\}$: The integral of this function is transformed by the substitution $x = \cosh u$ into the type 4.6.3, when we again introduce

$$t=\sqrt{\left(\frac{x-1}{x+1}\right)}=\tanh\frac{u}{2}$$

4.6.6 Integration of $R\left\{x,\sqrt{\left(x^2-1\right)}\right\}$: The integral of this function is transformed by the substitution $x = \sinh u$ into the type 4.6.3, whence it can be integrated in terms of elementary functions. Instead of the further reduction to an integral of a rational function by the substitution $e^u=\tau$ or $\tanh u/2 = t$, we could have reached the integral of a rational function by a single step by either of the substitutions

$$\tau=x+\sqrt{\left(x^2+1\right)},t=\frac{-1+\sqrt{\left(x^2+1\right)}}{x}$$

4.6.7 Integration of $R\left\{x,\sqrt{\left(ax^2+2bx+c\right)}\right\}$:The integral of this function is rational in terms of x and the square root of an arbitrry polynomial of the second degree in x; it can be reduced immediately to one of the types above. As in 4.5.1,we write

$$ax^2+2bx+c \quad = \quad \frac{1}{a}(ax+b)^2+\frac{ac-b^2}{a}$$

If $ac - b > 0$, we introduce a new variable ξ= by means of the transformation $\xi=\frac{ax+b}{\sqrt{\left(ac-b^2\right)}}$ whence the surd takes the form $\sqrt{\left\{\frac{ac-b^2}{a}\left(\xi^2+1\right)\right\}}$. Expressed in terms of ξ, we have now the type of 4.6.6. The constant a must be positive here, in order that the square root may have real values.

If $ac - b = 0$, $a > 0$, then we see from

$$\sqrt{\left(ax^2+2bx+c\right)} \quad = \quad \sqrt{a}\left(x+\frac{b}{a}\right)$$

that the integrand was rational from the start.

Finally, if $ac - b < 0$, we set $\xi = \frac{ax+b}{\sqrt{(b^2-ac)}}$ and obtain for the surd the expression $\sqrt{\left\{\frac{b^2-ac}{a}(\xi^2-1)\right\}}$. If a is positive, our integral is thus reduced to the type 4.6.5; on the other hand, if a is negative, we write the surd in the form $\sqrt{\left\{\frac{b^2-ac}{a}(\xi^2-1)\right\}}$ and see that the integral is thus reduced to the type of 4.6.4

Further Examples of Reduction to Integrals of Rational Functions: We shall briefly mention two other types of functions which can be integrated by reduction to rational functions:

1. Rational expressions involving two different surds of linear expressions

$$R\left\{x, \sqrt{(ax+b)}, \sqrt{(\alpha x+\beta)}\right\}$$

2. Expressions of the form

$$R\left\{x, \sqrt[n]{\left(\frac{ax+b}{\alpha x+\beta}\right)}\right\},$$

where a, b, α, β are constants. In (1), we introduce the new variable $\xi = \sqrt{(\alpha x+\beta)}$, so that $\alpha x+\beta = \xi$, whence

$$x = \frac{\xi^2-\beta}{\alpha} \text{ and } \frac{dx}{d\xi} = \frac{2\xi}{\alpha};$$

then

$$\int R\left\{x, \sqrt{(ax+b)}, \sqrt{(\alpha x+\beta)}\right\}dx$$

$$= \int R\left\{\frac{\xi^2-\beta}{\alpha}, \sqrt{\frac{1}{\alpha}\{a\xi^2-(a\beta-b\alpha)\}}, \xi\right\}\frac{2\xi}{\alpha}d\xi$$

which is of the type 4.6.7.

If we introduce in the second case the new variable

$$\xi = \sqrt[n]{\left(\frac{ax+b}{\alpha x+\beta}\right)}$$

we have

$$\xi^n = \frac{ax+b}{\alpha x+\beta}, x = \frac{-\beta\xi^n+b}{\alpha\xi^n-a}, \frac{dx}{d\xi} = \frac{a\beta-b\alpha}{(a\xi^n-\alpha)^2}\cdot n\xi^{n-1}$$

and immediately arrive at

$$\int R\left(x, \sqrt[n]{\left(\frac{ax+b}{\alpha x+\beta}\right)}\right)dx = \int R\left(\frac{-\beta\xi^n + b}{\alpha\xi^n - \alpha}, \xi\right)\frac{\alpha\beta - b\alpha}{\left(\alpha\xi^n - \alpha\right)^2} n\xi^{n-1} d\xi$$

which is an integral of a rational function.

Comments on Exercises: The preceding discussion is mainly of theoretical interest. In the case of complicated expressions, the actual calculations would be far too involved. It is therefore expedient to employ, whenever it is possible, a special form of the integrand, in order to simplify the work. For example, in order to integrate the expression

$1/(a\sin x + b\cos x)$,

it is better to use the substitution $t = \tan x$ instead of that given in 4.6.2, since $\sin x$ and $\cos x$ can be expressed rationally in terms of $\tan x$ and it is therefore unnecessary to go back to $t = \tan x/2$. The same is true for every expression formed rationally from $\sin x$, $\cos x$ and $\sin x \cos x$*, because $\sin x \cos x = \tan x \cos x$ can, of course, be expanded in terms of $\tan x$. Moreover, for the calculation of many integrals, a trigonometrical form is to be preferred to a rational one, provided that the trigonometrical form can be evaluated by some simple recurrence method.

For example, although the integrand in $\int x^n\left\{\sqrt{\left(1-x^2\right)}\right\}^m dx$ can be reduced to a rational form, it is simpler to write $x = \sin u$ and convert it into $\int \sin^n u \cos^{m+1} u du$ since this can be treated easily by the recurrence method in 4.4.3 (or by using the addition theorems to reduce the powers of the sine and cosine to sines and cosines of multiple angles).

For the evaluation of the integral

$$\int \frac{dx}{a\cos x + b\sin x}\left(a^2 + b^2 > 0\right)$$

we determine, instead of referring to the general theory, a number A and an angle θ in such a way that

$$a = A\sin\theta, b = A\cos\theta$$

that is, we write

$$A = \sqrt{a^2 + b^2}, \sin\theta = \frac{a}{A}, \cos\theta = \frac{b}{A}$$

The integral then becomes

$$\frac{1}{A}\int \frac{dx}{\sin(x+\theta)}$$

and, on introducing the new variable $x+\theta$, (4.2.3) yields the value of the integral

$$\frac{1}{A}\log\left|\tan\frac{x+\theta}{2}\right|$$

Definition of Functions by means of Integrals. Elliptic Integrals: The above examples of types of functions which can be integrated by reduction to rational functions practically exhaust the list of such functions. Attempts to express general integrals such as

$$\int\frac{dx}{\sqrt{\left(a_0+a_1^x+\ldots+a_nx^n\right)}},\ \int\sqrt{\left(a_0+a_1x+\ldots+a_nx^n\right)}dx$$

or $\int\frac{e^x}{x}dx$ in terms of elementary functions have always ended up in failure; and finally, in the Nineteenth Century, it has been proved that it is actually impossible to cany out integrations in terms of elementary functions. Hence, if the objective of the integral calculus were integration of functions in terms of elementary functions, we should have come to a definite halt. But such a restricted objective has no intrinsic justiflcation; indeed, it is of a somewhat artificial nature. We know that the integral of every continuous function exists and is itself a continuona function of the upper limit, and this fact has nothing to do with the question whether an integral can be expressed in terms of elementary functions or not. The distinguishing features of the elementary functions are based on the fact that their properties are easily recognized, that their application to numerical problems is often facilitated by convenient tables or, as in the case of the rational functions, that they are readily calculated with as great a degree of accuracy as we please.

Whenever an integral of a function cannot be expressed in terms of functions with which we are already acquainted, there is nothing to stop us from introducing such an integral as a new higher function into analysis, which really means no more than giving it a name. Whether the introduction of such a new function is convenient or not depends on the its properties, the frequency at which it occurs and the ease with which it can be manipulated in theory and in practice. Hence, in this sense, the process of integration forms a base for the generation of new functions.

After all, we are already acquainted with this principle from our dealings with the elementary functions. Thus, we found ourselves obliged in 3.6 to introduce the previously unknown integral of $1/x$ as a new function, which we called the logarithm and the properties of which we could easily determine. We could have introduced the trigonometric functions in a similar way, only making use of the rational functions, the process of integration, and the process of inversion. For this purpose, we only need introduce one or the other of the equations

$$\operatorname{arc}\tan x = \int_0^x\frac{dt}{1+t^2}\quad\text{or}\quad\operatorname{arc}\sin x=\int_0^x\frac{dt}{\sqrt{\left(1-t^2\right)}}$$

as the definitions of the function artan x or arsin x, respectively, in order to arrive at the trigonometric functions by inversion. By this process, the definition of these functions becomes separated from geometry, but we are naturally left with the task of also developing their properties independently of geometry.

We shall not go here into the development of these ideas. The essential step is to prove the addition theorem for the inverse functions, i,e. for sin and tan.

The first and most important example, which leads us beyond the region of elementary functions, is are the elliptic integrals. These are integrals in which the integrand is formed in a rational way from the variable of integration and the square root of an expression of the third or fourth degree. Among these integrals, the function

$$u(s) = \int_0^s \frac{dx}{\sqrt{(1-x^2)(1-k^2x^2)}}$$

turns out to be of special importance. Its inverse function $s(u)$ has a correspondingly important role. In particular, for $k = 0$, we obtain $u(s) = \text{arsin}\, x$ and $s(u) = \sin u$, respectively. The function $s(u)$ has been examined as thoroughly and tabulated as the elementary functions. However, this leads us away from the line of the present discussioa and into the realm of the so-called elliptic functions, which occupy a central position in the theory of functions of a complex variable. We shall merely note here that the name elliptic integral arises from the fact that such integrals enter into the problem of the determination of the length of an arc of an ellipse.

Moreover, we may note that integrals which at first sight have quite a different appearance turn out after a simple substitution to be elliptic integrals. For example, the integral

$$\int \frac{dx}{\sqrt{(\cos\alpha - \cos x)}}$$

is transformed by the substitution $u = \cos x/2$ into the integral

$$-k\sqrt{2}\int \frac{du}{\sqrt{(1-u^2)(1-k^2u^2)}}, \qquad k = \frac{1}{\cos\alpha/2}$$

the integral

$$\int \frac{dx}{\sqrt{(\cos 2x)}}$$

by the substitution $u = \sin x$ into the integral

$$\int \frac{du}{\sqrt{(1-u^2)(1-2u^2)}}$$

and, finally, the integral

$$\int \frac{dx}{\sqrt{(1-k^2\sin^2 x)}}$$

by the substitution $u = \sin x$ into the integral

$$\int \frac{dx}{\sqrt{(1-u^2)(1-k^2u^2)}}$$

On Differentiation and Integration: We insert here another remark on the relationship between differentiation and integration. Differentiation may be viewed to be a more elementary process than integration, since it does not lead us away from the domain of known functions. On the other hand, we must remember that the differentiability of an arbitrary, continuous function is by no means a foregone conclusion, but a very stringent additional assumption. In fact, we have seen that there are continuous functions which are not differentiable at isolated points and we may mention without proof that, since the time of Weierstrass, many examples have been constructed of continuous functions which do not possess a derivative anywhere . Hence, there is much less in the mathematical definition of continuity than simple intuition would lead us to suspect. In contrast, even though integration in terms of elementary functions is not always possible, we are under all circumstances at least certain that the integral of a continuous function exists.

Altogether, we see that integration and differentiation cannot be simply classified as more elementary and less elementary, but that from some points of view the one and from other points of view the other should be considered to be more elementary.

In as far as the concept of integral is concerned, we shall see in the next section that it is not closely tied to the assumption that the integrand is continuous, but that it may be extended to wider classes of functions with discontinuities.

Extension of the Concept of Integral. Improper Integrals

Functions with Jump Discontinuities: To start with, we see that there arises no difficulty with an extension of the concept of integral to the case where the integrand has jump discontinuities at one or more points in the interval of integration. In fact, we need only take the integral of the function as the sum of the integrals over the separate subintervals in which the function is continuous. The integral then retains its intuitive meaning as an area.

We should really note that, in our previous definition of integral, we assumed that the interval is closed and tbe function is continuous in the closed interval. This presents no difficulties, since we can extend in each closed subinterval the function so that it is continuous by taking for the value of the function at the end-point the limit of the function as x approaches the end-point from inside the interval.

Functions with Infinite Discontinuities: It is quite a different matter when a function has an infinite discontinuity inside the interval or at one of its ends. In order to gain even a notion of an integral in this case, we must introduce a further limiting process. Before stating the general definition, we shall illustrate some of the possibilities by means of examples.

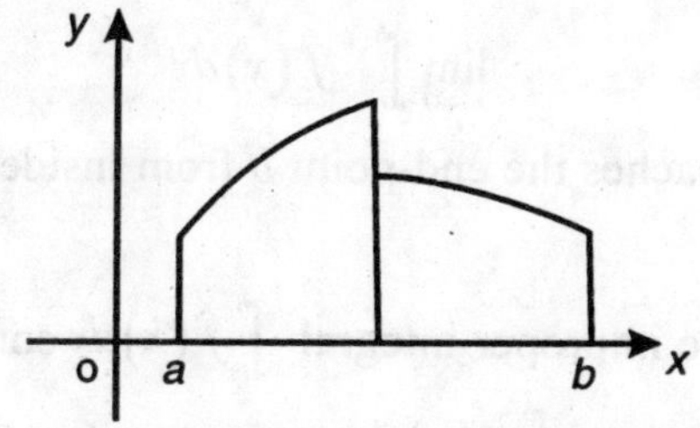

Fig. The integral of a discontinuous function

$$\int \frac{dx}{x^{\alpha}}$$

where α is a positive number. The integrand $1/x^{\alpha}$ becomes infinite as $x \to 0$, whence we cannot extend the integral to the lower limit 0. However, we can try to find what happens as we take the integral from the positive limit ε to the limit 1, say, and finally let $\varepsilon \to 0$. According to the elementary rules of integration, we obtain, provided $\alpha \neq 1$,

$$\int_{\varepsilon}^{1} \frac{dx}{x^{\alpha}} = \frac{1}{1-\alpha}\left(1-\varepsilon^{1-\alpha}\right)$$

We recognize immediately that the following possibilities arise: (1) $\alpha > 1$, when, as $\varepsilon \to 0$, the right hand side tends to ∞; (2) $\alpha < 1$, when the right hand side tends to the limit $1/(1-\alpha)$. hence, in the second case, we shall simply take this limiting value as the integral between the limits 0 and 1. In the first case, we shall say that the integral from 0 to 1 does not exist. (3) In the third case, where $\alpha = 1$, the integral will be equal to $-\log \varepsilon$ and therefore, as $\varepsilon \to 0$, it approaches no limit, but tends to ∞, that is, the integral from 0 to 1 does not exist.

Another example of the extension of the integral of a function up to an infinite discontinuity is given by the integrand $\dfrac{1}{\sqrt{(1-x^2)}}$. We find that

$$\int_{0}^{1-\varepsilon} \frac{dx}{\sqrt{(1-x^2)}} = \arcsin(1-\varepsilon)$$

If we let ε tend to 0, the right hand side converges to the definite limit $\pi/2$; we therefore call this the value of the integral $\displaystyle\int_{0}^{1} \frac{dx}{\sqrt{(1-x^2)}}$, even though the integrand becomes infinite at the point $x = 1$.

In order to extract a perfectly general concept from these examples, we note, in the first place, that it clearly makes no essential difference whether the discontinuity of the integrand lies at the upper or the lower end of the interval of integration. We now make the following statement:

If in an interval $a \leq x \leq b$ the function $f(x)$ is continuous with the single exception of the end-point b, we define $\int_{a}^{b} f(x)\,dx$ as the limit

$$\lim_{\varepsilon\to 0}\int_a^{b-\varepsilon} f(x)\,dx$$

when the point $b-\varepsilon$ approaches the end-point b from inside the interval provided such a limit exists.

In this case, we say that the improper integral $\int_a^b f(x)\,dx$ converges. However, if no such limit exists, we say that the integral $\int_a^b f(x)\,dx$ does not exist or does not converge or that it diverges.

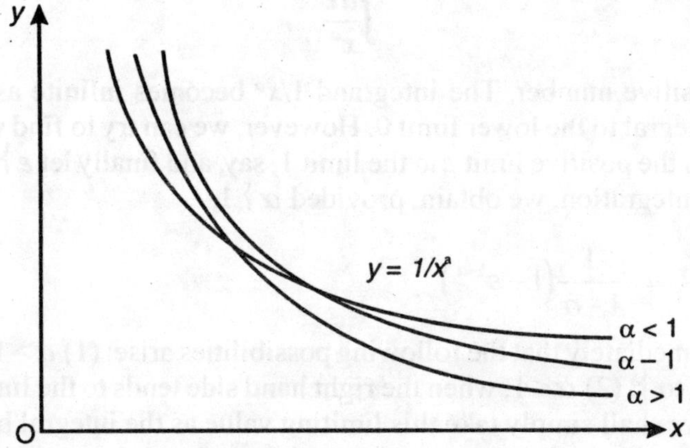

An analogous definition holds for the case where the lower limit of the interval of integration instead of the upper one is the exceptional point.

Even improper integrals can be intepreted as areas. In the first instance, of course, there is no sense in speaking of the area of a region which estends to infinity; yet, one may attempt to define such an area by means of a passage to the limit from a bounded region with a finite area. For example, the above results for the function $1/x^{\alpha}$ imply that the area, bounded by x axis, the line $x=\varepsilon$ and the curve $y = 1/x^{\alpha}$ tends to a finite limit as ε 0, provided that $\alpha < 1$, and that it tends to infinity if $\alpha < 1$. This fact may be simply expressed as follows: The area between the x-axis, the y-axis, the curve and the line $x = 1$ is finite or infinite according to whether $\alpha < 1$ or $\alpha < 1$.

Naturally, intuition can give us no precise information about the finitenees or infiniteness of the area of a region stretching to infinity. We can only say about such a region that the more closely its sides approach one another, the more likely it is to have a finite area. In this sense, Figure.illustrates the fact that for $\alpha < 1$, the area under our curve remains finite, while it is infinite for $\alpha < 1$.

In order to discover whether a function $f(x)$, which has an infinite discontinuity at the point $x = b$, can be integrated up to b, we can often save ourselves a special investigation by means of the criterion:

Let the function $f(x)$ be positive (A8.3); we will show that this restriction of sign can be removed) in the interval $a < x < b$ and let $\lim_{x\to b} f(x) = \infty$ Then the integral $\int_a^b f(x)\,dx$ converges, if there exist both a positive number μ less than 1 and a fixed number M independent of x such that everywhere in the interval $a < x < b$ the inequality $f(x) = M/(b-x)^{\mu}$ is true, in other words, if at the point $x = b$, the function $f(x)$ becomes infinite at a lower order at least than the first. On the other hand, the integral diverges, if there exist both a number $\nu < 1$ and a fixed number N such that everywhere in the interval $a < x < b$ the inequality $f(x) = N/(b-x)^{\nu}$ is true, in other words, if at the point $x = b$ the function $f(x)$ becomes infinite to the first order at least.

The proof follows almost immediately from a comparison with the very simple special case discussed above. In order to prove the first part of the theorem, we observe that for $0 < \varepsilon < b - a$, we have

$$0 \leq \int_a^{b-c} f(x)\,dx \leq \int_a^{b-c} \frac{M}{(b-x)^{\mu}}\,dx$$

As $\varepsilon < 0$, the integral on the right hand side, which is obtained from the integral $\int \frac{dx}{x^{\alpha}}$, by a simple change of notation, has a limit and therefore remains bounded. Moreover, the values of $\int_a^{b-c} f(x)\,dx$ increase monotonically as $\varepsilon < 0$; since they are also bounded, they must possess a limit and therefore the integral $\int_a^{b} f(x)\,dx$ converges.

The parallel proof of the second part of the theorem is left as an exercise for the reader.

We likewise see at once that exactly analogous theorems hold when the lower limit of the integral is a point of infinite discontinuity. If a point of infinite discontinuity lies in the interior of the interval of integration, we merely use this point to subdivide the interval into two parts and then apply the above considerations separately to each of these.

As a further example, we consider the elliptic integral

$$\infty$$

We conclud at oncee from the identity $1 - x = (1 - x)(1 + x)$ that, as $x < 1$, the integrand becomes only infinite at order , whence the improper integral exists.

Infinite Interval of Integration: Another important extension of the concept of integral occurs when one of the limits of integration as infinite. In order to make this extension precise, we introduce the notation: If the integral

$$\int_a^A f(x)\,dx$$

where a is fixed, tends to a definite limit when A increases positively beyond all bounds, we denote the limit by

$$\int_a^\infty f(x)\,dx$$

and call it the integral of the function $f(x)$ from a to ∞. Of course, such an integral does not necessarily exist or, as we often will say, converge.

Simple examples of the various possibilities are again given by the function $f(x)=1/x^\alpha$:

$$\int_1^A \frac{dx}{x^\alpha} = \frac{1}{1-\alpha}\left(A^{1-\alpha}-1\right)$$

We see here that, if we again exclude the case $\alpha = 1$, the integral to infinity exists for the case $\alpha > 1$, and, in fact,

$$\int_1^\infty \frac{dx}{x^\alpha} = \frac{1}{\alpha-1}$$

in contrast, when $\alpha < 1$, the integral no longer exists. In the case $\alpha = 1$, the integral again clearly fails to exist, since $\log x$ tends to infinity with x. Hence we see that, as regards integration over an infinite interval, the functions $1/x^\alpha$ do not behave in the same way as for integration up to the origin. This statement also is made plausible by a glance at fiigure. In , we see that, the larger is α, the more slowly do the curves draw in towards the x-axis when x is large, so that we can readily assume that the area under consideration tends to a definite limit for sufficiently large values of α.

The following criterion for the existence of an integral with an infinite limit is often useful. We again assume that for sufficiently large valued of x, say, for $x<\alpha$, the integrand has always the same sign, which, without loss of generality, we can choose to be positive. Then we have the statement:

The integral $\int_a^\infty f(x)\,dx$ converges if the function $f(x)$ vanishes at infinity to a higher order than the first, that is, if there is a number $\nu > 1$ such that for all values of x, no matter how large, the relation $0<f(x)\}M/x^\nu$ is true, where M is a fixed number independent of x. Again, the integral diverges, if the function remains positive and vanishes at infinity to an order not higher than the first, that is, if there is a fixed number $N > 0$ such that $xf(x)= N$.

The proof of these criteria, which run exactly parallel to the previous argument, is left to the reader.

A very simple example is the integral $\int_a^\infty \frac{1}{x^2}dx\,(a>0)$. The integrand vanishes at infinity to the second order. As a matter of fact, we see at once that the integral does converge, because $\int_a^A \frac{1}{x^2}dx = \frac{1}{a} - \frac{1}{A}$ whence

$$\int_a^\infty \frac{1}{x^2}dx = \frac{1}{a}$$

Another equally simple example is

$$\int_0^\infty \frac{1}{1+x^2}dx \;=\; \lim_{A\to\infty}(\arctan A-\arctan 0)=\frac{\pi}{2}$$

The Gamma Function: Another example of particular importance in analysis is offered by the socalled Γ–function

$$\Gamma(n) \;=\; \int_0^\infty e^{-x}x^{n-1}dx\;(n>0)$$

Here too the criterion of convergence is satisfied; for example, if we choose ν=2, we have since the exponential function e^x tends to zero to a higher order than any power $1/x^m$ $(m>0)$. This gamma function, which we can think of as a function of the number n (not necessarily an integer), satisfies a remarkable relation, which we can arrive at in the following way by integration by parts: To begin with, we have

$$\int e^{-x}x^{n-1}dx \;=\; -e^{-x}x^{n-1}+(n-1)\int e^{-x}x^{n-2}dx$$

If we take this formula between the limits 0 and A and then let A increase beyond all bounds, we immediately obtain

$$\Gamma(n) \;=\; (n-1)\int_0^\infty e^{-\alpha}x^{n-2}dx=(n-1)\Gamma(n-1)$$

and by this recurrence formula, provided μ is an integer and $0<\mu<n$,

$$\Gamma(n) \;=\; (n-1)(n-2)...(n-\mu)\int_0^\infty e^{-\alpha}x^{n-\mu-1}dx$$

In particular, if n is a positive integer, we have

$$\Gamma(n) \;=\; (n-1)(n-2)...3.2.1\int_0^\infty e^{-\alpha}dx$$

and, since

$$\int_0^\infty e^{-\alpha}dx \;=\; 1$$

it follows finally that

$$\Gamma(n) \;=\; (n-1)(n-2)...2.1=(n-1)!$$

This expression for a factorial by an integral is of importanoe in many applications.

The integrals

$$\int_0^\infty e^{-x^2}dx,\; \int_0^\infty x^n e^{-\alpha^2}dx$$

also converge, as we may easily verify by means of our criterion.

The Dirichlet Integral: A convergent integral, important in many applications, the convergence of which does not follow directly from our criterion and which is a simple case of a type investigated by Dirichlet, is

$$I = \int_0^{\infty} \frac{\sin x}{x} dx$$

This integral is easily seen to be convergent, if the upper limit is finite, because $\sin x/x \to 1$ as $x \to 0$. Its convergence in the infinite interval is due to the periodic change of sign of the integrand, which causes the contributions to the integral from neighbouring intervals of length π almost to cancel each another. In order to utilize this fact, we write the expression

$$D_{AB} = \int_A^B \frac{\sin}{x} dx$$

in the form

$$D_{AB} = \int_A^{A+\pi} \frac{\sin x}{x} dx - \int_B^{B+\pi} \frac{\sin x}{x} dx + \int_{A+\pi}^{B+\pi} \frac{\sin t}{t} dt$$

introduce in the last of the three integrals the new variable $x = t - p$, whence $\sin t = -\sin x$, and obtain

$$D_{AB} = \int_A^{A+\pi} \frac{\sin x}{x} dx - \int_B^{B+\pi} \frac{\sin x}{x} dx - \int_A^B \frac{\sin x}{x+\pi} dx$$

Addition of this expression to the original one for D_{AB} yields

$$2D_{AB} = \int_A^{A+\pi} \frac{\sin x}{x} dx - \int_B^{B+\pi} \frac{\sin x}{x} dx + \pi \int_A^B \frac{\sin x}{x(x+\pi)} dx$$

Hence, if we assume that $B > A > 0$, it follows that

$$|2D_{AB}| < \frac{2\pi}{A} + \pi \int_A^B \frac{dx}{x^2}$$

in fact, we may use the method of 2.2.6, while observing that

$$-\frac{1}{x} \leq \frac{\sin x}{x} \leq \frac{1}{x}$$

and

$$-\frac{1}{x^2} \leq \frac{\sin x}{x(x+\pi)} \leq \frac{1}{x^2}$$

for positive values of x. The integral on the right hand side converges, by our criterion and our formula shows that $|D_{AB}| \to 0$ as both A and B tend to infinity. Now

$$|D_{OB} - D_{OA}| = |D_{AB}|$$

and it follows from Cauchy's convergence test that D_{OB} tends to a definite limit as B . In other words, the integral I exists.

Substitution: It is obvioua that all rules for the substitution of new variables, etc. remains valid for convergent improper integrals. As an example, in order to calculate $\int_0^\infty xe^{-x^2}dx$, we introduce the new variable $u = x$ and obtain

$$\int_0^\infty xe^{-x^2}dx = \frac{1}{2}\int_0^\infty e^{-u}du = \lim_{A\to\infty}\frac{1}{2}\left(1-e^{-A}\right) = \frac{1}{2}$$

Other examples of the use of substitution in the investigation of improper integrals are the Fresnel integrals, which occur in the theory of diffraction of light:

$$F_1 = \int_0^\infty \sin\left(x^2\right)dx, F_2 = \int_0^\infty \cos\left(x^2\right)dx$$

The substitution $x = u$ yields

$$F_1 = \frac{1}{2}\int_0^\infty \frac{\sin u}{\sqrt{u}}du, F_2 = \frac{1}{2}\int_0^\infty \frac{\cos u}{\sqrt{u}}du$$

Integrating by parts, we have

$$\int_A^B \frac{\sin u}{\sqrt{u}}du = \frac{\cos A}{\sqrt{A}} - \frac{\cos B}{\sqrt{B}} - \frac{1}{2}\int_A^B \frac{\cos u}{u^{3/2}}du$$

The integral also tends to 0. Hence, by the same argument as for the Dirichlet integral, we see that the integral F_1 converges. The convergence of the integral F_2 is proved in exactly the same way.

These Fresnel integrals show that an improper integral may exist even although the integrand does not tend to zero as x. In fact, an improper integral can exist even when the integrand is unbounded, as is shown by the example

$$\int_0^\infty 2u\cos\left(u^4\right)du$$

When $u^4 = n\pi$, i.e. when $u = \sqrt[4]{n\pi}$, $n = 0, 1, 2, \ldots$ the integrand becomes $2\sqrt[4]{n\pi}\cos n\pi = \pm 2\sqrt[4]{n\pi}$, so that the integrand is unbounded. However, the substitution $u = x$ reduces the integral to

$$\int_0^\infty \cos\left(x^2\right)dx$$

which we have just shown to be convergent.

By means of a substitution, an improper integral may often be transformed into a proper one. For example, the transformation $x = \sin u$ yields

$$\int_0^1 \frac{dx}{\sqrt{\left(1-x^2\right)}} = \int_0^{\pi/2} du = \frac{\pi}{2}$$

On the other hand, integrals of continuous functions may be transformed into improper integrals; this oocurs if the transformation $u = \phi(x)$ is such that at the end of the interval of integration the derivative $\phi'(x)$ vanishes, so that dx/du is infinite.

The Second Mean Value Theorem of the Integral Calculus

The method of integration by parts yields us an easy way for proving an important theorem on the estimation of integrals, usually called the second mean value theorem of the integral calculus.

Let the function $\phi(x)$ be monotonic and continuous in the interval $a \rangle x \rangle b$, the derivative $\phi'(x)$ be continuous and moreover $f(x)$ be an arbitrary function which is continuous in the same interval. Then one has the second mean value theorem of the integral calculus:

There exists a number ξ such that $a \rangle \xi \rangle b$ for which

$$\int_a^b f(x)\phi(x)dx = \phi(a)\int_a^\xi f(x)dx+\phi(b)\int_\xi^b f(x)dx$$

We prove this by noting first that we can assume that $\phi(b) = 0$; in fact, replacement of $\phi(x)$ by $\phi(x) - \phi(b)$ changes both sides of the equation by the same amount and yields a function which vanishes at $x = b$. Moreover, we can assume that $\phi(a) > 0$; in fact, if $\phi(a) < 0$, we need only replace $\phi(x)$ by $-\phi(x)$, which changes the sign of both sides of the equation. (The case $\phi(a) = 0$ is trivial; in fact, if both $\phi(a)$ and $\phi(b)$ vanish, $\phi(x)$ must be identically zero and our equation becomes $0 = 0$.) Hence, we must only prove that, if $\phi(x)$ is continuous and monotonic decreasing and $\phi(b) = 0$, then

$$\int_a^b f(x)\phi(x)dx = \phi(a)\int_a^\xi f(x)dx$$

We now set $F(x) = \int_a^x f(x)dx$ and apply the formula for integration by parts to the left hand side of the last equation; we then have

$$\int_a^x f(x)\phi(x)dx = F(x)\phi(x)\Big|_a^b + \int_a^b F(x)\{-\phi'(x)\}dx$$

The integrated part vanishes, since $F(a)$ and $\phi(b)$ are zero. The expression $-\phi'(a)$ is everywhere positive, so that we can apply the first mean value theorem of the integral calculus. We thus find that the integral on the right hand side has the value

$$F(\xi)\int_a^b \{-\phi'(x)\}dx, a \leq \xi \leq b$$

However,

$$F(\xi) = \int_a^\xi f(x)dx \text{ and } \int_a^b \{-\phi'(x)\}dx = \phi(a) - \phi(b) = \phi(a)$$

and our theorem is established.

This theorem can be extended (although we shall not carry out the proof) to more general classes of functions. The theorem remains true for all continuous monotonic functions $\phi(x)$ whether they have derivatives or not. In fact, it is true for any discontinuous monotonic function for which we are in a position to integrate $f(x)\phi(x)$.

Chapter 6

Applications

Representation of Curves

Parametric representation: When we represent a curve by means of an equation $y = f(x)$, we must always restrict ourselves to a single-valued branch. Hence it is often more convenient - when dealing, in particular, with a closed curve - to introduce other analytical methods of representation. The most general and at the same time the most useful representation of a curve is parametric representation. Instead of considering one of the rectangular co-ordinates as a function of the other, we think of both the co-ordinates x and y as functions of a third independent variable t, the socalled parameter; the point with the co-ordinates x and y then describes the curve as t traverses a definite interval. Such parametric representations have already been encountered. For example, for the circle $x + y = a$, we obtain a parametric representation in the form

$$x = a\cos t, y = a\sin t.$$

Here, as we know already, t has the geometrical meaning of an angle at the centre of the circle. For the ellipse $x/a + y/b = 1$ we likewise have the parametric representation $x = a\cos t$, $y = b\sin t$, where t is the solaced eccentric angle, that is, the angle at the centre corresponding to the point of the circumscribed circle lying vertically above or below the point P ($a\cos t$, $b\sin t$) of the ellipse. In both these cases, the point with the co-ordinates x, y describes the complete circle or ellipse as the parameter t traverses the interval from 0 to 2π.

In general, we can seek to represent a curve parametrically by taking $x = f(t) = x(t)$, $y = y(t) = y(t)$ that is, by considering two functions of a parameter t; the shorter notation $x(t)$ and $y(t)$ will henceforth be used whenever there is no danger of confusion. For a given curve, these two functions $\phi(t)$ and $\psi(t)$ must be determined in such a way that the totality of pairs of functional values $x(t)$ and $y(t)$ corresponding to a given interval of values of t yields all the points on the curve and no points which are not on the curve. If a curve, in the first instance, is given in the form $y=f(x)$, we can arrive at a representation of this kind by first writing $x=\phi(t)$, where $\phi(t)$ is any continuous monotonic function which in a definite interval passes exactly once through each of the values of x in question; it then follows that $y = f\{\phi(t)\}$, that is, the second function $\psi(t)$ is determined by compounding f and ϕ. We thus see that, owing to the arbitrariness in the choice of the function ϕ, we have a great deal of freedom in representing a given curve parametrically; in particular, we may actually take $t = x$ and thus think of the original representation $y = f(x)$ as a parametric representation with the parameter $t = x$.

The advantage of the parametric representation is that this arbitrariness may be utilized for purposes of simplification. For example, we represent the curve $y = \sqrt[3]{x^2}$ by taking $x = t,\ y = t$, so that $\phi(t) = t,\ \psi(t) = t$. The point with the co-ordinates x,y will then describe the entire curve (semi-cubical parabola) as t varies from $-\}$ to $+\}$.

On the other hand, if a curve is originally given in parametric representation $x = \phi(t), y = \psi$ (t) and we wish to obtain the equation of the curve in non-parametric form, that is, in the form $y = f(x)$, we have only to eliminate the parameter t from the two equations. In the case of the above parametric representations of the circle and ellipse, we can do this at once by squaring and using the equation sin$t$ + cos$t$ = 1. (Another example is given below.) In general, we should have to find an expression for t from the equation $x = \phi(t)$ by means of the inverse function $t = \Phi(x)$ and substitute this into $y = \psi(t)$, in order to obtain the representation $y = \psi\{\Phi(x)\} = f(x)$. Naturally, during such an elimination, we must ordinarily restrict ourselves to a portion of the curve; in fact, to a portion which is not intersected twice by any line parallel to the y-axis.

However, it may happen that the equation $y = f(x)$ obtained in this way represents more than the original parametric representation. For example, the equations $x = a \sin t, y = b \sin t$ represent only the finite portion of the line $y = bx/a$ between the points $x = -a, y = -b$ and $x = a, y = b$, whereas the equation $y = bx/a$ represents the entire line.

The parametric representation has associated with it a definite sense in which a curve is described, corresponding to the direction in which the values of the parameter increase; we shall call this direction the positive sense. For example, if the point $x = x(t), y = y(t)$ describes a curve C as t traverses an interval $t_0 \} t \} t_1$ and the end-points P_0 and P_1 of the curve correspond to t_0 and t_1, respectively, then the curve is traversed positively in the direction from P_0 to P_1. If we introduce $\tau = -t$ as a new parameter, the curve C will correspond to the values $-t_1 \} \tau \} -t_0$ of the variable τ, and the points P_0 and P_1 will correspond to $\tau = -t_0$ and $\tau = -t_1$, respectively. If we now traverse the curve from P_0 to P_1, we proceed in the direction in which the values of the parameter τ decrease, that is, in the negative sense. In general, a change of parameter $t = t(\tau)$ preserves the sense in which a curve is described, if the function $t(\tau)$ is monotonic increasing, but reverses it, if the function $t(\tau)$ is monotonic decreasing.

Interpretation of the Parameter: Change of Parameter: In many cases, we can give an immediate physical interpretation to the parameter t, that is time. Any motion of a point in the plane may be expressed mathematically by the fact that the co-ordinates x and y appear as functions of the time. Hence, these two functions determine the motion along a path or trajectory in parametric form.

An example of this are the cycloids which arise when a circle rolls along a straight line on another circle. We limit ourselves here to the simplest case, in which a circle of radius a rolls along the x-axis and we consider a point on its circumference. This point then describes a common cycloid. If we choose the origin of the co-ordinate system and the initial time in such a way that for time $t = 0$ the corresponding point of the curve coincides with the origin, we obtain the parametric representation for the cycloid

$$x = a(t - \sin t),\ y = a(1 - \cos t),$$

where t denotes the angle through which the circle has turned from its original position; in the case when the velocity of rolling is uniform, it is proportional to the time.

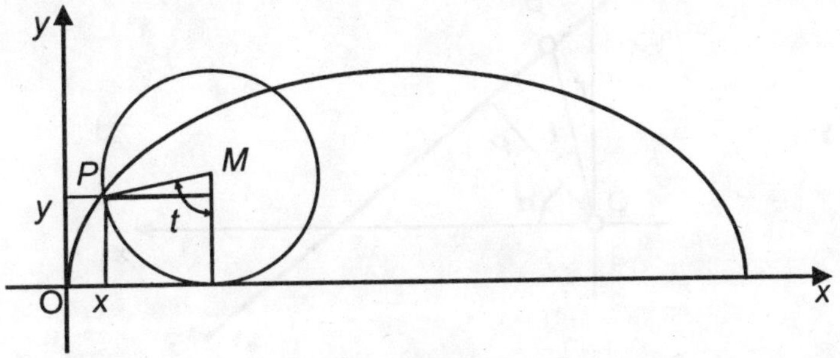

Fig. Cycloid

By elimination of the parameter t, we can obtain the equation of the curve in non-parametric form, however, at the cost of the neatness of the expression. We have

$$\cos t = \frac{a-y}{a},\ \text{arc}\cos\frac{a-y}{a},\ \sin t = \pm\sqrt{\left\{1-\frac{(a-y)^2}{a^2}\right\}}$$

$$\text{whence } x = a\ arc\ cos\ \frac{a-y}{a} \mp \sqrt{\{(2a-y)y\}},$$

and we obtain x as a function of y.

In the parametric representation of a given curve, we have a great deal of freedom in the choice of the parameter. For example, we could take instead of the time t the quantity $\tau = t$ as parameter or, indeed, any arbitrary quantity t which is related to the original parameter t by an arbitrary equation of the form $\tau = \omega(t)$, where we assume that for the entire interval of values of t under consideration this function has a unique inverse $t = \kappa(\tau)$. If increasing values of t correspond to increasing values of t, the positive sense of description remains the same; otherwise it is reversed.

Naturally, parametric representation is not limited to rectangular co-ordinates; for example, it can just as well be used with the polar co-ordinates r and θ, which are linked to the rectangular co-ordinates by the well-known equations

$$x = r\cos\theta,\ r\sin\theta \text{ or } r = \sqrt{\left(x^2+y^2\right)},\ \sin\theta = y/r,\ \cos\theta = x/r;$$

the equations of the curve would then be $r = r(t)$, $\theta = \theta(t)$.

As an example, the straight line may be represented parametrically by

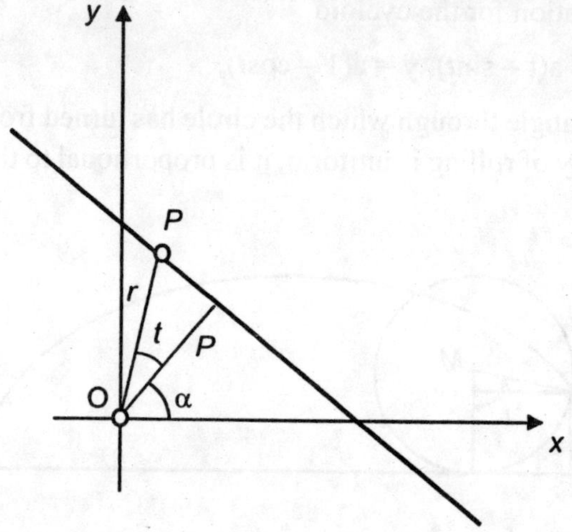

$$r = \frac{p}{\cos t},\ \theta = \alpha + t$$

(p and α being constants), from which we immediately obtain the equation of the line in polar co-ordinates

$$r = \frac{p}{\cos(\theta-\alpha)^9}$$

by eliminating the parameter t.

The Derivatives for a Parametrically Represented Curve: If, on the one hand, a curve is given by an equation $y = f(x)$ and, on the other hand, parametrically by $x = x(t)$, $y = y(t)$, then we must have $y = f\{x(t)\}$. By the chain rule for differentiation, it follows that

$$\frac{dy}{dt} = \frac{dy}{dx}\frac{dx}{dt}$$

or

$$y' = \frac{dy}{dx} = \frac{t}{2'}$$

where we use as an abbreviation for differentiation with respect to the parameter t a dot over the variable (Newton's notation) instead of the dash ' ; we shall reserve the latter for differentiation with respect to x.

For example, for the cycloid, we have

$$\dot{x} = a(1 - \cos t) = 2a \sin^2 \frac{t}{2},$$

$$\dot{y} = a \sin t = 2a \sin \frac{t}{2} \cos \frac{t}{2},$$

These formulae show that the cycloid has a cusp with a vertical tangent at the points $i = 0$, $2\pi, 4\pi, \ldots$, at which it meets the x-axis, because, on approaching these points, the derivative y' $= \dot{y}/\dot{x} = \cot(t/2)$ becomes infinite. At these points, y is equal to 0, while everywhere else $y > 0$.

The equation of the tangent to the curve is $(\xi - x)\dot{y} - (\eta - y)\dot{x} = 0$

where ξ and η are the current co-ordinates, that is, the variable co-ordinates corresponding to an arbitrary point on the tangent. For the equation of the normal, i.e., the straight line through a point of the curve, perpendicular to the tangent at that point, we likewise obtain

$$(\xi - x)\dot{x} + (\eta - y)\dot{y} = 0$$

The direction cosines of the tangent, that is, the cosines of the angles a, b which the tangent makes with the x and y axes, respectively, are given by

$$\cos\alpha = \frac{\dot{x}}{\pm\sqrt{(\dot{x}^2 + \dot{y}^2)}}, \quad \cos\beta = \frac{\dot{y}}{\pm\sqrt{(\dot{x}^2 + \dot{y}^2)}},$$

as we may verify by elementary methods. The corresponding direction cosines of the normal are given by

$$\cos\alpha' = \frac{-\dot{y}}{\pm\sqrt{(\dot{x}^2 + \dot{y}^2)}}, \quad \cos\beta' = \frac{\dot{y}}{\pm\sqrt{(\dot{x}^2 + \dot{y}^2)}},$$

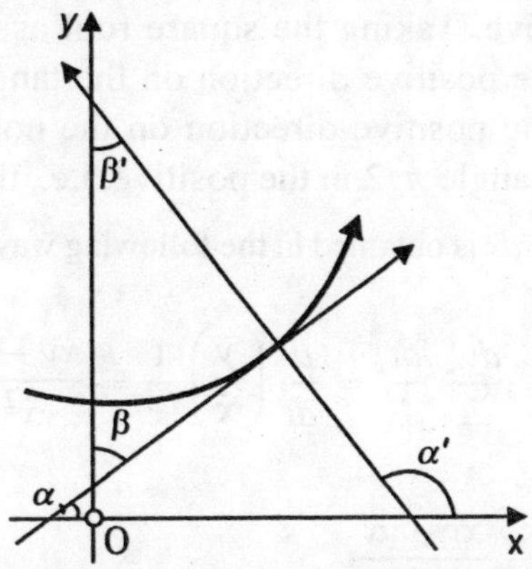

Fig. Direction cosines of the tangent and the normal

These formulae show us that at every point, at which $\dot{x}$ and $\dot{y}$ are continuous and $\dot{x}^2 + \dot{y}^2 \neq 0$, the direction of the tangent varies continuously with t. This is the most important case for us; however, it is interesting to illustrate by examples the various possibilities which arise when our assumptions are not fulfilled and we cannot state directly that the tangent keeps on turning continuously. At a point, at which $\dot{x} = \dot{y} = 0$, the tangent may or may not turn continuously. As one example, we have the curve $x = t, y = t$, discussed in 2.3.5, which has a cusp at the origin even though $\dot{x}$ and $\dot{y}$ are continuous everywhere. Consider as another example the curve $x = t, y = t$, which is the straight line $y = x$. This curve has the same tangent direction everywhere; the latter is therefore continuous, although the derivatives $\dot{x}$ and $\dot{y}$ both vanish for $t = 0$. Moreover, at a point at which $\dot{x}$ and $\dot{y}$ are discontinuous, the direction of the tangent may or may not be continuous. In fact, let $\phi(t)$ be any continuous monotonic increasing function, defined for $t_1 < t < t_2$, which has a sharp corner at $t = t_3$, $t_1 < t < t_2$. Then the curve $x = t$, $y = \phi(t)$, which is the same curve as $y = \phi(x)$, has a sharp corner at $x = t_3$; while the curve $x = \phi(t)$, $y = \phi(t)$, which is a segment of the straight line $y = x$, has a constant tangent direction even though the derivatives $\dot{x}$ and $\dot{y}$ do not exist at $t = t_3$. This indicates that, if we wish to investigate the behaviour of the tangent at a point where our theorem does not apply, we should first use the formulae to find $\cos\alpha$ or $\cos\beta$ as functions of t and then investigate these direction cosines themselves.

From a well-known formula in trigonometry or analytical geometry, we find that the angle between the two curves represented parametrically by $x=x_1(t), y=y_1(t)$ and $x=x_2(t), y=y_2(t)$, respectively, (that is, the angle between their tangents or normals) is given by the expression

$$\cos\delta = \frac{\dot{x}_1\dot{x}_2 + \dot{y}_1\dot{y}_2}{\pm\sqrt{\left(\dot{x}_1^2 + \dot{y}_1^2\right)}\sqrt{\left(\dot{x}_2^2 + \dot{y}_2^2\right)}},$$

The indeterminacy of the signs of the square roots in the last few formulae suggests that the angles are not completely determined, since we can still specify either sense of direction on the tangent or normal as positive. Taking the square root as positive, as it is usually done, corresponds to choosing for the positive direction on the tangent the direction in which the parameter increases, and for the positive direction on the normal the direction obtained by rotating the tangent through an angle $\pi/2$ in the positive, i.e., the counter-clockwise sense.,

The second derivative $y'' = dy/dx$ is obtained in the following way by means of the chain rule and the rule for differentiating a quotient:

$$y'' = \frac{dy'}{dx} = \frac{dy'}{dt}\frac{dt}{dx} = \frac{d}{dt}\left(\frac{\dot{y}}{\dot{x}}\right)\frac{1}{\dot{x}} = \frac{\dot{x}\ddot{y} - \dot{y}\ddot{x}}{\dot{x}^2}\frac{1}{\dot{x}},$$

whence

$$y'' = \frac{d^2y}{dx^2} = \frac{\dot{x}\ddot{y} - \dot{y}\ddot{x}}{\dot{x}^3}$$

Change of Axes for Parametrically represented Curves: If we rotate the axes through an angle α in the positive direction, the new rectangular co-ordinates ξ, η and the old ones x,y are interrelated by the equations

$$x = \xi \cos\alpha - \eta \sin\alpha, \qquad \xi = x \cos\alpha + y \sin\alpha,$$
$$y = \xi \sin\alpha - \eta \cos\alpha, \qquad \eta = -x \sin\alpha \sin\alpha + y \cos\alpha$$

Thus, the new co-ordinates ξ and η are specified along with x and y as functions of the parameter t. We obtain at once by differentiation

$$\dot{x} = \dot{\xi} \cos\alpha - \dot{\eta} \sin\alpha, \qquad \dot{\xi} = \dot{x} \cos\alpha + \dot{y} \sin\alpha,$$

$$\dot{y} = \dot{\xi} \sin\alpha + \dot{\eta} \cos\alpha, \qquad \dot{\eta} = -\dot{x} \sin\alpha \sin\alpha + \dot{y} \cos\alpha$$

Let the curve be given in polar co-ordinates and both polar and rectangular co-ordinates be given as functions of a parameter t. Then, by differentiation with respect to t, we obtain from the equations $x = r \cos\theta$, $y = r \sin\theta$ the formulae

$$\left.\begin{aligned} \dot{x} &= \dot{r}\cos\theta - r\sin\theta.\dot{\theta}, \\ \dot{y} &= \dot{r}\sin\theta + r\cos\theta.\dot{\theta}, \end{aligned}\right\} \qquad \text{...(a)}$$

which are frequently used in passing from rectangular to polar co-ordinates. As an example, consider the polar equation of a curve, $r=f(\theta)$ which might arise from a parametric representation $r=r(t)$, $\theta=\theta(t)$ by elimination of the parameter t. The angle ψ between the radius vector to a point on the curve and the tangent to the curve at that point is then given by

$$\tan\psi = \frac{f(\theta)}{f'(\theta)}$$

We can convince ourselves of this in the following way. If we think of the curve as being given by an equation $y = F(x)$ and use θ as a parameter, so that $\dot{\theta} = 1$ and $\dot{r} = f'(\theta)$ we have

$$\tan\alpha = y' = \frac{\dot{y}}{\dot{x}} = \frac{\dot{r}\tan\theta + r}{\dot{r} - r\tan\theta}$$

In addition, $\psi = \alpha - \theta$, whence

$$\tan\psi = \frac{y' - \tan\theta}{1 + y'\tan\theta} = \frac{r + r\tan^2\theta}{\dot{r} + \dot{r}\tan^2\theta} = \frac{r}{\dot{r}}$$

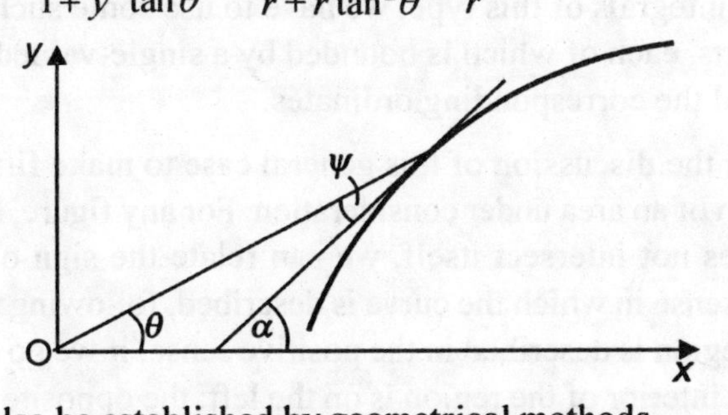

This formula can also be established by geometrical methods.

General Remarks:. In discussing given curves, we sometimes consider properties which do not assert anything about the form of a curve itself, but merely something about the position of the curve with respect to the co-ordinate system; for example, the occurrence of a horizontal tangent, expressed by the equation $\dot{y} = 0$ or the occurrence of a vertical tangent, expressed by $\dot{x} = 0$. Such properties do not persist when the axes are rotated.

In contrast to this, a point of inflection will still be a point of inflection after the axes have been rotated. According to equation , the condition for a point of inflection is

$$\dot{x}\ddot{y} - \ddot{x}\dot{y} = 0$$

If we replace on the left hand side the expressions $\dot{x}$, $\dot{y}$, $\ddot{x}$, $\ddot{y}$ by their values in terms of the new co-ordinates ξ, η, we readily obtain

$$\dot{x}\ddot{y} - \ddot{x}\dot{y} = \dot{\xi}\ddot{\eta} - \ddot{\xi}\dot{\eta}$$

Hence it follows from the equation $\dot{x}\ddot{y} - \ddot{x}\dot{y} = 0$ that $\dot{\xi}\ddot{\eta} - \ddot{\xi}\dot{\eta} = 0$ so that our equation expresses a property of the point of the Curve which is independent of the co-ordinate system.

We shall often see later on that properties which are truly geometrical are expressed by formulae the form of which is not altered by rotation of the axes.

Applications to the Theory of Plane Curves

We shall consider two different kinds of geometrical properties or quantities associated with curves. The first type consists of properties or quantities which depend only on the behaviour of the curve in the small, i.e. in the immediate neighbourhood of a point, and which can be expressed analytically by means of the local derivative. Properties of the second type depend on the entire course or a portion of the curve and are expressed analytically by means of the concept of integral. We shall begin by considering properties of the second type.

Orientation of Area: The idea of area was our starting point for the definition of the integral; but the connection between the definite integral and area is still somewhat incomplete. The areas, with which we are concerned in geometry, are bounded by given closed curves; on the other hand, the area measured by the integral $\int_{x_1}^{x_2} f(x)dx$ is bounded only partly by the given curve $y = f(x)$, the rest of the boundary consisting of lines which depend on the choice of the co-ordinate system. If we wish to determine the area interior to a closed curve, such as a circle or an ellipse, by means of integrals of this type, we have to use some such device as breaking up the area into several parts, each of which is bounded by a single-valued branch of the curve as well as by the x-axis and the corresponding ordinates.

It is convenient for the discussion of this general case to make first some remarks on the determination of the sign of an area under consideration. For any figure, bounded by an arbitrary closed curve which does not intersect itself, we can relate the sign of its area to the purely geometrical idea of the sense in which the curve is described, following the convention: We say that the boundary of a region is described in the positive sense, if we go around the boundary in such a direction that the interior of the region is on the left; the opposite sense we call negative. If we then consider a region, the boundary of which is traversed in an assigned sense - a solaced oriented region, we consider the area to be positive if this sense is positive, and negative if this sense is negative.

If we wish to avoid the words right and left in such a context, we say that the triangle, the ordered vertices of which are the origin, the point $x = 1$, $y = 0$ and the point $x = 0$, $y = 1$, is described in the positive sense, if the vertices are passed in the order mentioned. For every other region, we say that the boundary is positively described. if it is described in the same sense as this triangle, otherwise it is described negatively

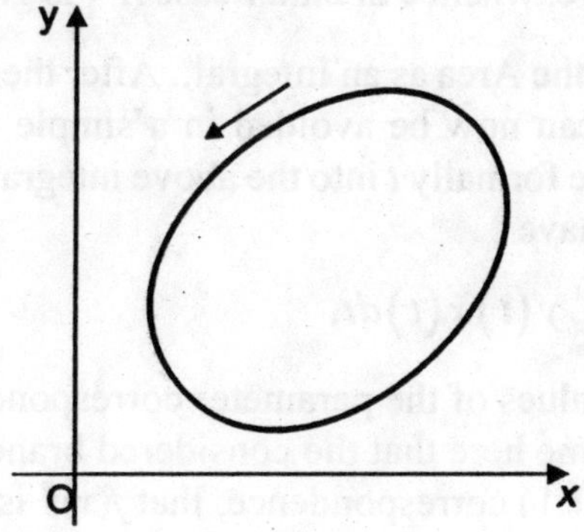

Fig. A positive area

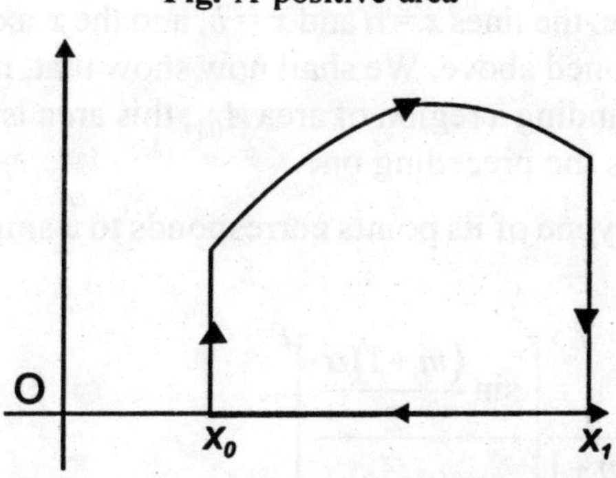

In particular, let in the interval $a < x < b$ the function $f(x)$ be everywhere positive. We consider the closed curve obtained by starting at the point $x = b = x_1, y = 0$, traversing the x-axis back to the point $x = a = x_0, y = 0$, then proceeding along the ordinate to the curve $y = f(x)$, then along the curve to the ordinate $x = b$, and finally along the ordinate to the x-axis. The absolute value of the area interior to this curve - the number of square units contained in it - is, as we know, $\int_a^b f(x)dx$. Hence, denoting by $A^0{}_1$ the area with its sign as determined above, the integral yields the value $A^0{}_1$ except for its sign. In order to determine the sign, we need only observe that the boundary of the region is traversed in the negative sense, so that $A^0{}_1$ is negative; hence

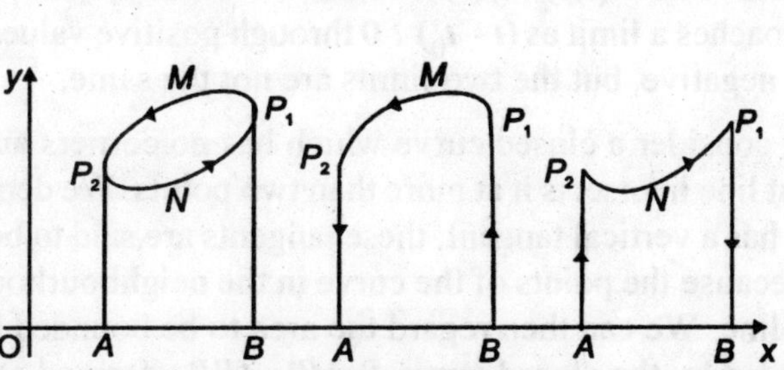

Fig. Area of a closed curve

$$A^0{}_1 = -\int_a^b f(x)dx$$

Similarly, if $a > b$, we find that, according to our convention, $A^0{}_1$ is positive, while the integral $\int_a^b f(x)dx$ is negative, whence in either case $A^0{}_1$ is given by the above equation.

The General formula for the Area as an Integral:. After these preliminaries, the difficulties mentioned at the beginning can now be avoided in a simple way by representing our curve parametrically. If we introduce formally t into the above integral as a new independent variable, writing $x = x(t)$, $y = y(t)$, we have

$$A_{01} = -\int_{t_0}^{t_1} y(t)\dot{x}(t)dt,$$

where t_0 and t_1 are the values of the parameter corresponding to the abscissae $x_0 = a$ and $x_1 = b$, respectively. We assume here that the considered branch of the curve $y=f(x)$ is related to an interval $t_0 \} t \} t_1$ by a (1,1) correspondence, that $f(x)$* is everywhere positive and that) never vanishes in this interval. As we have seen, our expression then yields the area of the region bounded by the curve, the lines $x = a$ and $x = b$, and the x-axis. It is, of course, still subject to the disadvantages mentioned above. We shall now show that, if the curve $x = x(t)$, $y = y(t)$, $t_0 \} t \} t_1$ is a closed curve bounding a region of area A_{01}, this area is given by an integral which in form is exactly the same as the preceding one.

*** i.e., is such that everyone of its points corresponds to a single value of t in the interval $t_0 \} t \} t_1$ and conversely.**

$$s_m(\alpha) = \frac{1}{m+1}\left[\frac{\sin\frac{(m+1)\alpha}{2}}{\sin\frac{\alpha}{2}}\right]^2$$

Consider now a closed curve which is represented parametrically by the equations $x=x(t)$, $y = y(t)$, the curve being described just once as t describes the interval $t_0 \} t \} t_1$. In order that the curve may be closed, it is essential that $x(t_0) = x(t_1)$ and $y(t_0) = y(t_1)$. We shall assume that the derivatives are continuous except at most for a finite number of jump-discontinuities, and that $-\int_{t_0}^{t_1} y\dot{x}dt$ differs from zero except perhaps at a finite number of points which may be corners of the curve.

A continuous curve $x = x(t)$, $y = y(t)$ is said to have a corner at $t = t_0$, if the positive direction of the tangent approaches a limit as $(t - t_0) \} 0$ through positive values and approaches a limit as $(t - t_0) \} 0$ through negative, but the two limits are not the same.

We shall first consider a closed curve which has no corners and is convex and of such a type that no straight line intersects it at more than two points. We denote by P_1 and P_2 the points at which the curve has a vertical tangent; these tangents are said to be lines of support at P_1 and P_2, respectively, because the points of the curve in the neighbourhood of P_1 and P_2 lie entirely on one side of the line. We can then regard the area to be bounded by the curve as the sum of the area A_{12}, bounded by the closed curve $P_1MP_2ABP_1$, formed as in the preceding section,

and the area A_{21}, bounded by the closed curve $P_2NP_1BAP_2$. We assume here that the curve is described in the positive sense, as in the figure; by our sign convention, A_{12} is then positive and A_{21} negative. Let the point $x(t)$, $y(t)$ describe the upper part of the curve from P_1 to P_2 as t moves from t_0 to t, and the lower part from P_2 to P_1 as t moves from τ to t_1. We then obtain immediately

$$A_{12} = -\int_{t_0}^{t} y(t)\dot{x}(t)dt, \quad A_{21} = -\int_{t}^{t_1} y(t)\dot{x}(t)dt$$

whence the total area bounded by the convex curve is

$$A = -\int_{t_0}^{t_1} y(t)\dot{x}(t)dt$$

If we denote by the absolute area of a region the number of square units contained in it, - which is, of course, never negative - then the above expression always yields the absolute area bounded by the curve except perhaps for the sign. In order to see what happens when we reverse the sense in which the curve is described, we simply take the same integral from t_1 to t_0 instead of from t_0 to t_1; then our integral becomes $-\int_{t_1}^{t_0} y\dot{x}\,dr$

which is equal to $-A$. We thus recognize the truth of the statement:

The area represented by our formula is positive or negative according to the sense in which the boundary is described.

In drawing the figure, we have assumed that $y > 0$ for all points of the curve. This really does not restrict the generality of the result. In fact, if we displace the curve by a distance a parallel to the y-axis, without rotating it, in other words, replace y by $y + a$, the area is unchanged; the value of the integral is likewise unaltered, for the above integral is replaced by $-\int_{t_0}^{t_1} (y+a)\dot{x}(t)dt$ and, since the curve is closed,

$$\int_{t_0}^{t_1} a\dot{x}\,dt = a\{x(t_1) - x(t_0)\} = 0$$

Two simple observations enable us to extend our results. Firstly, our formula remains valid for closed curves which do not intersect themselves, even if they are not convex, but have a more general form. Secondly, the derivatives may have jump discontinuities or may both vanish at a finite number of points, which may represent corners; according to equation , the function $y\dot{x}$ remains integrable. We assume that the curve has only a finite number of lines of support, corresponding to the points $P_1, P_2, \ldots, P_n$ and subdivide the curve into the single-valued branches $P_1P_2 \ldots\ldots\ldots P_{n-1}P_nP_1$.

Then, as in fiigure., we obtain the area bounded by the curve in the form $A = A_{12} + A_{23} + \ldots + A_{n-1,n} + A_{n1}$. If we express each of these portions of area parametrically and combine the expressions into a single integral, we find that the area bounded by the curve is given by $-\int_{t_n}^{t_1} y\dot{x}\,dt$ which, as before, has the same sign as the sense in which the boundary curve is traversed.

In a certain sense, our formula even gives us the area in the case where the curve intersects itself. But we shall not enter into such a discussion here; readers who wish to do so may turn to A5.2,

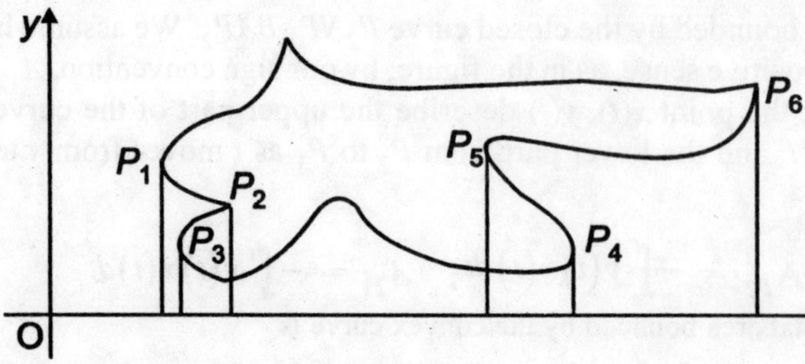

We can express our formula for the area in a more elegant symmetrical form if we first apply to the integral integration by parts:

$$\int_{t_0}^{t_1} y\dot{x}\,dt = -\int_{t_0}^{t_1} x\dot{y}\,dt + xy\Big|_{t_0}^{t_1}$$

Since the curve is closed,

$$x(t_0) = x(t_1), \qquad y(t_0) = y(t_1)$$

whence

$$A = -\int_{t_0}^{t_1} y\dot{x}\,dt = \int_{t_0}^{t_1} x\dot{y}\,dt$$

If we form the arithmetic mean of the two expressions, we obtain the symmetrical form

$$A = -\frac{1}{2}\int_{t_0}^{t_1} (y\dot{x} - x\dot{y})\,dt$$

Instead of finding for the area the second expression above by integration by parts, we could have derived it by using the fact that, as regards the definition of area, the x-axis and the y-axis are interchangeable, except that the sense of rotation which brings the x-axis into the y-axis along the shortest way is opposite to the sense which brings the y-axis into the x-axis along the shortest way.

Remarks and an Example: In connection with these expressions, we must make a remark of a fundamental nature. Both the proof and the statement of the formulae depend on a particular system of rectangular co-ordinates. But the value of the area - a purely geometrical quantity - cannot depend on the chosen co-ordinate system. It is therefore important to show that a change of co-ordinates does not affect our integrals.

If the axes are merely displaced without rotation, obviously the integrals are unaltered. Now let us assume that the axes are rotated through an angle α; instead of x and y, we now have new variables ξ and η, defined by the equations

$$x = \xi\cos\alpha - \eta\sin\alpha, \qquad y = \xi\sin\alpha + \eta\cos\alpha$$

the new variables being also functions of the parameter t. If we recall that

$$\dot{x} = \dot{\xi}\cos\alpha - \dot{\eta}\sin\alpha, \qquad \dot{y} = \xi\sin\alpha + \eta\cos\alpha$$

a short calculation yields

$$y\dot{x} - x\dot{y} = \eta\dot{\xi} - \xi\dot{\eta}$$

whence

$$A = -\frac{1}{2}\int_{t_2}^{t_1}(y\dot{x} - x\dot{y})dt = -\frac{1}{2}\int_{t_2}^{tl}\left(\eta\dot{\xi} - \xi\dot{\eta}\right)dt$$

This equation expresses the fact that the area does not depend on the co-ordinate system.

Our integral expression for the area also does not depend on the choice of parameter. In fact, let us introduce a new parameter t by the equation $t = t\,(t)$; we have

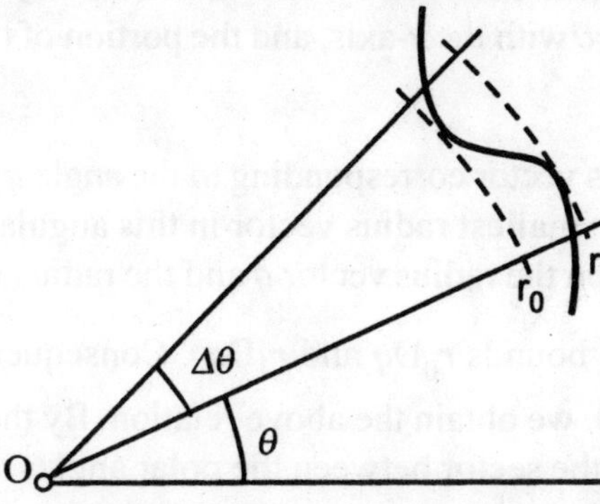

Fig. Element of area in polar co-ordinstes

$$\frac{dx}{dt} = \frac{dx}{d\tau}\frac{d\tau}{dt}, \qquad \frac{dy}{dt} = \frac{dy}{d\tau}\frac{dr}{dt}$$

so that

$$-\int_{t_0}^{t_1}\left(y\frac{dx}{dt} - x\frac{dy}{dt}\right)dt = -\int_{t_2}^{t_1}\left(y\frac{dx}{d\tau} - x\frac{dt}{d\tau}\right)\frac{d\tau}{dt}dt$$

$$= -\int_{t_2}^{t_1}\left(y\frac{dx}{d\tau} - x\frac{dy}{d\tau}\right)d\tau$$

where τ_0 and τ_1 are the initial and final values of the new parameter, corresponding to the parametric values t_0 and t_1, respectively.

In this section, we have based the definition of area on the concept of the integral and have shown that this analytical definition has a truly geometrical character, since it yields a quantity independent of the co-ordinate system. However, it is easy to give a direct geometrical definition of the area bounded by a closed curve which does not intersect itself, as follows: The area is the upper bound of the areas of all polygons lying interior to the curve. The proof that the two definitions are equivalent is quite simple, but will not be given here.

As an example of the application of our formulae for the area, we consider the ellipse $y = \frac{b}{a}\sqrt{\left(a^2 - x^2\right)}$. In order to find its area, we take the upper and lower halves of the ellipse separately and in this way express its area by the integral $2\frac{b}{a}\int_{-a}^{+a}\sqrt{\left(a^2 - x^2\right)}dx$

However, if we use the parametric representation $x = a \cos t, y = b \sin t$, we find immediately that its area is given by $ab \int_0^{2\pi} \sin^2 t \, dt$

This can be integrated as in 4.2.3; its value is $ab\pi$.

Areas in Polar Co-ordinates: For many purposes, it is important to be able to calculate areas using polar co-ordinates. Let $r = f(\theta)$ be the equation of a curve in polar co-ordinates. Let $A(\theta)$ be the area of the region which is bounded by the x-axis (that is, the line $\theta = 0$), the line through the origin forming an angle θ with the x-axis, and the portion of the curve between these two Lines. Then $ab \int_0^{2\pi} \sin^2 t \, dt$

In fact, if we consider the radius vector corresponding to the angle q and that corresponding to the angle $q + \mathrm{D}q$, and denote the smallest radius vector in this angular interval by r_0 and the largest by r_1, the sector lying between the radius vector q and the radius vector $q + \mathrm{D}q$ will have an area DA which lies between the bounds $r_0\mathrm{D}q$ and $r_1\mathrm{D}q$). Consequently, $\frac{1}{2}r_0^2 \leqq \frac{\Delta A}{\Delta\theta} \leqq \frac{1}{2}r_1^2$ and on passing to the limit as $\Delta\theta$ ⟩ 0, we obtain the above relation. By the fundamental theorem of the integral calculus, the area of the sector between the polar angles a and b is then given by $\frac{1}{2}\int_\alpha^\beta r^2 d\theta$

If $\beta > \alpha$, this expression cannot be less than zero. Since we readily see that, as θ increases, the point with co-ordinates (r, θ) describes the boundary of the region in the positive sense, this is in agreement with our previous sign convention.

As an example, consider the area bounded by one loop of a lemniscate/. Its equation is $r = 2\ a\cos 2\theta$ and we obtain one loop by letting θ vary from $-\pi/4$ to $+\ \pi/4$. This gives us the expression $a^2 \int_{-\pi/4}^{\pi/4} \cos 2\theta \, d\theta$ for the area. This can be integrated at once by introducing the new variable $u = 2q$; we find the value of the integral to be a.

Length of a Curve: Another important geometrical concept - the length of arc - leads to integration.

To start with, we shall explain geometrically how we are led to a definition of the length of an arbitrary curve. The elementary process of measuring a length consists of comparing the length to be measured with rectilinear standards of length. The simplest method is to apply our standard length to the curve, with its ends on the curve, and count the number of times that we have to repeat the process in order to pass from the beginning to the end of the curve; we can refine the method as required by using smaller and smaller standards of length. By analogy with this elementary intuitive idea, we set up the definition of the length of a curve as follows: We assume that our curve is given by the equations $x = x(t), y = y(t), \alpha < t < \beta$ (This includes curves in the form $y = f(x)$, since these can be written as $y = f(t), x = t$.) In the interval between α and β, we choose points $t_0 = \alpha, t_1, t_2, \ldots, t_n = \beta$ in that order. We join the points on the curve, corresponding to these values of t, in order of the line segments, thus obtaining part of a polygon inscribed in the curve; we now measure the perimeter of this polygon. This length will depend on the way in which the points t_n, or, as we may also say, the vertices of the polygon are chosen. We now let the number of the points t_n increase beyond all bounds in such a way that the length of the longest subinterval in the interval $\alpha < t < \beta$ at the same time tends to 0; this causes the number of sides of our polygon to increase without limit, while the length of the longest side tends to 0. The length of the curve is then defined to be the limit of the perimeters of these inscribed polygons, provided that such a limit exists and is independent of the particular way in which the polygons are chosen. It is only when this assumption that the limit exists (the assumption of rectifiability is fulfilled that we can speak of the length of the curve. We shall soon see that very wide classes of curves can be proved to be rectifiable.

In order to express the length analytically by an integral, we think, in fact, of the curve as being represented in the first instance by a function $y = f(x)$ with a continuous derivative y'. We subdivide by the points $a = x_1, x_2, \ldots, x_n = b$ the interval $a < x < b$ of the x-axis, above which lies our curve, into $(n-1)$ intervals of lengths $\Delta x_1, \ldots, \Delta x_{n-1}$. We inscribe in the curve a polygon the vertices of which lie vertically above these points. By Pythagoras' theorem, the total length of this inscribed polygon is given by

$$\sum_{v=1}^{n-1} \sqrt{\left(\Delta x_v^2 + \Delta y_v^2\right)} = \sum_{v=1}^{n-1} \sqrt{\left\{1 + \left(\frac{\Delta y_v}{\Delta x_v}\right)^2\right\}} \Delta x_v$$

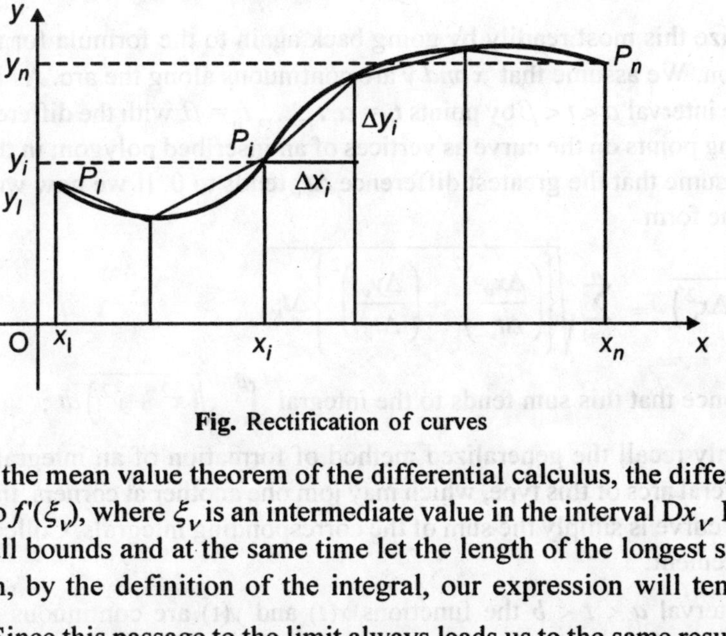

Fig. Rectification of curves

However, by the mean value theorem of the differential calculus, the difference quotient $\Delta y_v / \Delta x_v$ is equal to $f'(\xi_v)$, where ξ_v is an intermediate value in the interval Dx_v. If we now let n increase beyond all bounds and at the same time let the length of the longest sub-interval Δx tend to zero, then, by the definition of the integral, our expression will tend to the limit $\int_a^b \sqrt{\left(1 + y'^2\right)} dx$. Since this passage to the limit always leads us to the same result, namely, the integral, no matter how the subdivision of the interval is made, we have established the theorem:

Every curve $y = f(x)$, for which the derivative $f'(x)$ is continuous, is a rectifiable curve and its length between $x = a$ and $x = b$ $(b < a)$ is given by

$$s(a,b) = \int_a^b \sqrt{\left(1+y'^2\right)}\,dx$$

If we denote by *s* the length of arc measured from an arbitrary fixed point to the point with abcissa *x*, the above equation yields for the derivative of the length of arc with respect to *x*:

$$\frac{ds}{dx} = \sqrt{1+y'^2}$$

Our expression for the length of arc is still subject to the special and artificial assumption that the curve consists of one single-valued branch above the *x*-axis. Parametric representation removes this restriction. If a curve of the kind under consideration is given in parametric form by the equations $x = x(t)$, $y = y(t)$, then we obtain by introduction of the parameter *t* into the above expression the parametric form of the length of arc

$$s(\alpha,\beta) = \int_\alpha^\beta \sqrt{\left(\dot{x}^2 + \dot{y}^2\right)}\,dt$$

where *a* and *b* are the values of *t* which correspond to the points $x=a$ and $x=b$ of the curve, respectively.

This parametric expression for the length of a curve has a considerable advantage over the previous form in that it is not restricted to single-valued branches of the curves, represented by the equation $y = f(x)$, but instead it holds for any arbitrary arcs of curves, including closed curves, provided that the derivatives $\dot{x}$ *and* $\dot{y}$ are continuous along the arcs.

We recognize this most readily by going back again to the formula for the length of the inscribed polygon. We assume that $\dot{x}$ *and* $\dot{y}$ are continuous along the arc. As in the definition, we subdivide the interval $\alpha < t < \beta$ by points $t_0 = \alpha, t_1, \ldots, t_n = \beta$, with the differences Δt_ν and use the corresponding points on the curve as vertices of an inscribed polygon; in the passage to the limit $n\}\}$, we assume that the greatest difference Δt_ν tends to 0. If we now write the length of the polygon in the form

$$\sum_{\nu=1}^{n} \sqrt{\left(\Delta x_\nu^2 + \Delta x_\nu^2\right)} = \sum_{\nu=1}^{n} \sqrt{\left\{\left(\frac{\Delta x_\nu}{\Delta t_\nu}\right)^2 + \left(\frac{\Delta y_\nu}{\Delta t_\nu}\right)^2\right\}}\,\Delta t_\nu$$

We see at once that this sum tends to the integral $\int_\alpha^\beta \sqrt{\left(\dot{x}^2 + \dot{y}^2\right)}\,dt$;

We need only recall the generalized method of formation of an integral. If the curve is composed of several arcs of this type, which may join one another at corners, the expression for the length of the curve is simply the sum of the corresponding integrals. Collecting the results, we have the statement:

If in the interval $a < t < b$ the functions $x(t)$ and $y(t)$ are continuous as well as their derivatives $\dot{x}$ *and* $\dot{y}$, except perhaps for a finite number of jump discontinuities, the arc of $x = x(t)$, $y = y(t)$ has the length $\int_\alpha^\beta \sqrt{\left(\dot{x}^2 + \dot{y}^2\right)}\,dt$ where this integral, if necessary, is to be taken as an improper integral in the sense of 4.8.

By virtue of this formula, in which α must be less than β, there is a meaning in ascribing a negative length, given by the same formula, to an arc of a curve traversed in the direction in which the value of the parameter t decreases. The sign of the length of arc therefore depends on the choice of the parameter. If we introduce a new parametric expression for the same curve which does not reverse the sense of description, that is, if we introduce a new parameter by the equation $\tau = \tau(t)$, where $d\tau/dt > 0$, we see *a priori* that our integral formula should give the same value no matter whether t or τ is used as parameter, because the two integrals yield the length of the same curve and must therefore be equal. However, this may also be verified directly, because

$$\int\sqrt{\left(\dot{x}^2+\dot{y}^2\right)}dt = \int\sqrt{\left\{\left(\frac{dx}{d\tau}\right)^2\left(\frac{d\tau}{dt}\right)^2+\left(\frac{dy}{d\tau}\right)^2\left(\frac{d\tau}{dt}\right)^2\right\}}dt$$

$$= \int\sqrt{\left\{\left(\frac{dx}{d\tau}\right)^2+\left(\frac{dy}{d\tau}\right)^2\right\}}d\tau$$

We now present the expression for the length of arc when the curve is expressed in polar co-ordinates. In the last expression, we need only substitute for $\dot{x}$ *and* $\dot{y}$ their values as given by the formula (*a*) in order to obtain $\dot{x}^2+\dot{y}^2 = \dot{r}^2+r^2\dot{\theta}^2$

whence

$$s(\alpha,\beta) = \int_\alpha^\beta\sqrt{\left(\dot{r}^2+r^2\dot{\theta}^2\right)}dt$$

If we now step over from the parametric expression to the equation in the form $r = f(\theta)$ by introducing as parameter $t = \theta$ itself, so that $\dot{\theta}=1$ we find for the length of arc.

$$s(\theta_0, \theta_1) = \int_{\theta_0}^{\theta_1}\sqrt{\left(\dot{r}^2+r^2\right)}d\theta$$

A simple example of the explicit calculation of the length of an arc is given by the parabola $y = x/2$; we obtain immediately for its length of arc the integral $\int_a^b\sqrt{\left(1+x^2\right)}dx$. Which with the substitution $x = \sinh u$ becomes

$$\int_{ar\sinh a}^{ar\sinh b}\cosh^2 u\,du = \frac{1}{2}\int_{ar\sinh a}^{ar\sinh b}(1+\cosh 2u)du = \frac{1}{2}(u+\sinh u\cosh u)\Big|_{ar\sinh a}^{ar\sinh b}$$

So that the length of arc of the parabola between the abscissae $x=a$ and $x=b$ is given by

$$s(a,b) = \frac{1}{2}\left\{ar\sinh b + b\sqrt{\left(1+b^2\right)} - ar\sinh a - a\sqrt{\left(1+a^2\right)}\right\}$$

For the catenary $y = \cosh x$, we find

$$s(a,b) = \int_a^b\sqrt{\left(1+\sinh^2 x\right)}dx = \int_a^b\cosh x\,dx, \quad or\ s(a,b) = \sinh b - \sinh a$$

Finally, note that it is convenient in many cases to introduce as parameter the length of arc reckoned from some fixed point P_0 on the curve, that is, to take $x = x(s)$ and $y = y(s)$. Points of the curve on opposite sides of P_0 will correspond to values of s with opposite signs. In this case, we have

$$\dot{x}^2 + \dot{y}^2 = \left(\frac{ds}{dt}\right)^2 = 1$$

whence by differentiation $\dot{x}\ddot{x} + \dot{y}\ddot{y} = 0$ these two relations are applied frequently.

Curvature of a Curve: The axes and the length of arc of a curve depend on its complete course. We now insert a discussion of a concept which has reference only to the behaviour of a curve in the neighbourhood of a point - its curvature.

If we think of a curve as being described uniformly in the positive sense in such a way that equal lengths of arc are passed over in equal periods of time, the direction of the curve will vary at a definite rate, which we take as a measure of its curvature. Hence, denoting the angle between the positive direction of the tangent (5.1.3) and the positive x-axis by α and thinking of a as a function of the length of arc s, we shall define the curvature k at the point corresponding to the length of arc s by the equation $k = da/ds$. We know that a = artan y', whence, by the chain rule,

$$\frac{d\alpha}{ds} = \frac{d\alpha}{dx} \div \frac{ds}{dx} = \frac{y''}{1+y'^2} \cdot \frac{1}{\sqrt{(1+y'^2)}}$$

(where the positive sign of the square root means that increasing values of x correspond to increasing values of s). Hence, the curvature is given by

$$k = \frac{y''}{\left(1+y'^2\right)^{3/2}}$$

Using the parametric formulae for y' and y'' we obtain the simple expression for the curvature of a parametrically represented curve:

$$k = \frac{\dot{x}\ddot{y} - \dot{y}\ddot{x}}{\left(\dot{x}^2+\dot{y}^2\right)^{3/2}}$$

which, of course, can also be found directly from the equation

$$\alpha = \text{arc tan}\frac{\dot{y}}{\dot{x}} = \text{arc cot}\frac{\dot{x}}{\dot{y}}$$

In contrast to the previous expression, which depends on the equation $y = f(x)$ and consequently involves a special assumption about the position of the arc with respect to the x-axis, the parametric expression for the curvature holds for all arcs along which $\dot{x}, \dot{y}, \ddot{x}$, *and* $\ddot{y}$ are continuous functions of t and $\dot{x}^2 + \dot{y}^2 \neq 0$. In particular, it holds for points where $\dot{x} = 0$, i.e., where dy/dx becomes infinite.

If we introduce the length of arc s as parameter and recall that $\dot{x}^2 + \dot{y}^2 = 1$ and $\dot{x}\ddot{x} + \dot{y}\ddot{y} = 0$ we find

$$k = \dot{x}\ddot{y} - \dot{y}\ddot{x} = \ddot{y}\left(\dot{x} + \dot{y}\frac{\dot{y}}{\dot{x}}\right) = \frac{\ddot{y}}{\dot{x}} = -\frac{\ddot{x}}{\dot{y}}$$

We thus obtain a particularly simple expression for the curvature.

The sign of the curvature is changed, if we reverse the sense of description of the curve, that is, if we replace the parameter t or s by the new parameter $\tau = -t$ or $\sigma = -s$, because then $\dot{x}$ *and* $\dot{y}$ change their sign, but not $\ddot{x}, \ddot{y}, \dot{x}^2$ *or* $\dot{y}^2$ as follows from a simple calculation:

$$\frac{d}{d\tau}x\{t(\tau)\} = \frac{dx}{dt}\frac{dt}{d\tau} = (\dot{x})(-1);$$

$$\frac{d^2}{d\tau^2}x\{t(\tau)\} = \frac{d}{d\tau}\left[-\dot{x}\{t(\tau)\}\right] = -\frac{d\dot{x}}{dt}\frac{dt}{d\tau} = (-\ddot{x})(-1)$$

(A similar calculation can be made for y.) In the case of the expression $k = \dfrac{y''}{\left(1 + y'^2\right)^{3/2}}$ found first, this fact is concealed, since it is natural and customary to think of a curve as described from the left to the right hand side, in which case the square root can only be positive.

As an example, consider the curvature of a positively described circle with radius a. If we start from the parametric representation $x = a\cos t$, $y = a\sin t$, we obtain immediately

$$k = \frac{1}{a}$$

Hence the curvature of a positively described circle is the reciprocal of its radius. This result assures us that our definition of curvature is really suitable, because, in the case of a circle, we naturally think of the reciprocal of the radius as a measure of its curvature.

Let us set $\rho = 1/k$. In general, the quantity $|\rho| = 1/|k|$ is called the radius of curvature of a curve at the point in question. For a given point on a curve, that circle which touches the curve at the point and has there the same sense of description and the same curvature as the curve and, moreover, has its centre on the positive or negative side of the normal according to whether k is positive or negative, is called the circle of curvature, corresponding to the point. Let us think of the equation of the circle as being written in the form $y=g(s)$. Then, at the point in question, we do not only have $f(x) = g(x)$ and $f'(x) = g'(x)$, as follows from the fact that the circle and curve touch, but, by virtue of the relation

$$\frac{f''(x)}{\sqrt{\left\{1 + f'(x)^2\right\}^3}} = k = \frac{g''(x)}{\sqrt{\left\{1 + g'(x)^2\right\}^3}}$$

we also have $f''(x) = g''(x)$

The centre of the circle of curvature is called the centre of curvature, corresponding to the given point. Its co-ordinates are expressed parametrically by

$$\xi = x - \frac{\rho \dot{y}}{\sqrt{\dot{x}^2 + \dot{y}^2}}, \qquad \eta = y + \frac{\rho \dot{x}}{\sqrt{\dot{x}^2 + \dot{y}^2}}$$

In order to prove this, we need only employ the formulae for the direction cosines of the normal, on which the centre of curvature lies at a distance $1/|k|=|\rho|$ from the tangent. These formulae yield an expression for the centre of curvature in terms of the parameter t. As t describes its range, the centre of curvature describes a curve, the socalled evolute of the given curve and since, together with x and y, we must regard $\dot{x}, \dot{y}, and\, \rho$ as known functions of t, the formulae above yield parametric equations for the evolute.

Centre of Mass and Moment of a Curve: We now come to some applications which take us into mechanics. We consider a system of n particles in a plane. Let m_1, m_2, m_n be the masses of these particles, and y_1, y_2, y_n their respective ordinates.

$$T = \sum_{\nu=1}^{n} m_\nu y_\nu = m_1 y_1 + m_2 y_2 + \ldots + m_n y_n$$

We then call the moment of the system of particles with respect to the x-axis. The expression $\eta = T/M$, where M denotes the total mass $m_1 + m_2 + \ldots m_n$ of the system, gives us the height of the centre of mass of the system of particles above the x-axis. We define the moment about the y-axis and the abscissa of the centre of mass in a corresponding way.

We shall now see that this idea is readily extended to yield a definition of the moment of a curve along which a mass is distributed uniformly, and of the co-ordinates ξ and η of the centre of mass of such a curve. Merely for the sake of brevity, we assume that the density along the curve is constant, say m; any continuous distribution could equally well be discussed in the same manner.

In order to arrive at this extension, we return to the consideration of a system of a finite number of particles and then pass on to the limit. For this purpose, we assume that the length of arc s is introduced as a parameter on the curve and that the curve is subdivided by $(n—1)$ points into arcs of lengths $\Delta s_1, \Delta s_2, \ldots, \Delta s_n$. We represent as concentrated the mass $\mu \Delta s_i$ of each arc Δs_i at an arbitrary point s of the arc, say, with the co-ordinate y_i.

By definition, the moment of this system of particles with respect to the x-axis has the value

$$T = \mu \sum y_i \Delta s_i$$

If now the largest of the quantities Δs_i tends to 0, this sum tends to a definite limit given by

$$T = \mu \int_{s_0}^{s_1} y\, ds = \mu \int_{x_0}^{x_1} y \sqrt{\left(1 + y'^2\right)}\, dx$$

which we shall therefore naturally accept as the definition of the moment of a curve with respect to the x-axis. Since the total mass of the curve equals its length, multiplied by μ, is

$$\mu \int_{s_0}^{s_1} ds = \mu(s_1 - s_0)$$

we are immediately led to the expressions for the co-ordinates of the centre of mass of the curve:

$$\eta = \frac{\int_{s_0}^{s_1} yds}{s_1 - s_0}, \qquad \xi = \frac{\int_{s_0}^{s_1} xds}{s_1 - s_0}$$

These statements are actually definitions of the moment and centre of mass of a curve, but they are such straightforward extensions of the simpler case of a number of particles, so that we naturally expect that - as is actually the case - any statement in mechanics, which involves the centre of mass or the moment of a system of particles, will also apply to curves. In particular, the position of the centre of mass with respect to a curve does not depend on the co-ordinate system.

Area and Volume of a Surface of Revolution:. If we rotate the curve $y = f(x)$, for which $f(x) \geq 0$, about the x-axis, it describes a so-called surface of revolution. The area of this surface, the abscissae of which we assume to lie between the bounds x_0 and $x_1 > x_0$, can be obtained by a discussion analogous to the preceding work. In fact, if we replace the curve by an inscribed polygon, we shall have instead of the curved surface a figure composed of a number of thin truncated cones. Following these intuitive suggestions, we define the area of a surface of revolution as the limit of the areas of these conical surfaces as the length of the longest side of the inscribed polygon tends to zero. We know from elementary geometry that the area of each truncated cone is equal to its slant height multiplied by the circumference of the circular section of mean radius. If we add these expressions and then carry out the passage to the limit, we obtain for the area the expression

$$A = 2\pi \int_{x_2}^{x_1} y\sqrt{(1+y'^2)}\,dx = 2\pi \int_{s_0}^{s_1} y\,ds$$

Expressed in words, this result states that the area of a surface of revolution is equal to the length of the generating curve multiplied by the distance, traversed by the centre of mass (Guldin's rule).

In the same way, we find that the volume interior to a surface of revolution, which is bounded at the ends by the planes $x = x_0$ and $x = x_1 > x_0$, is given by

$$V = \pi \int_{x_0}^{x_1} y^2 dx$$

This formula is obtained by following the intuitive suggestion that the volume in question is the limit of the volumes of the above-mentioned figures consisting of truncated cones. The remainder of the proof is left to the reader.

Moment of Inertia: In the study of rotatory motion in mechanics, an important role is played by a certain quantity called a moment of inertia which will now be discussed briefly.

We suppose that a particle m at a distance y from the x-axis rotates uniformly about that axis with angular velocity ω (that is, it rotates in unit time through an angle ω). The kinetic energy of the particle, expressed by half the product of the mass and the square of its velocity, is obviously

$$\frac{m}{2}(y\omega)^2$$

We call the coefficient of $\omega/2$, that is the quantity my, the moment of inertia of the particle about the x-axis.

Similarly, if we have n particles with masses $m_1, m_2, \dots, m_n$ and ordinates $y_1, y_2, \dots, y_n$, we call the expression

$$T = \sum_i m_i y_i^2$$

the moment of inertia of the system of masses about the x-axis. The moment of inertia is a quantity which belongs to the system of masses itself, without any reference to its state of motion. Its importance lies in the fact that, if the entire system is set in constant rotation about an axis, without change of the distances between pairs of particles, the kinetic energy is obtained by multiplying the moment of inertia about that axis by half the square of the angular velocity. Thus, the moment of inertia about an axis has the same role in rotation about an axis as has mass in rectilinear motion.

Suppose now that we have an arbitrary curve $y = f(x)$, lying between the abscissae x_0 and $x_1 > x_0$, along which a mass is distributed uniformly with unit density. In order to define the moment of inertia of this curve, we just proceed as we did in the 5.2.7; as before, we arrive at an expression for the moment of inertia about the x-axis, namely,

$$T_x = \int_{s_0}^{s_1} y^2 ds = \int_{x_2}^{x_1} y^2 \sqrt{\left(1+y'^2\right)} dx$$

We have for the moment of inertia about the y-axis the corresponding expression

EXAMPLES

The theory of plane curves with its great variety of special forms and properties offers us a rich store of examples of these abstract concepts. But in order to avoid being lost in a mass of detail, we must limit ourselves to a few typical applications.

The Common Cycloid: We obtain at once from the equations $x = a(1 - \cos t)$, $y = a(1 - \sin t)$ the equations $\dot{x} = a(1-\cos t)$, $\dot{y} = a\sin t$, whence the length of arc is

$$s = \int_0^{a} \sqrt{\left(\dot{x}^2 + \dot{y}^2\right)} dt = \int_0^{a} \sqrt{\left\{2a^2(1-\cos t)\right\}} dt$$

However, since $1 - \cos t = 2\sin t/2$, the integrand is equal to $2a \sin t/2$, whence for $0 < \alpha < 2\pi$

$$s = 2a\int_0^{a} \sin\frac{t}{2} dt = -4a\cos\frac{t}{2}\bigg|_0^{a} = 4a\left(1-\cos\frac{\alpha}{2}\right) = 8a\sin^2\frac{\alpha}{4}$$

In particular, if we consider the length of arc between two successive cusps, we must set $\alpha = 2\pi$, since the interval $0 \rangle\, t \,\rangle\, 2\pi$ of values of the parameter corresponds to one revolution of the rolling circle. We thus obtain the value $8a$, that is, the length of arc of the cycloid between successive cusps is equal to four times the diameter of the rolling circle.

Similarly, we calculate the area bounded by one arch of the cycloid and the x-axis:

$$\begin{aligned} I &= \int_0^{2\pi} y\dot{x}\,dt = a^2 \int_0^{2\pi} (1-\cos t)^2\,dt \\ &= a^2 \int_0^{2\pi} \left(1 - 2\cos t + \cos^2 t\right) dt \\ &= a^2 \left(t - 2\sin t + \frac{t}{2} + \frac{\sin 2t}{4} \right) \Bigg|_0^{2\pi} = 3a^2\pi \end{aligned}$$

This area is therefore three times the area of the rolling circle.

For the radius of curvature $\rho = 1/k$, we have

$$\rho = \frac{\left(\dot{x}^2 + \dot{y}^2\right)^{3/2}}{\dot{x}\ddot{y} - \dot{y}\ddot{x}} = -2a\sqrt{\{2(1-\cos t)\}} = -4a\left|\sin\frac{t}{2}\right|;$$

at the points $t = 0$, $t = 2p$, ..., this expression has the value zero. These are actually the cusps, where the cycloid meets the x-axis at right angles.

The area of the surface of revolution formed by rotation of an arch of the cycloid about the x-axis is given by our formula as

$$\begin{aligned} A &= 2\pi \int_0^{8a} y\,ds = 2\pi \int_0^{2n} a(1-\cos t)\,2a\sin\frac{t}{2}\,dt \\ &= 8a^2\pi \int_0^{2\pi} \sin^3\frac{t}{2}\,dt = 16a^2\pi \int_0^{\pi} \sin^3 u\,du \\ &= 16a^2\pi \int_0^{\pi} \left(1-\cos^2 u\right)\sin u\,du \end{aligned}$$

The last integral can be evaluated by means of the substitution $\cos u = v$; we find

$$A = 16a^2\pi \left(-\cos u + \frac{1}{3}\cos^3 u \right) \Bigg|_0^{\pi} = \frac{64a^2\pi}{3}$$

As an exercise, the reader should find the height h of the centre of mass of the cycloid above the x-axis as well as its moment of inertia

The Catenary: The length of arc of the catenary has already been found as an example

$$\begin{aligned} A &= 2\pi \int_a^b \cosh^2 x\,dx = 2\pi \int_a^b \frac{1+\cosh 2x}{2}\,dx \\ &= \pi\left(b - a + \frac{1}{2}\sinh 2b - \frac{1}{2}\sinh 2a \right) \end{aligned}$$

For the area of the surface of revolution obtained by rotating the catenary about the x-axis - the solaced catenoid - , we find

$$A = 2\pi \int_a^b \cosh^2 x dx = 2\pi \int_a^b \frac{1+\cosh 2x}{2} dx$$
$$= \pi\left(b-a+\frac{1}{2}\sinh 2b-\frac{1}{2}\sinh 2a\right)$$

Moreover, this yields the height of the arc's centre of mass between a and b

$$\eta = \frac{A}{2\pi s} = \frac{b-a+\frac{1}{2}\sinh 2b-\frac{1}{2}\sinh 2a}{2(\sinh b-\sinh a)}$$

and, finally, the curvature

The Ellipse and the Lemniscate: The length of arc of these two curves cannot be reduced to elementary functions; they belong to the class of elliptic integrals.

For the ellipse, $y=\frac{b}{a}\sqrt{a^2-x^2}$ we obtain

$$s = \frac{1}{a}\int\sqrt{\left\{\frac{a^4-(a^2-b^2)x^2}{a^2-x^2}\right\}}dx = a\int\frac{1-x^2\xi^2}{\sqrt{(1-\xi^2)(1-x^2\xi^2)}}$$

where we have set $x/a = \xi$, $1 - b/a = \kappa$. By the substitution $\xi = \sin\phi$, this integral becomes

$$s = \int\sqrt{\left\{a^2-(a^2-b^2)\sin^2\varphi\right\}}d\varphi = a\int\sqrt{(1-x^2\sin^2\varphi)}d\varphi$$

In order to obtain the semi-perimeter of the ellipse, we must let here x cover the interval from $-a$ to +a, which corresponds to the interval $-1 \leqq \xi \leqq +1$ *or* $-\pi/2 \leqq \varphi \leqq +\pi/2$

For the lemniscate with the equation in polar co-ordinates $r = 2a\cos 2t$, we obtain in a similar manner

$$s = \int\sqrt{(r^2+\dot{r}^2)}dt = \int\sqrt{\left(2a^2\cos 2t + 2a^2\frac{\sin^2 2t}{\cos 2t}\right)}dt$$
$$= a\sqrt{2}\int\frac{dt}{\sqrt{\cos 2t}} = a\sqrt{2\int\frac{dt}{\sqrt{1-2\sin^2 t}}}$$

If we introduce the independent variable $u = \tan t$ into the last integral, we have

$$\sin^2 = \frac{u^2}{1+u^2}, \qquad dt = \frac{du}{1+u^2}$$

whence

$$s = a\sqrt{2\int\frac{du}{\sqrt{(1-u^4)}}}$$

In a complete loop of the lemniscate, u ranges from -1 to $+1$, whence the length of arc equals $a\sqrt{2}\int_{-1}^{+1}\frac{du}{\sqrt{(1-u^4)}}$, a special elliptic integral which occupied an important place in the research of Gauss.

Some Very Simple Problems in the Mechanics of a Particle

Next to geometry, the differential and integral calculus are especially indebted to the science of mechanics for their early development. Mechanics rests upon certain basic principles which were first laid down by Newton; the statement of these principles involves the concept of the derivative and their application requires the theory of integration. Without analyzing these basic principles in detail, we shall illustrate by some simple examples how the integral and differential calculus are applied in mechanics.

The Fundamental Hypotheses of Mechanics: We shall restrict ourselves here to the consideration of a single particle, that is of a point at which a mass m is imagined to be concentrated. Moreover, we shall assume that motion can only occur along a certain fixed curve, on which the position of the particle is specified by the length of arc s measured from a fixed point on the curve; in particular, the curve may be a straight line, in which case we shall use the abscissa x as the coordinate of the point instead of s. The motion of the point is determined by expressing the co-ordinate $s = \phi(t)$ as a function of the time. We shall mean by the velocity of motion the derivative $f\phi'(t)$ or, as we shall also write,

$$\frac{ds}{dt} = \phi'(t) = \dot{s}$$

We shall call the second derivative

$$\frac{d^2s}{dt^2} = \phi''(t) = \ddot{s}$$

the acceleration.

In mechanics, we start from the assumption that the motion of a point can be explained by means of forces of definite direction and magnitude. In the case of motion on a given curve, Newton's second fundamental law may be expressed as follows:

The mass multiplied by the acceleration is equal to the force acting on the particle in the direction of the curve; in symbols: $m\ddot{s} = F$

Thus, the direction of the force is always the same as that of the acceleration; its direction is that of increasing values of s, if the velocity in that direction is increasing, otherwise it is opposed to the direction of increasing values of s.

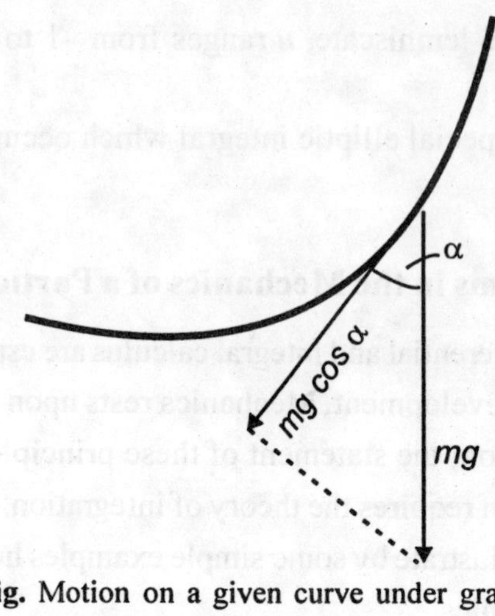

Fig. Motion on a given curve under gravity

The law of Newton is in the first instance nothing more than a definition of the concept of force. The left hand side of our equation is a quantity, which can be determined by observation of the motion, by means of which we measure the force. But this equation has a far deeper meaning. As a matter of fact, it turns out that in many cases we can determine the acting force from other physical assumptions without any consideration of the corresponding motion. The above fundamental law of Newton is then no longer a definition of the force, but instead it is a relation from which we can draw important conclusions about the motion.

The most important example of a known force is given to us by gravity. We know from direct measurements that the force of gravity acting on a mass m is directed vertically downwards and is of magnitude mg, where the constant g, the so-called gravitational acceleration, is approximately equal to 981 if the time is measured in seconds and the lengths in centimetres. If a mass moves along a given curve, we learn by experiments that the force of gravity in the direction of this curve is equal to $mg\cos\alpha$, where α denotes the angle between the vertical and the tangent to the curve at the point under consideration.

In the case of motion on our given curve, the basic problem of mechanics is: If we know the force acting on the particle (e.g., the force of gravity), we have to determine the position of the point, that is, its co-ordinate s or x, as a function of the time.

If we restrict ourselves to the simplest case, in which this force * $mf(s)$ is known at the outset as a function of the length of arc - so that the force is independent of the time - we shall show how the course of the motion along the curve can be found from the equation

$$\ddot{s} = \frac{1}{m}F = f(s)$$

* The separation of the factor m in the expression for the given force is not essential, but makes the formula simpler.

We have to deal here with a differential equation, that is, an equation from which an unknown function - here $s(t)$ - is to be determined and in which the derivative of this function occurs as well as the function itself .

Freely Falling Body. Resistance of Air: In the case of the free fall of a particle along the vertical x-axis, Newton's law yields the differential equation $\ddot{x} = g$

Hence we have the equation $\dot{x}(t) = gt + v_0$ where v_0 is a constant of integration. Its meaning is easily found by setting $t = 0$. We then find $\dot{x}(0) = v_0$, that is, v_0 is the velocity of the particle at the instant from which the time is reckoned - the initial velocity. Another integration yields

$$x(\mathrm{t}) = \frac{1}{2}gt^2 + v_0^t + x_0$$

Where x_0 is also a constant of integration, the value of which is again found by setting $t =$ 0; we thus find that x_0 is the initial position, that is, the co-ordinate of the point at the beginning of the motion.

Conversely, we can choose the initial position x_0 and the initial velocity v_0 arbitrarily and then obtain the complete representation of the motion from the equation $x = \frac{1}{2}gt^2 + v_0^t + x_0$

If we wish to take account of the effect of the friction or air resistance acting on the particle, we have to consider this as a force the direction of which is opposite to the direction of motion and concerning which we must make definite physical assumptions.

These assumptions must be chosen to suit the particular system under consideration; for example, the law of resistance for low speeds is not the same as that for high speeds (e.g., bullet velocities).

We shall now work out the results of different physical assumptions: (*a*) the resistance is proportional to the velocity, being given by an expression of the form $-r\dot{x}$ where r is a positive constant; (*b*) the resistance is proportional to the square of the velocity, given as $-r\dot{x}^2$. In accordance with Newton's law, we obtain for the equations of motion

$$\text{(a) } m\ddot{x} = mg - r\dot{x}, \quad \text{(b) } m\ddot{x} = mg - r\dot{x}^2,$$

If we at first consider $\dot{x} = u(t)$. as the function sought, we have $\ddot{x}(t) = \dot{u}(t)$ so that

$$\text{(a) } m\dot{u} = mg - ru, \quad \text{(b) } m\dot{u} = mg - ru^2,$$

Instead of determining by these equations u as a function of t, we determine t as a function of u, writing our differential equations in the form

$$\text{(a) } \frac{dt}{du} = \frac{1}{g - ru/m}, \quad \text{(b) } \frac{dt}{du} = \frac{1}{g - ru^2/m}$$

With the methods, we can immediately carry out the integrations and obtain

$$\text{(a) } t(u) = -\frac{m}{r}\log\left(1 - \frac{r}{mg}u\right) + t_0,$$

$$\text{(b) } t(u) = -\frac{1}{2}k\log\frac{kg - u}{kg + u} + t_0,$$

where we have set $\sqrt{m/rg} = k$ and t_0 is a constant of integration. Solving these equations for u, we find

$$\text{(a) } u(t) = -\frac{mg}{r}\left(e^{-r(t-t_0)/m} - 1\right),$$

$$\text{(b) } u(t) = -gk\frac{e^{-2(t-t_0)/k} - 1}{e^{-2(t-t_0)/k} + 1}$$

These equations at once reveal an important property of the motion. The velocity does not increase with time beyond all bounds, but tends to a definite limit which depends on the mass m. We have for

$$\text{(a) } \lim_{t\to\infty} u(t) = \frac{mg}{r}, \qquad \text{(b) } \lim_{t\to\infty} u(t) = \sqrt{\frac{mg}{r}}$$

A second integration, performed on our expressions for $u(i) = \dot{x}$ by the methods yields the results (which may be verified by differentiation)

$$\text{(a) } x(t) = \frac{m}{r} g e^{-r(t-t_0)/m} + \frac{mg}{r} t + c,$$

$$\text{(b) } x(t) = \frac{m}{r} \log\cosh\sqrt{\frac{rg}{m}}\left(t - t_0\right) + c,$$

where c is a new constant of integration. The two constants of integration t_0 and c are readily determined, if we know the initial position $x(0) = x_0$ and the initial velocity $\dot{x}(0) = u(0) = v_0$ of the falling particle.

The Simplest Type of Elastic Vibration: As a second example, we consider the motion of a particle which moves along the x-axis and is pulled back towards the origin by an elastic force. As regards this force, we assume that it is always directed towards the origin and that its magnitude is proportional to the distance from the origin. In other words, we assume the force to be equal to $-kx$, where the coefficient k is a measure of the stiffness of the elastic link. Since k is assumed to be positive, the force is negative when x is positive and positive when x is negative. Newton's law now says $m\ddot{x} = -kx$

We cannot expect that this differential equation will determine the motion completely, but it is plausible to assume that for a given instant of time, say $t = 0$, we can arbitrarily assign the initial position $x(0) = x_0$ and the initial velocity $\dot{x}(0) = v_0$ in physical language, it says that we can launch the particle from an arbitrary position with an arbitrary velocity and that thereafter the motion is determined by the differential equation. Mathematically speaking, this is expressed by the fact that the general solution of our differential equation contains two constants of integration, at first undetermined, the values of which we find by means of the initial conditions. We shall prove this fact immediately.

We can easily state directly such a solution. If we set $\omega = \sqrt{k/m}$, we may at once verify by differentiation that our differential equation is satisfied by the above expression. This expression is readily written in the form

$$x(t) = a \sin \omega (t-\delta) = -\alpha \sin \omega\delta \cos \omega t + a \cos \omega\delta \sin \omega t;$$

we need only write $-a \sin \omega\delta = c_1$ and $a \cos \omega\delta = c_2$, thus introducing instead of c_1 and c_2 the new constants a and δ. Motions of this type are said to be sinusoidal or simple harmonic. Such motions are periodic; any state, i.e. position $x(t)$ and velocity $\dot{x}(t)$, is repeated after the time $T=2p/w$, which is called the period, since the functions $\sin \omega t$ and $\cos \omega t$ have the period T. The number a is called the maximum displacement or amplitude of the oscillation. The number $1/T = \omega/2\pi$ is called the frequency of the oscillation; it measures the number of oscillations per unit time.

Motion on a Given Curve: Finally, we shall discuss the most general form of the problem stated above, namely, the problem of the motion along a given curve under an arbitrary pre-assigned force $mf(s)$.

The objective here is the determination of the function $s(t)$ as a function of t by means of the differential equation

$$\ddot{s} = f(s),$$

where $f(s)$ is a given function. This differential equation in s can be solved completely by the following device.

We begin by considering any primitive function $F(s)$ of $f(s)$, so that $F'(s)=f(s)$, and multiply both sides of the equation $\ddot{s} = f(s) = F'(s)$ by $\dot{s}$. We can then write the left hand side in the form $\frac{d}{dt}\left(\frac{1}{2}\dot{s}^2\right)$ as we see at once by differentiating the expression $\dot{s}^2$; however, he right hand side $F'(s)\dot{s}$ is the derivative of $F(s)$ with respect to the time t, if we regard in $F(s)$ the quantity s as a function of t. Hence, we have immediately $\frac{d}{dt}\left(\frac{1}{2}\dot{s}^2\right) = \frac{d}{dt}F(s)$, or, by integration, $\frac{1}{2}\dot{s}^2 = F(s) + c$, where s denotes a constant yet to be determined.

Let us write this equation in the form $\frac{ds}{dt} = \sqrt{2(F(s)+c)}$ We see that we cannot find s immediately from this equation as a function of t by integration. However, we arrive at a solution of the problem, if we at first content ourselves with finding the inverse function $t(s)$, that is, the time taken by the particle to reach a definite position s. We have for this the equation $\frac{dt}{ds} = \frac{1}{\sqrt{2\{F(s+c)\}}}$; thus, the derivative of the function $t(s)$ is known and we have

$$t = \int \frac{ds}{\sqrt{2\{F(s)+c\}}} + c_1,$$

where c_1 is another constant of integration. As soon as we have performed the last integration, we have solved the problem, for which we have not the position s as a function of t; we have found inversely the time t as a function of the position s. The fact that the two constants of integration c and c_1 are still available, allows us to fit the general solution to special initial conditions.

In the above example of elastic motion, we have to identify x with s; we have $f(s) = -\omega s$ and correspondingly, say, $F(s) = -\omega s$. We therefore obtain

$$\frac{dt}{ds} = \frac{1}{\sqrt{(2c - \omega^2 s^2)}}$$

whence

$$t = \int \frac{ds}{\sqrt{(2c - \omega^2 s^2)}} + c_1$$

However, this Integral is easily evaluated by introducing $\omega s / \sqrt{2c}$ as a new variable; we thus obtain

$$t = \frac{1}{\omega} \arcsin \frac{\omega s}{\sqrt{2c}} + c_1$$

or, forming the inverse function

$$s = \frac{\sqrt{2c}}{\omega} \sin \omega (t - c_1)$$

Thus, we are led to exactly the same statement of the solution as before.

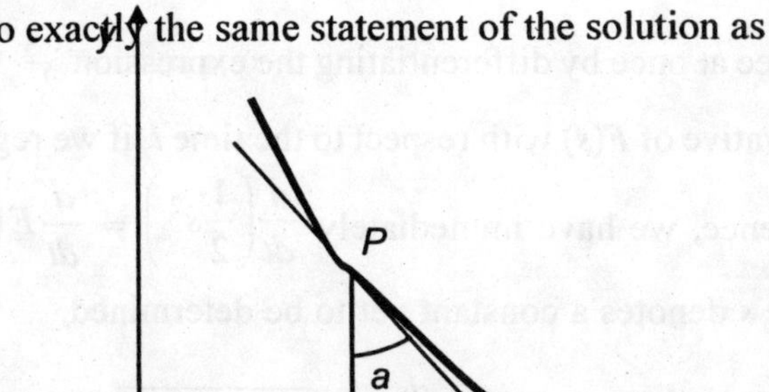

From this example, we also see what the constants of integration mean and how they are to be determined. For example, if we require that at time $t = 0$ the particle shall be at the point $s = 0$ and at that instant have the velocity $\dot{s}(0) = 1$. we obtain the two equations

$$0 = \frac{\sqrt{2c}}{\omega} \sin \omega c_1, \qquad 1 = \sqrt{2c} \cos \omega e_1,$$

from which we find that the constants have the values $c_1 = 0$, $c =$ ýÿ. The constants of integration c and c_1 can be determined in exactly the same way when the initial position s_0 and the initial velocity $\dot{s}_0$ are prescribed arbitrarily.

Further Applications. A Particle sliding down a Curve

General Remarks: The case of a particle sliding along a frictionless curve under the influence of gravity can be treated very simply by the method just described. We shall first discuss the motion in general and then with special reference to the case of the ordinary pendulum and the cycloidal pendulum. We choose axes in such a way that the *y*-axis points vertically upwards, i.e., opposite to the direction of the force of gravity, and consider the curve as given in terms of a parameter θ by the parametric equations

$$x = \phi(\theta) = x(\theta),\ y = \psi(\theta) = y(\theta).$$

A portion of the curve, for which the motion will be studied. At every point of the curve, the force of gravity acts downwards (i.e., in the direction of decreasing *y*) on the particle with magnitude *mg*. If we denote the angle between the negative *y*-axis and the tangent to the curve by α, then, according to the hypothesis in 5.4.1, the force acting along the direction of the curve is

$$mg \cos \alpha = -mg\frac{y'}{\sqrt{x'^2 + y'^2}},$$

where

$$x' = \frac{d\varphi}{d\theta} = \varphi'(\theta), \qquad y' = \frac{d\Psi}{d\theta} = \Psi'(\theta),$$

In particular, if we introduce the length of arc *s* as parameter in place of θ, we obtain the expression - $mg\,dy/ds$ for the force along the curve. Hence, by Newton's law, the function $s(t)$ satisfies the differential equation

$$\ddot{s} = -g\frac{dy}{ds}$$

The right hand side of this equation is a known function of *s*, since we know the curve and must therefore regard the quantities *x* and *y* to be known functions of *s*.

As in 5.4.4, we multiply both sides of this equation by $\dot{s}$ The left hand side then becomes the derivative of $\frac{1}{2}\dot{s}^2$ with respect to *t*. If we regard in the function $y(s)$ the variable *s* to be a function of *t*, the right hand side of our equation is the derivative of $-gy$ with respect to *t*. Hence we find on integrating

$$\frac{1}{2}\dot{s}^2 = -gy + c,$$

where *c* is a constant of integration. In order to fix the meaning of this constant, we assume that at the time $t = 0$ our particle is at the point of the curve for which the value of the parameter is θ_0 and the co-ordinates are $x=\phi(\theta_0)$, $y_0=\psi(\theta_0)$, and that at this instant its velocity is zero, i.e.,

$\dot{s}(0) = 0$. Then, setting $t = 0$, we find immediately $-gy_0 + c = 0$, whence

$$\frac{1}{2}\dot{s}^2 = -g(y - y_0)$$

Now, instead of regarding s as a function of t, we shall consider the inverse function $t(s)$. Then we obtain at once

$$\frac{dt}{ds} = \pm\frac{1}{\sqrt{\{2g(y_0 - y)\}}},$$

which is equivalent to

$$t = c_1 \pm \int\frac{ds}{\sqrt{\{2g(y_0 - y)\}}},$$

where c_1 is a new constant of integration. As regards the sign of the square root, which is the same as the sign of $\dot{s}$, we note that, if the particle moves along an arc, which is lower than y_0 everywhere except at the ends, the sign cannot change.

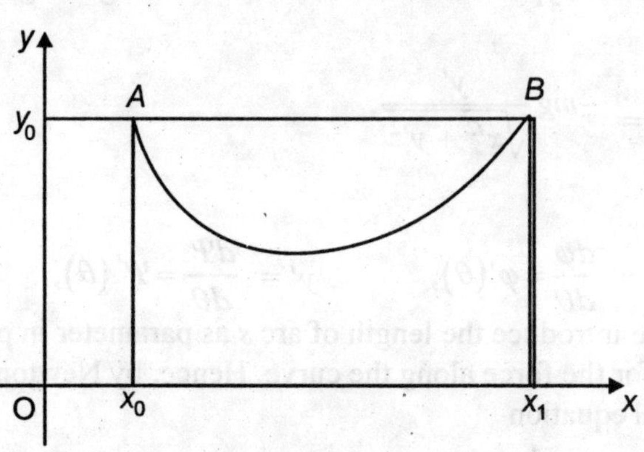

In fact, the sign of $\dot{s}$ can change only when $\dot{s} = 0$, that is, where $y - y_0 = 0$. The integrand on the right hand side is known in terms of the parameter θ, since the curve is known. Introducing θ as independent variable, we obtain

$$t = c_1 \pm \int\frac{ds}{d\theta}\frac{d\theta}{\sqrt{\{2g(y_0 - y)\}}} = c_1 \pm \int\sqrt{\left(\frac{x'^3 + y'^2}{2g(y_0 - y)}\right)}d\theta$$

where the functions $x' = \phi'(\theta)$, $y' = \psi(\theta)$, $y = \psi(\theta)$ are known. In order to determine the constant of integration c_1, we note that for $t = 0$ the value of the parameter must be θ_0. This yields immediately our solution in the form

$$t = \pm\int_{\theta_0}^{\theta}\sqrt{\left(\frac{x'^2 + y'^2}{2g(y_0 - y)}\right)}d\theta$$

When integrated, the equation represents the time taken by the particle to move from the parameter value θ_0 to the parameter value θ. The inverse function $\theta(t)$ of this function $t(\theta)$ enables us to describe the motion completely, because we can determine at each instant t the point $x = \phi\{\theta(t)\}, y = \psi\{\theta(t)\}$, which the particle is then passing.

Discussion of the Motion: We can deduce the general nature of the motion by simple intuitive reasoning from the equations just found, without an explicit expression for the result of the integration. We assume that our curve is of the type, that it consists of an arc which is convex downwards; we take s as increasing from the left hand side to the right hand side. If we initially release the particle at the point A with the co-ordinates $x=x_0, y=y_0$, corresponding to $\theta=\theta_0$, the velocity increases, because the acceleration $\ddot{s}$ is positive. The particle travels from A to the lowest point with increasing velocity. However, after it reaches the lowest point, the acceleration is negative, since the right hand side $-gdy/ds$ of the equation of motion is negative. The velocity therefore decreases. From the equation $s = -2g(y - y_0)$, we see at once that the velocity reaches the value 0 when the particle reaches the point B, the height of which is the same as that of the initial position A. Since the acceleration is still negative, the motion of the particle must be reversed at this point, so that the particle will swing back to the point A; this action will repeat itself indefinitely. In this oscillatory motion, the time which the point takes to return from B to A must obviously be the same as the time taken to move from A to B. If we denote the time required for a complete journey from A to B and back by T, the motion will obviously be periodic with the period T. If θ_0 and θ_1 are the values of the parameter corresponding to the points A and B, respectively, the half-period is given by

$$\frac{T}{2} = \frac{1}{\sqrt{2g}}\left|\int_{\theta_0}^{\theta_1}\sqrt{\left(\frac{x'^2+y'^2}{y_0-y}\right)}d\theta\right|$$

$$= \frac{1}{\sqrt{2g}}\left|\int_{\theta_0}^{\theta_1}\sqrt{\left(\frac{\varphi'^2(\theta)+\Psi'^2(\theta)}{\Psi(\theta_0)-\Psi(\theta)}\right)}d\theta\right|$$

If θ_2 is the value of the parameter corresponding to the lowest point of the curve, the time which the particle takes to fall from A to this lowest point is

$$\frac{1}{\sqrt{2g}}\left|\int_{\theta_0}^{\theta_2}\sqrt{\left(\frac{x'^2+y'^2}{y_0-y}\right)}d\theta\right|$$

The Ordinary Pendulum:. The simplest example is given by the so-called ordinary pendulum, when the curve under consideration is a circle of radius l:

$$x = l\sin\theta, \qquad y = -l\cos\theta,$$

where the angle θ is measured in the positive sense from the position of rest. We obtain at once from the general expression above

$$T = \sqrt{\frac{2l}{g}}\int_{-a}^{\alpha}\frac{d\theta}{\sqrt{(\cos\theta-\cos\alpha)}} = \sqrt{\frac{l}{g}}\int_{a}^{a}\frac{d\theta}{\sqrt{\left(\sin^2\frac{\alpha}{2}-\sin^2\frac{\theta}{2}\right)}}$$

where $\alpha\,(0 < \alpha < \pi)$ denotes the amplitude of oscillation of the pendulum, i.e., the angular position from which the particle is released at time $t=0$ at velocity 0. The substitution

$$\alpha = \frac{\sin(\theta/2)}{\sin(\alpha/2)}, \qquad \frac{du}{d\theta} = \frac{\cos(\theta/2)}{2\sin(\alpha/2)}$$

yields for the period of oscillation of the pendulum

$$T = 2\sqrt{\frac{l}{g}}\int_{-1}^{1}\frac{du}{\sqrt{(1-u^2)\left(1-u^2\sin^2(\alpha/2)\right)}}$$

We have thus expressed the period of oscillation of the pendulum by an elliptic integral. If we assume that the amplitude of the oscillation is small, so that we may with sufficient accuracy replace the second factor under the square root by 1, we obtain $2\sqrt{\frac{l}{g}}\int_{-1}^{1}\frac{du}{\sqrt{1-u^2}}$ as an approximation for the period of oscillation. We can evaluate this last integral by using 13. in our table of integrals and obtain for T the approximate expression $2\pi\sqrt{\frac{l}{g}}$.

The Cycloidal Pendulum: The fact that the period of oscillation of the ordinary pendulum is not strictly independent of the amplitude of oscillation caused Christian Huygens, in his prolonged efforts to construct accurate clocks, to look for a curve for which the period of oscillation is strictly independent of the particular position on the curve at which the oscillating particle begins its motion; the oscillations are then said to be isochronous. Huygens discovered that the cycloid is such a curve.

In order that a particle may actually be able to oscillate on a cycloid, its cusps must point in the direction opposite to that of the force of gravity, i.e., we must rotate the cycloid, considered earlier equation about the x-axis, whence we write the equations of the cyloid in the form

$$x = a(\theta - \sin\theta)$$
$$y = a(1 + \cos\theta)$$

which also involves a translation of the curve by a distance $2a$ in the positive y-direction. The time which the particle takes to travel from a point at the height

$$y_0 = a(1 + \cos\alpha) \quad (0 < \alpha < \pi)$$

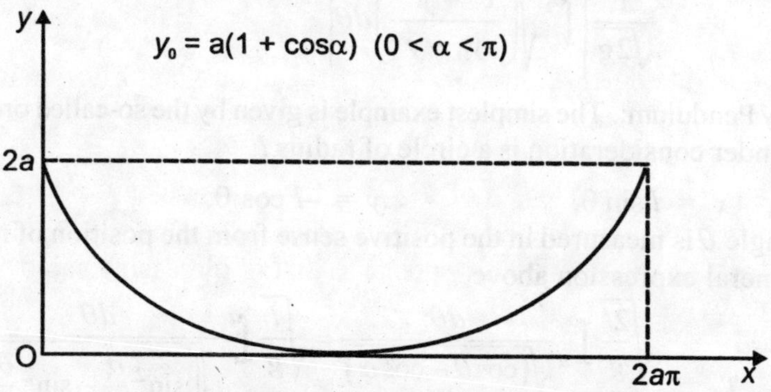

Fig. Path described by a cycloidal pendulum

down to the lowest point is, by the formula worked out in ,

$$\frac{T}{4} = \sqrt{\frac{1}{2g}}\int_{\alpha}^{\pi}\sqrt{\left(\frac{x'^2+y'^2}{y_0-y}\right)}d\theta = \sqrt{\frac{a}{g}}\int_{\alpha}^{\pi}\sqrt{\left(\frac{1-\cos\theta}{\cos\theta-\cos\theta}\right)}d\theta$$

We now use the equation

$$\cos\alpha - \cos\theta = 2\left(\cos^2\frac{\alpha}{2} - \cos^2\frac{\theta}{2}\right);$$

it yields

$$\frac{T}{4} = \sqrt{\frac{a}{g}}\int_{\alpha}^{\pi}\frac{\sin\dfrac{\theta}{2}}{\sqrt{\left(\cos^2\dfrac{\alpha}{2} - \cos^2\dfrac{\theta}{2}\right)}}dq$$

We then work out the definite integral, employing the substitution

$$\cos\frac{\theta}{2} = u\cos\frac{\alpha}{2}, \qquad \sin\frac{\theta}{2}d\theta = -2\cos\frac{\alpha}{2}du$$

$$\int\frac{\sin\dfrac{\theta}{2}}{\sqrt{9\cos^2\dfrac{\theta}{2} - \cos^2\dfrac{\theta}{2}}}d\theta = -2\int\frac{du}{\sqrt{1-u^2}} = -2 \text{ arc sin } u,$$

whence

$$T = -8\sqrt{\frac{a}{g}}\text{arcsin}\frac{\cos\dfrac{\theta}{2}}{\cos\dfrac{\theta}{2}}\Bigg|_a^{\pi} = 4\pi\sqrt{\frac{a}{g}},$$

The period of oscillation T is therefore actually independent of the amplitude.

Work

General Remarks: The concept of work throws new light on the considerations of the last section and on many other questions of mechanics and physics.

Let us think again of a particle, moving on a curve under the influence of a force acting along the curve, and assume that its position is specified by the length of the arc measured from any fixed initial point. The force itself will then, as a rule, be a function of s. We assume that it is a continuous function $f(s)$ of the length of arc. This function will have positive values, where the direction of the force is the same as the direction of increasing values of s, and negative values where the direction of the force is opposite to that of increasing value of s.

If the magnitude of the force is constant along the path, we mean by the work done by the force the product of the force by the distance $(s_1 - s_0)$ traversed, where s_1 denotes the final point and s_0 the initial point of the motion. If the force is not constant, we define the work by means of a limiting process. We subdivide the interval from s_0 to s_1 into n equal or unequal sub-intervals and note that, if the sub-intervals are small, the force in each one is nearly constant; if σ_n is an arbitrarily chosen point in the n-th subinterval, then throughout this subinterval the force will be approximately $f(\sigma_n)$. If the force throughout the ν-th subinterval were exactly $f(\sigma_n)$, the work done by our force would be exactly

$$W = \int_{s_0}^{s_1} f(s)\,ds,$$

which we naturally call the work done by the force.

If the direction of the force and that of the motion on the curve are the same, the work done by the force is positive; we then say that the force does work. On the other hand, if the direction of the force and that of the motion are opposed, the work done by the force is negative; we then say that work is done against the force.

Note that we must here carefully distinguish the forces uder consideration. For example, in lifting a weight, the work done by the force of gravity is negative: Work is done against gravity. But from the point of view of the person doing the lifting, the work done is positive, for the person must exert a force opposed to gravity.

If we regard the co-ordinate of the position s as a function of the time t, so that the force $f(s) = p$ is also a function of t, then we can plot the point in a plane with rectangular co-ordinates s and p with the co-ordinates $s = s(t), p = p(t)$ as a function of the time. This point will describe a curve, which may be called the work diagram of the motion. If we are dealing with a periodic motion, as in the case of any machine, then, after a certain time T (one period), the moving point $s = s(t), p = p(t)$ will return to the same position, i.e., the work diagram will he a closed curve. In this case, the curve may consist simply of one and the same arc, traversed first forwards and then backwards, as happens, for instance, in elastic oscillations. But it is also possible for the curve to be a more general closed curve, enclosing an area, for example, in the case of machines in which the pressure on a piston is not the same during the forward and backward stroke. The work done in one cycle, i.e., in time T, will then be given simply by the negative area of the work diagram or, in other words, by the integral $\int_{t_0}^{t_0+T} p(t)\frac{ds}{dt}di$, where the interval of time from t_0 to $t_0 + T$ represents exactly one period of the motion. If the boundary of the area is positively traversed, the work done is negative, if negatively traversed, it is positive. If the curve consists of several loops, some traversed positively and others negatively, the work done is given by the sum of the areas of the loops, each with its sign changed.

These considerations are illustrated in practice by the indicator diagram of a steam engine. By a suitably designed mechanical vice, a pencil is made to move over a sheet of paper; the horizontal motion of the pencil relative to the paper is proportional to the distance s of the piston from its extreme position, while the vertical motion is proportional to the steam pressure, and hence proportional to the total force p of the steam on the piston. The piston therefore describes the work diagram for the engine on a known scale. The area of this diagram is measured (usually by means of a planimeter) and the work done by the steam on the piston is thus found. Here we also see that our convention for the sign of an area. In fact, it does happen sometimes, when an engine is running light, that the highly expanded steam at the end of the stroke has a pressure lower than that required to expel it on the return stroke; on the diagram, this is shown by a positively traversed loop; the engine itself is drawing energy from the flywheel instead of providing energy.

The Mutual Attraction of Two Masses: Let a particle attract another particle according to Newton's law of attraction; as a first example, we shall consider the work done by the force of attraction as the second particle moves along the line joining the two particles. By Newton's law of gravitation, the attracting force is inversely proportional to the square of the distance. If we imagine the first particle to be at rest at the origin and the second particle at the distance r from the origin, the attracting force is given by

$$f(r) = -\mu \frac{1}{r^2}$$

where μ is a positive constant. Hence the work done by this force, as the particle moves from the distance r to the distance $r_1 > r$, is positive and equal to the integral

$$-\mu \int_r^{r_1} \frac{ds}{s^2} = \mu\left(\frac{1}{r_1} - \frac{1}{r}\right)$$

If the particle is moved further away from the origin from r to the distance $r_1 > r$ by means of an opposing force, the work done by the force of attraction will, of course, still be given by this integral (now negative). The work done by the opposing force has the same numerical value, but the opposite sign, whence it is equal to $\mu(1/r - 1/r_1)$. If we think of the final position as being chosen further and further away, this approaches the limiting value μ/r, which we may call the work which must be done against the force of attraction in order to move the particle from the distance r to infinity. This important expression is called the mutual potential of the two particles. Hence, the potential here is defined as the work required to separate completely two attracting masses; for example, it is the work required to tear an electron completely away from an atom (ionization potential)

The Stretching of a Spring: As a second example, we consider the work done in stretching a spring. Usually, in the theory of elasticity, we assume that the force needed to stretch the spring is proportional to x - the increase in the length of the spring, i.e., $p = kx$, where h is a constant. The work which must be done in order to stretch the spring from the unstressed position $x = 0$ to the final position $x = x_1$ is therefore given by the integral

$$\int_0^{x_1} kx\,dx = \frac{1}{2}kx_1^2$$

The Charging of a Condenser:. The concept of work in other branches of physics can be treated in a similar manner. For example, we may consider the charging of a condenser. If we denote the quantity of electricity in the condenser by Q, its capacity by C and the potential difference (voltage) across the condenser by V, then we know from physics that $Q = CV$. Moreover, the work done in moving a charge Q through a potential difference V is equal to QV. Since during the charging of the condenser the difference of potential V is not constant, but increases with Q, we perform a passage to the limit exactly analogously to that and obtain for the work done during charging a condenser

$$\int_0^{Q_1} V\,dQ = \frac{1}{C}\int_0^{Q_1} Q\,dQ = \frac{1}{2}\frac{Q_1^2}{C} = \frac{1}{2}Q_1V_1,$$

where Q_1 is the total quantity of electricity fed into the condenser and V_1 is the difference of potential across the condenser at the end of the charging process.

TAYLOR'S THEOREM AND THE APPROXIMATE EXPRESSION OF FUNCTIONS BY POLYNOMIALS

In many respects, the rational functions are the simplest functions of analysis. They are formed by a finite number of applications of the rational operations of calculation, while in the last resort the formation of every other function involves a more or less concealed passage to the limit from rational functions. The questions whether and how a given function can be expressed approximately by rational functions, in particular, by polynomials are therefore of great importance in theory as well as in practice.

The Logarithm and the Inverse Tangent

The Logarithm: We begin with some special cases in which the integration of the geometrical progression leads almost at once to the desired approximations. We first remind the reader of the following fact. For $q \neq 1$ and for positive integers n, we have

$$\frac{1}{1-q} = 1+q+q^2+\ldots+q^{n-1}+r_n,$$

where

$$r_n = \frac{q^n}{1-q}$$

If $q<1$, the remainder r_n tends to 0 as n increases and we then obtain the infinite geometric series $1+q+q^2+\ldots$ with the sum $\frac{1}{1-q}$

We take as our starting-point the formula

$$\log(1+x) = \int_0^x \frac{dt}{1+t}$$

and expand the integrand in accordance with the above formula, setting $q=-t$. Then we obtain at once by integration

$$\log(1+x) = x-\frac{x^2}{2}+\frac{x^3}{3}-\frac{x^4}{4}+-\ldots+(-1)^{n-1}\frac{x^n}{n}+R_n,$$

where

$$R_n = \int_0^x r_n\,dt = (-1)^n\int_0^x \frac{t^n\,dt}{1+t}$$

Hence, for any positive integer n, we have expressed the function $\log(1+x)$ approximately by a polynomial of the n-th degree, namely $x-\frac{x^2}{2}+\frac{x^3}{3}-+...+(-1)^{n-1}\frac{x^n}{n}$; at the same time, the quantity R_n - the remainder - specifies the size of the error made in the approximation.

In order to estimate the accuracy of this approximation, we need only have an estimate for the remainder R_n; such an estimate is given to us immediately by the integral estimates in 2.7. If we assume at first that $x \succ 0$, then the integrand is nowhere negative in the entire interval of integration and nowhere exceeds t^n. Consequently,

$$|R_n| \leqq \int_0^x t^n dt = \frac{x^{n+1}}{n+1},$$

and we therefore see that, for every value of x in the interval $0<x<1$, this remainder can be made as small as we please by choosing n large enough. If, on the other hand, the quantity x is in the interval $-1<x<0$, the integrand will not change sign and its absolute value will not exceed $|t|^n/(1+x)$, and we thus obtain the remainder estimate

$$|R_n| \leqq \frac{1}{1+x}\int_0^{|x|} t^n dt = \frac{|x|^{n+1}}{(1+x)(n+1)}$$

Hence, we see that here again the remainder is arbitrarily small when we make n sufficiently large. Of course, our estimate has no meaning for $x=-1$.

Summing up, we can say that

$$\log(1+x) = x-\frac{x^2}{2}+\frac{x^3}{3}-+...+(-1)^{n-1}\frac{x^n}{n}+R_n$$

where the remainder R_n tends to zero as n increases, provided that x lies in the interval $-1<x<0$; note that this interval is open on the left hand side and closed on the right hand side. In fact, we can find from the above inequalities an estimate for the remainder, independent of x, which is valid for all values of x in the interval $-1+h<x<1$, where h is a number such that $0<h<1$. In fact, then $|R_n| \leqq \frac{1}{h}\frac{1}{n+1}$ and this formula shows that in the entire interval the function $\log(1+x)$ is expressed approximately by our polynomial of the n-th degree, the error being nowhere greater than $1/h(n+1)$. We leave it to the reader to convince himself that for all values of x for which $|x|>1$, the remainder not only fails to approach zero, but, in fact, increases numerically beyond all bounds as n increases, so that for such values of x our polynomial does not yield an approximation to the logarithm.

The fact that in the above interval the remainder R_n tends to zero may be expressed by saying that we have in this interval for the logarithm the infinite series.

$$\log(1+x) = x-\frac{x^2}{2}+\frac{x^3}{3}-\frac{x^4}{4}+-...$$

If we insert in this series the particular value $x=1$, we obtain the remarkable formula

$$\log 2 = 1 - \frac{1}{2} + \frac{1}{3} - \frac{1}{4} + - \ldots$$

This is one of the relations the discovery of which left a deep impression on the minds of the first pioneers of the differential and integral calculus.

The above approximation for the logarithm leads us to another formula which is useful for many purposes, particularly in numerical calculations. Provided that $-1 < x < 1$, we have only to write $-x$ in place of x in the above formula to obtain

$$\log(1-x) = -x - \frac{x^2}{2} - \frac{x^3}{3} - \frac{x^4}{4} - \ldots - S_n$$

Assuming that n is even and subtracting, we find

$$\frac{1}{2}\log\frac{1+x}{1-x} = x + \frac{x^3}{3} + \frac{x^5}{5} + \ldots + \frac{x^{n-1}}{n-1} + R_n,$$

where R_n is given by

$$R_n = \frac{1}{2}(R_n + S_n) = \frac{1}{2}\int_0^x t^n\left(\frac{1}{1+t} + \frac{1}{1-t}\right)dt = \int_0^x \frac{t^n}{1-t^2}dt$$

Because $|R_n| \leq \frac{|x^{n+1}|}{n+1}\frac{1}{1-x^2}$, the remainder tends to zero as n increases, a fact which we again express by writing the expansion as an infinite series:

$$\frac{1}{2}\log\frac{1+x}{1-x} = ar\tanh x = x + \frac{x^3}{3} + \frac{x^5}{5} + \frac{x^7}{7} + \ldots,$$

for all values of x such that $|x| < 1$.

An advantage of this formula is that as x traverses the interval -1 to $+1$, the expression $(1 + x)/(1 - x)$ ranges over all positive numbers, whence, if the value of x is chosen suitably, this series enables us to calculate the value of the logarithm of any positive number with an error not exceeding the above estimate for R.

The Inverse Tangent: We can treat the inverse tangent in a similar way, if we begin with the formula, true for every positive integer n,

$$\frac{1}{1+t^2} = 1 - t^2 + t^4 - + \ldots + (-1)^{n-1}t^{2n-2} + r_n,$$

where

$$r_n = (-1)^n \frac{t^{2n}}{1+t^2}$$

By integration, we obtain

$$\text{ar tan}x = x - \frac{x^3}{3} + \frac{x^5}{5} - + ... + (-1)^{n-1}\frac{x^{2n-1}}{2n-1} + R_n,$$

$$= (-1)^n \int_0^x \frac{t^{2n}}{1+t^2} dt,$$

and see at once that in the interval -1 } x } 1 the remainder R_n tends to zero as n increases, because, by the mean value theorem of the integral calculus,

$$|R_n| \leq \int_0^{|x|} t^{2n} dt = \frac{|x|^{2n+1}}{2n+1}$$

Using this formula for the remainder, we can also readily show that for $|x| > 1$ the absolute value of the remainder increases beyond all bounds as n increases. We have thus derived the infinite series

$$\text{arc tan}x = x - \frac{x^3}{3} + \frac{x^5}{5} - + ...,$$

valid for $|x|$ 1. For $x = 1$, since arc tan 1 = $\pi/4$, we have

$$\frac{\pi}{4} = 1 - \frac{1}{3} + \frac{1}{3} - + ...,$$

as remarkable a formula as the one found previously for log 2.

Taylor's Theorem

An approximate representation by rational functions, as in the special cases above, can also be obtained in the case of an arbitrary function $f(x)$ for which we only assume that for all values of the independent variable in an assigned closed interval the function has continuous derivatives at least up to the (n+l)-th order . In most of the cases, which actually occur, the existence and continuity of all the derivatives of the function are known to begin with, so that we can choose for n any arbitrary integer. The approximation formula, which we shall now derive, was discovered in the early days of the differential and integral calculus by Newton's student Taylor and is known as Taylor's theorem.

A special case of this theorem is often referred to, without historical justification, as Maclaurin's theorem. We will not follow this practice.

Taylor's Theorem for Polynomials: In order to get a clear idea of the problem, we shall consider first the case where $f(x) = a_0 + a_1x + a_2x^2 + ... + a_nx^n$ is itself a polynomial of the n-th degree. We can then easily express the coefficients of this polynomial in terms of the derivatives of $f(x)$ at the point x=0. In fact, differentiating both sides of the equation once, twice, etc., with respect to x and setting $x = 0$, we at once find that the coefficients are $a_0 = f(0)$, $a_1 = f'(0)$, $a_0 = f''(0)$, ..., $a_n = \frac{1}{n!} f^{(n)}(0)$.

Any polynomial $f(x)$ of the n-th degree can therefore be written in the form

$$f(x) = f(0) + xf'(0) + \frac{x^2}{2!}f''(0) + \frac{x^3}{3!}f'''(0) + \ldots + \frac{x^n}{n!}f^{(n)}(0)$$

This formula merely states that the coefficients a_ν can be expressed in terms of the derivatives at $x = 0$ and gives the expressions for them.

We can generalize this Taylor series for the polynomial slightly, if we replace x by $x = \xi + h$ and consider the function $f(\xi) = f(x + h) = g(h)$ as a function of h, for moment thinking of x as fixed and h as the independent variable. It then follows that

$$g'(h) = f'(\xi), \ldots, g^{(n)}(h) = f^{(n)}(\xi)$$

and hence, if we set $h = 0$,

$$g'(0) = f'(x), \ldots, g^{(n)}(0) = f^{(n)}(x)$$

If we apply the previous formula to the function $f(x + h) = g(h)$, which is itself a polynomial of the n-th degree in h, we immediately obtain the Taylor series

$$f(\xi) = f(x + h) = f(x) + hf'(x) + \frac{h^2}{2!}f''(x) + \frac{h^3}{3!}f'''(x) + \ldots + \frac{h^n}{n!}f^{(n)}(x)$$

Taylor's Theorem for an Arbitrary Function: These formulae suggest that we should seek a similar formula in the case of an arbitrary function $f(x)$, which is not necessarily a polynomial; however, in this case, the formula can lead only to an approximation to the function by a polynomial.

We wish to compare the values of the function f at the point x and at the point $\xi = x+h$, so that $h = \xi - x$. If now n is any positive integer, as a rule, the expression $f(x) + (\xi - x)f'(x) + \ldots + \frac{(\xi - x)^n}{n!}f^{(n)}(x)$ will not be an exact expression for the functional value $f(\xi)$, whence we must set

$$f(\xi) = f(x) + (\xi - x)f'(x) + \frac{(\xi - x)^2}{2!}f''(x) + \ldots \frac{(\xi - x)^n}{n!}f^{(n)}(x) + R_n,$$

where the expression R_n denotes the remainder when $f(\xi)$ is replaced by the expression $f(x) + f'(x)(\xi - x) + \ldots$ In the first instance, this equation is nothing but a formal definition of the expression R_n. Its significance lies in the fact that we can easily find a neat and useful expression for R_n. For this purpose, we think of the quantity ξ as the fixed and the quantity x as the independent variable. The remainder is then the function $R_n(x)$. By the above equation, this function vanishes for $x = \xi$:

$$R_n(x) = 0$$

Moreover, by differentiation, we obtain

$$R_n'(x) = \frac{(\xi - x)^n}{n!}f^{(n+1)}(x)$$

In fact, if we differentiate this equation for the remainder with respect to x, we obtain 0 on the left hand side, since $f(\xi)$ does not depend on x and is therefore to be regarded as a constant. On the right hand side, we differentiate each term by the rule for products and find that all the terms cancel except the last one, which is written above with a minus sign.

Now, by the fundamental theorem of the integral calculus,

$$R_n(x) = R_n(x) - R_n(\xi) = \int_\xi^x R_n'(t)\,dt = -\int_x^\xi R_n'(t)\,dt,$$

so that we obtain the formula

$$R_n(x) = \int_x^{x+h} \frac{(x+h-t)^n}{n!} f^{(n+1)}(t)\,dt$$

If we introduce a new integration variable τ by means of the equation $\tau = t - x$, this becomes

$$R_n = \frac{1}{n!}\int_0^h (h-\tau)^n f^{(n+1)}(x+\tau)\,d\tau$$

Collecting these results, we can make the statement:

If the function $f(x)$ has continuous derivatives up to order $(n+1)$ in the interval under consideration, then

$$f(x+h) = f(x) + hf'(x) + \frac{h^2}{2!}f''(x) + \frac{h^3}{3!}f'''(x) + \dots + \frac{h^n}{n!}f^{(n)}(x) + R_n,$$

or (the equivalent expression for $h = \xi - x$)

$$f(\xi) = f(x) + (\xi - x)f'(x) + \frac{(\xi-x)^2}{2!}f''(x) + \dots \frac{(\xi-x)^n}{n!}f^{(n)}(x) + R_n,$$

where the remainder R_n is given by

$$R_n = \frac{1}{n!}\int_0^h (h-\tau)^n f^{(n+1)}(x+\tau)\,d\tau$$

In particular, if we set $x = 0$ and then replace h by x, we obtain

$$f(x) = f(0) + \frac{x}{1!}f'(0) + \frac{x^2}{2!}f''(0) + \dots + \frac{x^n}{n!}f^{(n)}(0) + R_n$$

with the remainder

$$R_n = \frac{1}{n!}\int_0^x (x-\tau)^n f^{(n+1)}(\tau)\,d\tau$$

These formulae are known as Taylor's theorem. They give expressions for the functions $f(x+h)$ and $f(x)$, respectively, in terms of polynomials of degree n in h and in x, respectively (the so-called polynomial of approximation), and a remainder. The polynomial of approximation is characterized by the fact that, when $h = 0$ (or $x = 0$, as the case may be), its value and that of its first n derivatives are the same as those of the given function and its first n derivatives. In contrast to the Taylor series for polynomials, which do not require a remainder, the remainder and the expression for it are here essential. The significance of the formula lies in the fact that the remainder, even though it has a more complicated form than the other terms of the formula, nevertheless provides useful means for estimating the accuracy with which the sum of the first $n+1$ terms $f(0)+\frac{x}{1!}f'(0)+\frac{x^2}{2!}f''(0)+...+\frac{x^n}{n!}f^{(n)}(0)$, represents the function $f(x)$.

Estimation of the Remainder: Whether the first $n+1$ terms of Taylor's series actually yield a sufficiently good approximation to the function depends naturally on whether the remainder is sufficiently small. Hence we turn now to the estimation of this remainder. Such an estimate can most easily be made by means of the mean value theorem of the integral calculus.

We employ this theorem in the form

$$\int_0^h p(\tau)\phi(\tau)d\tau = \phi(\theta h)\int_0^h p(\tau)d\tau,$$

where $p(\tau)$ is a continuous function which is negative nowhere in the interval of integration and $\phi(\tau)$ is merely a continuous function there, while q is a number in the interval $0 < q < 1$. In fact, we may assume that $0 < \theta < 1$, but this is not important here. If we set in the expression for the remainder $p(\tau)=(h-\tau)$, we obtain

$$R_n = \frac{h^{n+1}}{(n+1)!}f^{(n+1)}(x+\theta h),$$

while if we set $p(\tau) = 1$, we obtain

$$R_n = \frac{h^{n+1}}{n!}(1-\theta)^n f^{(n+1)}(x+\theta h),$$

which is less important for us and is stated here only for the sake of completeness. In these formulae, θ denotes a certain number in the interval $0 \rangle \quad \theta \rangle 1$, the value of which, in general, cannot be specified more accurately; of course, as a rule, this value is different in the two formulae for the remainder and depends, in addition, on n, x and h. The first form of the remainder was given by Lagrange, the second by Cauchy, whence they received their names.

These expressions for the remainder, as well as others, can be derived from the mean value theorem of the differential calculus and from the generalized mean value theorem, respectively. We apply these theorems to the function $R_n = R_n(x) - R_n(\xi)$ and to the pair of functions $R_n(x)$ and $(x-\xi)^{n+1}$, respectively, where we consider ξ to be fixed and employ the formula

$$R_n'(x) = -\frac{(\xi-x)^n}{n!}f^{(n+1)}(x)$$

These methods of deriving the formulae for the remainder emphasize more the fact that Taylor's theorem is a generalization of the mean value theorem; they also offer the advantage, which for many theoretical purposes is important, that we need only assume the existence and not the continuity of the $(n+1)$-th derivative. However, on the other hand, we lose the advantage of having an exact expression for the remainder in the form of an integral.

Our interest will be directed chiefly towards discovering whether the remainder R_n tends to zero as n increases; if this is the case, the larger we choose n the more accurately is $f(x+h)$

represented by the corresponding polynomial in h. In this case, we say that we have expanded the function in an infinite Taylor series

$$f(x+h) = f(x)+\frac{h}{1!}f'(x)+\frac{h^2}{2!}f''(x)+\frac{h^3}{3!}f'''(x)+...,$$

or, in particular, if we first set $x = 0$ and then write x in place of h,

$$f(x+h) = f(0)+\frac{x}{1!}f'(0)+\frac{x^2}{2!}f''(0)+\frac{x^3}{3!}f'''(0)+....$$

We shall encounter examples in the next section. However, first of all, we wish to point out the second important point of view arising from the consideration of Taylor series. If we think in the first formula of the quantity h as becoming smaller and tending to zero, then, in the terminology of, the various terms of the series will tend to zero at different orders of magnitude; we accordingly call the expression $f(x)$ the term of zero order in the Taylor series, the expression $hf'(x)$ the term of first order, the expression $hf''(x)/2!$ the term of second order, etc. We gather from the form of our remainder the fact:

In expanding a function as far as the term of n-th order, we make an error which tends to zero at order $(n+1)$ as $h < 0$.

Many important applications depend on this fact. It shows us that the nearer the point $x + h$ lies to the point x, the better is the representation of the function $f(x+h)$ by the polynomial approximation and that, in a given case, the approximation can be improved in the immediate neighbourhood of the point x by increasing the value of n.

Applications. Expansions of the Elementary Functions

We shall now use the general results of the preceding section in order to express the elementary functions approximately by polynomials and to expand them in Taylor series. However, we shall retrict ourselves to those functions for which the coefficients of the expansion in series are given by simple laws of formation. The series for certain other functions will be discussed in 8.6.

The Exponential function: The simplest example is offered by the exponential function $f(x) = e^x$, all the derivatives are which identical with $f(x)$ and therefore have the value 1 for $x = 0$. Hence, using Lagrange's form for the remainder.

$$e^{\alpha} = 1 + \frac{x}{1!}+\frac{x^2}{2!}+\frac{x^3}{3!}+...+\frac{x^n}{n!}+\frac{x^{n+1}}{(n+1)!}e^{\theta x}$$

If we now let n increase beyond all bounds, the remainder will tend to zero, no matter what fixed value of x has been chosen, because $|e^{\theta x}| \succ e^{|x|}$ from the start. We now choose a fixed integer m large than $2|x|$, then for $n \succ m$

$$\frac{|x|}{n}<\frac{1}{2};\left|\frac{x^{n+1}}{(n+1)!}\right| = \frac{\left|x^m\right|}{m!}\cdot\frac{|x|}{m+1}\cdots\frac{|x|}{n+1}$$

$$\leq \frac{|x^m|}{m!}\frac{1}{2^{n+1-m}} \leq \frac{|2x|^m}{m!}\frac{1}{2^n},$$

whence $|R_n| \leq \frac{|2x|^m}{m!} e^{|x|} \frac{1}{2^n}$

Since the first two factors on the right hand side are independent of x, while the number $1/2^n$ tends to zero as n increases, our statement is proved. If we think of the number x as not being fixed, but free to vary in the interval $-a < x < a$, where a is a fixed positive number, there follows from the above, if we choose $m > 2a$, that the estimate $|R_n| \leq \frac{|2a|^m}{m!} e^a \frac{1}{2^n}$ is valid provided $n < m$. Thus, we have specified for the remainder a bound which holds for all values of x in the interval $-a < x < a$ and tends to zero as n. For the function e^x, we can therefore write the expansion as an infinite series

$$e^x = 1 + \frac{x}{1!} + \frac{x^2}{2!} + \frac{x^3}{3!} + \ldots = \sum_{\nu=0}^{\infty} \frac{x^\nu}{\nu!}$$

This expansion is valid for all values of x. Thus, we have again proved that the number e considered is the same as the base of the natural logarithm. Of course, we must employ for numerical calculations the finite form of Taylor's theorem with the remainder; for example, for $x = 1$, this yields

$$e = 1 + 1 + \frac{1}{2!} + \frac{1}{3!} + \ldots + \frac{1}{n!} + \frac{e^\theta}{(n+1)!}$$

If we wish to compute e with an error of at most 1/10,000, we need only choose n so large that the remainder is certainly less than 1/10,000; since this remainder is certainly less* than $3/(n+1)!$, it is sufficient to choose $n = 7$, since $8! > 30{,}000$. We thus obtain the approximate value $e = 2.71822$ with an error less than 0.0001. We do not take here into account the error due to neglect of the figures in the sixth decimal place.

* We have used here the fact that $e < 3$. This follows immediately from our series for e, because it is always true that $1/n < 1/2^{n-1}$, whence

$$e < 1 + 1 + \frac{1}{2} + \frac{1}{4} + \ldots = 1 + 1/(1 - \frac{1}{2}) = 3$$

Sin x, cos x, sinh x, cosh x: We find for the functions sin x, cos x,

sinh x and cosh x:

$$\begin{array}{lcllll}
f(x) & = & \sin x & \cos x & \sinh x & \cosh x, \\
f'(x) & = & \cos x & -\sin x & \cosh x & \sinh x, \\
f''(x) & = & -\sin x & -\cos x & \sinh x & \cosh x, \\
f'''(x) & = & -\cos x & \sin x & \cosh x & \sinh x, \\
f^{iv}(x) & = & \sin x & \cos x & \sinh x & \cosh x
\end{array}$$

If $f(x) = \sin x$ or $f(x) = \cos x$, the n-th derivative can always be represented by

$$f^{(n)}(x) = f(x + \frac{1}{2}n\pi)$$

Hence, the coefficients of the even powers of x vanish in the polynomial approximations for sin x and sinhx, while the coefficients of the odd powers vanish in those for cos x and cosh x. Thus, in the first case, the $(2n+1)$-th and the $(2n+2)$-th polynomial are identical, while, in the second case, the $2n$-th and the $(2n+1)$-th polynomials are identical. If we employ in each case the higher order ones of these polynomials, we obtain at once Lagrange's forms of the remainders

$$\sin x = x-\frac{x^3}{3!}+\frac{x^5}{5!}-+\ldots+(-1)^n\frac{x^{2n+1}}{(2n+1)!}+(-1)^{n+1}\frac{x^{2n+3}}{(2n+3)!}\cos(\theta x),$$

$$\cos x = 1-\frac{x^2}{2!}+\frac{x^4}{4!}-+\ldots+(-1)^n\frac{x^{2n}}{(2n)!}+(-1)^{n+1}\frac{x^{2n+2}}{(2n+2)!}\cos(\theta x),$$

$$\sinh x = x+\frac{x^3}{3!}+\frac{x^5}{5!}-+\ldots+\frac{x^{2n+1}}{(2n+1)!}+\frac{x^{2n+3}}{(2n+3)!}\cosh(\theta x),$$

$$\cosh x = 1+\frac{x^2}{2!}+\frac{x^4}{4!}+\ldots+\frac{x^{2n}}{(2n)!}+\frac{x^{2n+2}}{(2n+2)!}\cosh(\theta x),$$

where, of course, in each of the four formulae θ denotes a different number in the interval $0 < \theta < 1$, a number which, in addition, depends on n and on x. In these formulae, we can also make the approximation as exact as we please for each value of x, since the remainder tends to 0 as n increases. We thus obtain the four series

$$\sin x = x-\frac{x^3}{3!}+\frac{x^5}{5!}-+\ldots=\sum_{\nu=0}^{\infty}(-1)^{\nu}\frac{x^{2\nu}+1}{(2\nu+1)!},$$

$$\cos x = 1-\frac{x^2}{2!}+\frac{x^4}{4!}-+\ldots=\sum_{\nu=0}^{\infty}(-1)^{\nu}\frac{x^{2\nu}}{(2\nu)!},$$

$$\sinh x = x+\frac{x^3}{3!}+\frac{x^5}{5!}+\ \ldots=\sum_{\nu=0}^{\infty}\frac{x^{2\nu+1}}{(2\nu+1)!},$$

$$\cosh x = 1+\frac{x^2}{2!}+\frac{x^4}{4!}+\ \ldots=\sum_{\nu=0}^{\infty}\frac{x^{2\nu}}{(2\nu)!}$$

The last two series can also be obtained formally from the series for e^x in accordance with the definitions of the hyperbolic functions.

The Binomial Series: We may pass over the Taylor series for the functions log $(1+x)$ and artan x, which we have already dealt with directly in equation . However, we must take up the generalization of the binomial theorem for arbitrary indices, which is one of the most fruitful of

Newton's mathematical discoveries and which represents one of the most important cases of the expansion in Taylor series. Our objective is the expansion of the function

$$f(x) = (1+x)^{\alpha}$$

in a Taylor series, where $x > -1$ and α is an arbitrary number, positive or negative, rational or irrational. We have chosen the function $(1 + x)^{\alpha}$ instead of x^{α}, because it is not true at the point $x = 0$ that all the derivatives of x^{α} are continuous except in the trivial case of non-negative values of α. We calculate first the derivatives of $f(x)$:

$$f'(x) = \alpha(1+x)^{\alpha-1}, \qquad f''(x) = \alpha(\alpha-1)(1+x)^{\alpha-2}, \ldots,$$
$$f^{(\nu)}(x) = \alpha(\alpha-1) \ldots (\alpha-\nu+1)(1+x)^{\alpha-\nu}$$

In particular, we have for $x = 0$,

$$f'(0) = \alpha, f''(0) = \alpha(\alpha-1), \ldots, f^{(\nu)}(0) = \alpha(\alpha-1)\ \alpha(\alpha-\nu+1)$$

Taylor's theorem then yields

$$(1+x)^{\alpha} = 1 + ax + \frac{\alpha(\alpha-1)}{2!}x^2 + \ldots + \frac{\alpha(\alpha-1)(\alpha-2)\ldots(\alpha-n+1)}{n!}x^n + R_n$$

We must still discuss the remainder. This is not very difficult, but nevertheless is not quite as simple as in the cases treated previously. Here we shall pass over the remainder estimate, since the general binomial theorem will be proved completely in a somewhat different and simpler way in 8.6.2 and A6.3. The result, which we mention here in advance, is that whenever $|x| < 1$, the remainder tends to 0, whence the expression $(1 + x)^{\alpha}$ can be expanded in the infinite binomial series

$$(1+x)^{\alpha} = 1 + \frac{\alpha}{1!}x + \frac{\alpha(\alpha-1)}{2!}x^2 + \ldots = \sum_{\nu=0}^{\infty}\binom{\alpha}{\nu}x^{y}$$

where, for the sake brevity, we have introduced the general binomial coefficients

$$\binom{\alpha}{\nu} = \frac{\alpha(\alpha-1)\ldots(\alpha-\nu+1)}{\nu!}(for\,\nu > 0), \binom{\alpha}{0} = 1$$

Geometrical Applications

The behaviour of a function $f(x)$ in the neighbourhood of a point $x = a$ or of a given curve in the neighbourhood of a point can be studied with increased accuracy by means of Tailor's theorem, because this theorem resolves the increment of the function on passing to a neighbouring point $x = a + h$ into a sum of quantities of the first order, second order, etc.

Contact of Curves: We shall now use this method, in order to investigate the concept of contact of two curves. If at a point, say the point $x = a$, two curves $y = f(x)$ and $y = g(x)$ do not only intersect, but also have a common tangent, we shall say that at this point the curves touch one another or have contact of the first order. The Taylor expansions of the functions $f(a + h)$ and $g(a + h)$ then have the same terms of zero and first order in h. If at the point $x=a$ the second derivatives of $f(x)$ and $g(x)$ are also equal to each other, we say that the curves have contact of the second order. In the Taylor expansions, the terms of second order are then also the same and, if we assume that both functions have at least continuous derivatives of the third order, the difference $D(x) = f(x) - g(x)$ can be expressed in the form

$$D(a + h) = f(a + h) - g(a + h) = \frac{h^3}{3!} D'''(a + \theta h)$$

where the expression $F(h)$ tends to $f'''(a + h) - g'''(a)$ as h tends to zero, whence the difference $D(a + h)$ vanishes to at least the third order in h.

We can proceed in this way and consider the general case, when the Taylor series for $f(x)$ and $g(x)$ are the same up to terms of the n-th order, i.e., when

$$f(a) = g(a), f'(a) = g'(a), f''(a) = g''(a), \ldots, f^{(n)}(a) = g^{(n)}(a)$$

We assume here that the $(n + 1)$-th derivatives are also continuous. Under these conditions, we say that at this point the curves have contact of the n-th order. The difference of the two functions will then be

$$f(a + h) - g(a + h) = \frac{h^{n+1}}{(n+1)} f(h)$$

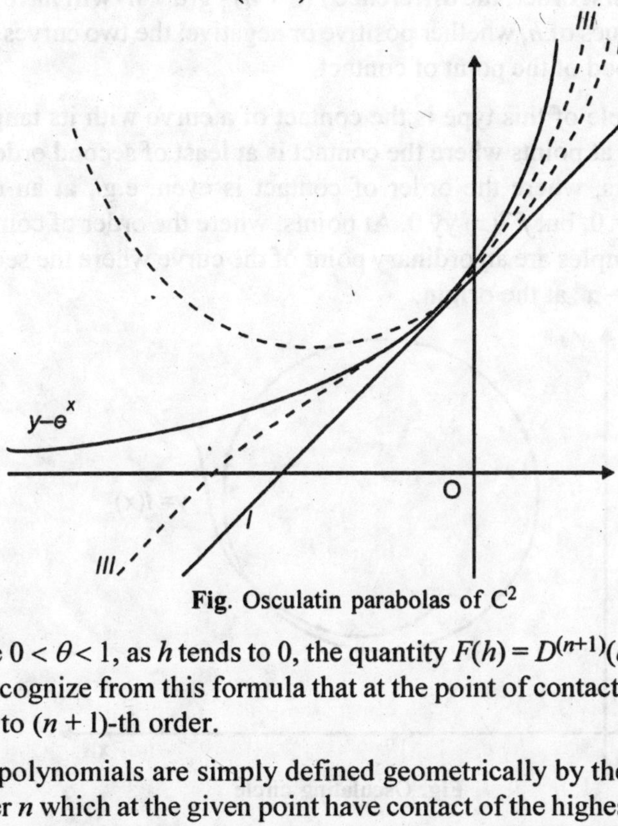

Fig. Osculatin parabolas of C^2

where, since $0 < \theta < 1$, as h tends to 0, the quantity $F(h) = D^{(n+1)}(a + q\,h)$ tends to $f^{(n+1)}(a) - g^{(n+1)}(a)$. We recognize from this formula that at the point of contact the difference $f(g) - g(x)$ vanishes at least to $(n + 1)$-th order.

The Taylor polynomials are simply defined geometrically by the fact that they are those parabolas of order n which at the given point have contact of the highest possible order with the graph of the given function. Hence they are sometimes called osculating parabolas. For the case $y = e^x$, figures gives the first three osculating parabolas at the point $x = 0$.

If two curves $y = f(x)$ and $y = g(x)$ have contact of order n, the definition does not exclude the possibility that the contact may be of a still higher order, i.e., that also the equation $f^{(n+1)}(a) = g^{(n+1)}(a)$ is true. If this is not the case, i.e., if $f^{(n+1)}(a)$ ýÿ $g^{(n+1)}(a)$, we speak of a contact of exactly the n-th order or say that the order of the contact is exactly n.

The fact that the order of contact of two curves is a genuine geometrical relationship, which is unaffected by a change of axes, is easily verified by means of the formulae for a change of axes.

From our formula as well as from our figures, we can at once state a remarkable fact which is often overlooked by beginners. If the contact of two curves is exactly of an even order, that is, if an even number n of derivatives of the two functions have the same value at the point in question, while the $(n+1)$-th derivatives differ, then, in conformity with the above formula, the difference $f(a+h) - g(a+h)$ will have different signs for small positive values and for numerically small negative values of h. The two curves will then cross at the point of contact. This case occurs, for example, in a contact of the second order, if the third derivatives have different values. However, if we consider in detail the case of an odd order contact, e.g., the case of an ordinary contact of the first order, the difference $f(a+h) - g(a+h)$ will have the same sign for all numerically small values of h, whether positive or negative; the two curves therefore will not cross in the neighbourhood of the point of contact.

The simplest example of this type is the contact of a curve with its tangent. The tangent can cross the curve only at points where the contact is at least of second order; it will actually cross the curve at points, where the order of contact is even, e.g., at an ordinary point of inflection, where $f''(x) = 0$, but $f'''(x)$ ýÿ 0. At points, where the order of contact is odd, it will not cross the curve; examples are an ordinary point of the curve where the second derivative is non-zero or the curve $y = x^4$ at the origin.

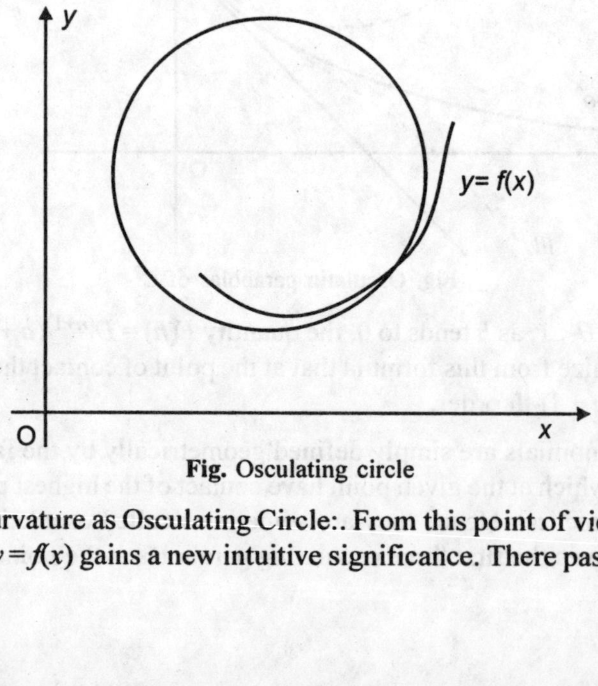

Fig. Osculating circle

The Circle of Curvature as Osculating Circle:. From this point of view, the concept of the curvature of a curve $y = f(x)$ gains a new intuitive significance. There pass through the definite point of the curve with co-ordinates $x = a$ and $y = b$ an infinite number of circles which touch the curve at the point. The centres of these circles lie on the normal to the curve and there corresponds to each point of this normal just one such tangent circle. We may expect that we can bring about by a proper choice of the centre of the circle a contact of second order between the curve and the circle.

As a matter of fact, we know from that for the circle of curvature at the point $x = a$, the equation of which is, say, $y = g(x)$, we do not only have $g(a)=f(a)$ and $g'(a) = f'(a)$, but also $g''(a) = f''(a)$, whence the circle of curvature is at the same time the osculating circle at the point of the curve under consideration, i.e., it is the circle which at that point has second order contact with the curve. In the limiting case of a point of inflection or, in general, of a point at which the curvature is zero and the radius of curvature infinite, the circle of curvature degenerates into the tangent. In ordinary cases, i.e., when the contact at the point in question does not happen to be of an order higher than the second, the circle of curvature will not merely touch the curve, but will also cross it.

ON THE THEORY OF MAXIMA AND MINIMA

A point $x = a$, at which $f'(a) = 0$, has a maximum of the function $f(x)$ if $f''(a)$ is negative, a minimum if $f''(a)$ is positive,whence these conditions are sufficient for the occurrence of a maximum or a minimum. They are by no means necessary, because there are three possibilities in the case where $f''(a)=0$; at the point in question, the function may have a maximum or a minimum or neither. Examples of the three possibilities are giver by the functions $y=-x^4$, $y=x^4$ and $y = x^3$ at the point $x = 0$. Taylor's theorem enables us immediately to make a general statement of sufficient conditions for a maximum or a minimum.We merely need expand the function $f(a + h)$ in powers of h; the essential point is then to discover whether the first non-vanishing term contains an even or an odd power of h. In the first case, we have a maximum or a minimum according to whether the coefficient of h is negative or positive; in the second case, we have a horizontal inflectional tangent and neither a maximum nor a minimum. The reader should complete the argument by using the formula for the remainder.

However, the earlier necessary and sufficient condition (3.5.2) is more general and more convenient in applications, i.e., provided the first derivative $f'(x)$ vanishes at only a finite number of points; a necessary and sufficient condition for the occurrence of a maximum or minimu,at one of these points is that the first derivative $f'(x)$ changes sign as the point is passed.

Chapter 7

Numerical Methods

PRELIMINARY REMARKS

Anyone who has to use analysis as an instrument for investigating physical or technical phenomena is faced by the question whether and how the theory can be adapted to yield useful practical methods for actual numerical calculations. Yet, even from the point of view of the theoretical person, who only desires to recognize the link between natural phenomena and not to conquer them, these questions are of no trifling interest. Here we can only discuss some particularly important points which are more or less closely linked to the preceding ideas. We wish to direct special attention to the fundamental fact that the meaning of an approximate calculation is not precise unless it is accompanied by a definite knowledge of the degree of accuracy attained.

Numerical Integration

We have seen that even relatively simple functions cannot be integrated in terms of elementary functions and that it is futile to make this unattainable goal the aim of the integral calculus. On the other hand, the definite integral of a continuous function does exist and this fact raises the problem of finding methods for calculating it numerically. Here we shall discuss the simplest and most obvious of these methods with the aid of geometrical intuition and thenconsider error estimation.

Our objective is to calculate the integral

$$I = \int_a^b f(x)dx,$$

where $a < b$. We imagine the interval of integration to have been subdivided into n equal parts of length $h = (b - a)/n$ and denote the points of subdivision by $x_1 = a, x_2 = a + h, \ldots, x_n = b$, the values of the function at the points of subdivision by $f_0, f_1, \ldots, f_n$ and, similarly, the values of the function at the midpoints of the intervals by $f_{1/2}, f_{3/2}, \ldots, f_{(2n-1)/2}$. We interpret our integral as an area and cut up the region under the curve into strips of width h in the usual manner. We must now obtain an approximation for each such strip of surface, that is, for the integral

$$I_\nu = \int_{x_y}^{x_y+h} f(x)dx,$$

The Rectangle Rule: The crudest and most obvious method of approximating the integral I is directly linked to the definition of the integral; we replace the area of the strip I_n by the rectangle of area $f_n h$ and obtain for the integral the approximate expression

$$I \approx h(f_0 + f_1 + ... + f_{n-1})$$

From here on, the symbol } means is approximately equal to.

$$x_\nu, x_{\nu+1} = x_\nu + h, \; and \; x_{\nu+2} = x_\nu + 2h$$

The Trapezoid and Tangent Formulae: We obtain a closer approximation with no greater trouble if we do not replace the area of the strip I_ν, as above, by a rectangular area, but by the trapezoid of area $(f_\nu + f_{\nu+1})h/2$. For the whole integral, this process yields the approximate expression

$$I \approx h(f_1 + f_2 + ... + f_{n-1}) + \frac{h}{2}(f_0 + f_n)$$

(trapezoid formula), since, when the areas of the trapezoids are added, each value of the function except the first and the last occurs twice.

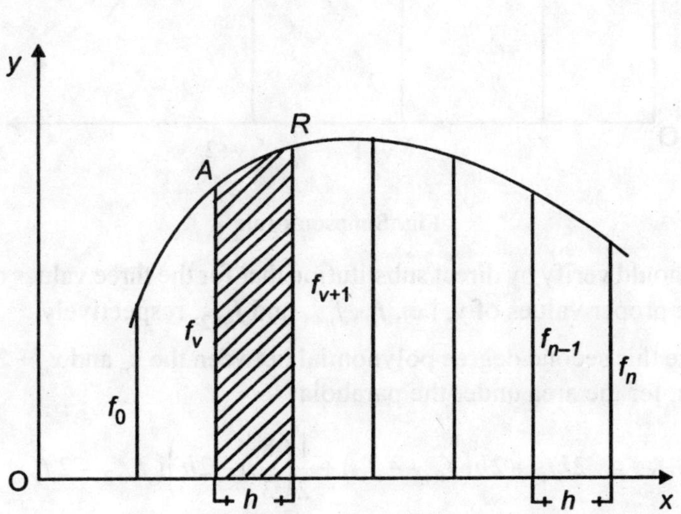

Fig. The Frapezoid Formula

As a rule, the approximation becomes even better if, instead of choosing the trapezoid under the chord AB as an approximation to the area of I, we select the trapezoid under the tangent to the curve at the point with the abscissa $x=x_v+h/2$. The area of this trapezoid is simply $hf_{v+\}}$, so that the approximation for the entire integral is

$$I \approx h\left(f_{1/2} + f_{3/2} + \dots + f_{2n-1/2}\right),$$

which is called the tangent formula.

Simpson's Rule: By means of Simpson's rule, we arrive with very little more trouble at a numerical result which is generally much more exact. This rule depends on estimating the area $I_v + I_{v+1}$ of the double strip between the abscissae $x = x_v$ and $x = x_v + 2 = x_{v+2}$ by considering the upper boundary to be no longer a straight line but a parabola - in order to be specific, that parabola which passes through the three points of the curve with the abscissae x_v, $x_{v+1}=x_v+h$, *and* $x_{v+2} = x_v + 2h$. The equation of this parabola is

$$y = f_v + (x - x_v)\frac{f_{v+1} - f_v}{h} + \frac{(x - x_v)(x - x_v - h)}{2}\frac{f_{v+2} - 2f_{v+1} + f_v}{h^2}$$

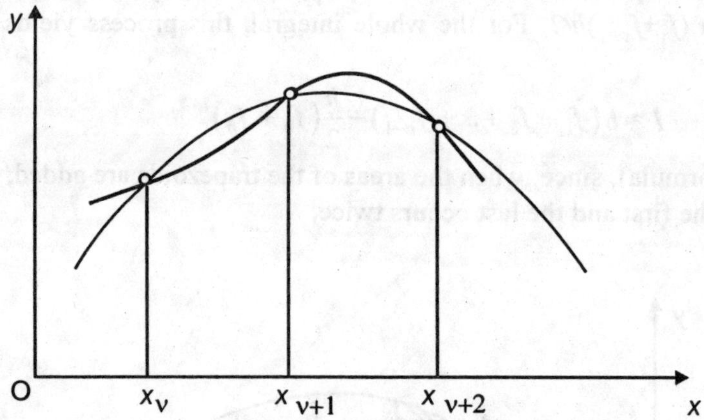

Fig. Simpson's rule

The student should verify by direct substitution that for the three values of x in question this eqnation gives the proper values of y, i.e., f_n, f_{n+1} and f_{n+2}, respectively.

If we integrate this second degree polynomial between the x_n and $x_n + 2h$, we obtain, after a brief calculation, for the area under the parabola

$$\int_{xv}^{xv+2h} y\,dx = 2hf_v + 2h\left(f_{v+1} - f_v\right) + \frac{1}{2}\left(\frac{8}{3}h - 2h\right)\left(f_{v+2} - 2f_{v+1} + f_v\right)$$

$$= \frac{h}{3}\left(f_v + 4f_{v+1} + f_{v+2}\right)$$

This represents the required approximation to the area of our strip $I_v + I_{v+1}$.

If we now assume that $n = 2m$, i.e., that n is an even number, we obtain, by addition of the areas of such strips, Simpson's rule

$$I \approx \frac{4h}{3}(f_1 + f_3 + +...f_{2m-1}) + \frac{2h}{3}(f_2 + f_4 + ... + f_{2m-2}) + \frac{h}{3}(f_0 + f_{2m})$$

Estimation of the Error:. It is easy to obtain an estimate of the error for each of our methods of integration, if the derivatives of the function $f(x)$ are known throughout the interval of the integration. We take M_1, M_2, ... as the upper bounds of the absolute values of the first, second, derivative, respectively, i.e., we assume that throughout the interval $|f^{(\nu)}(x)| < M_\nu$. Then the estimates are:

For the rectangle rule: $|I_\nu - hf_\nu| < \frac{1}{2}M_1h^2$ or $\left|I - h\sum_{\nu=0}^{n-1} f_\nu\right| < \frac{1}{2}M_1nh^2 = \frac{1}{2}M_1(b-a)h$

For the tangent rule: $\left|I_\nu - hf_{\nu+\frac{1}{2}}\right| < \frac{M}{24}h^3$ or $\left|I - h\sum_{\nu=0}^{n-1} f_{\nu+\frac{1}{2}}\right| < \frac{M_2}{24}(b-a)h^2$

For the trapezoid rule: $\left|I_\nu - \frac{h}{2}(f_\nu + f_{\nu+1})\right| < \frac{M_2}{12}h^3$

For Simpson's rule : $\left|I_\nu + I_{\nu+1} - \frac{h}{3}(f_\nu + 4f_{\nu+1} + f_{\nu+2})\right| < \frac{M_4}{90}h^5$

The last two estimates also lead to estimates for the entire integral I. We see that Simpson's rule has an error of much higher order in the small quantity h than the other rules, so that when M_4 is not too large, it is very advantageous for practical calculations. In order to avoid tiring the reader with the details of the proofs of these estimates, which are fundamentally quite simple, we shall restrict ourselves to the proof for the tangent formula. For this purpose, we expand the function $f(x)$ in the $(n + 1)$-th strip by Taylor's theorem:

$$f(x) = f_{\nu+\frac{1}{2}} + \left(x - x_\nu - \frac{h}{2}\right)f'\left(x_\nu + \frac{h}{2}\right) + \frac{1}{2}\left(x - x_\nu - \frac{h}{2}\right)^2 f''(\xi),$$

where ξ is a certain intermediate value in the strip. If we integrate the right hand side over the interval $x_n < x < x_n + h$, the integral of the middle term is zero. Since, as is easily verified,

$$\frac{1}{2}\int_{x_\nu}^{x_\nu+h}\left(x - x_\nu - \frac{h}{2}\right)^2 dx = \frac{h^3}{24},$$

it follows immediately that $\left|\int_{x_\nu}^{x_\nu+h} f(x)dx - hf_{\nu+\frac{1}{2}}\right| < M_2\frac{h^3}{24}$, **which proves the above assertion.**

Applications of the Mean Value theorem and of Taylor's Theorem

The Calculus of Errors: We now come to quite a different type of numerical calculations. These are applications of the mean value theorem or, more generally, of Taylor's theorem with remainder or, finally, of the infinite Taylor series. As an application which, although simple, is

quite important in practice, we shall consider the calculus of errors. This rests on the idea - which lies at the root of the entire differential calculus - that a function $f(x)$, which is differentiable a sufficient number of times, can be represented in the neighbourhood of a point by a linear function with an error of order less than the first, by a quadratic function with an error of order less than the second, etc. Consider the linear approximation to a function $y=f(x)$. If $y+\Delta y=f(x+\Delta x)=f(x+h)$, we find by Taylor's theorem

$$\Delta y = hf'(x) + \frac{h^2}{2} f''(\xi)$$

where $\xi = x+\theta h$ $(0<\theta<1)$ is an intermediate value, which need not be known more precisely. If $h=\Delta x$ is small, we obtain as a practical approximation $\Delta y \approx hf'(x)$.

In other words, we replace the difference quotient by the derivative to which it is approximately equal and the increment of y by the approximately equal linear expression in h.

We use this fairly obvious fact for practical purposes in the following way. Let two physical quantities x and y be linked by the relation $y = f(x)$. There arises then the question how an inaccuracy in the measurement of x affects the determination of y. If we happen to use instead of the true value of x the inaccurate value $x + h$, then the corresponding value of y differs from the true value $y = f(x)$ by $\Delta y = f(x + h) - f(x)$, whence the error is given approximately by the above relation.

Evaluation of π: Gregory's series, also called Leibnitz' series,

$$\frac{\pi}{4} = 1-\frac{1}{3}+\frac{1}{5}-\frac{1}{7}+-\ldots,$$

which we have obtained in 6.1.2 using the series for artan, is not suitable for the calculation of π due to the slowness of its convergence. However, we may calculate π with comparative ease by the artifice: From the addition theorem for the tangent

$$\tan(\alpha+\beta)=\frac{\tan\alpha+\tan\beta}{1-\tan\alpha\tan\beta},$$

changing to the inverse functions $\alpha=$ artan u, $\beta=$ artan v, we obtain

$$\text{arc tan}u + \text{arc tan}v = \text{arc tan}\left(\frac{u+v}{1-uv}\right)$$

If we now choose u and v in such a manner that $(u + v)/(1 - uv) = 1$, we obtain the value $\pi/4$ on the right hand side and, if u and v are small numbers, we can easily compute the left hand side by means of known series.

For example, if we set $u = 1/2$, $v = 1/3$, as Euler did, we obtain $\frac{\pi}{4} = \text{arc tan}\frac{1}{2} + \text{arc tan}\frac{1}{3}$

Moreover, if we note that $\left(\frac{1}{3}+\frac{1}{7}\right) \div \left(1-\frac{1}{21}\right) = \frac{1}{2}$

we have $\text{arc tan}\frac{1}{2} = \text{arc tan}\frac{1}{3} + \text{arc tan}\frac{1}{7}$, so that $\frac{\pi}{4} = 2 \text{ arc tan}\frac{1}{7}$ +arc tan.

Using this formula, Georg Vega (1756 - 1802) calculated the number π to 140 places.

Moreover, using the equation $\left(\frac{1}{5}+\frac{1}{8}\right)\div\left(1-\frac{1}{40}\right)=\frac{1}{3}$

we obtain $\text{arc tan}\frac{1}{3} = \text{arc tan}\frac{1}{5} + \text{arc tan}\frac{1}{8}$

or $\frac{\pi}{4} = 2 \text{ arc tan}\frac{1}{5} + \text{arc tan }\frac{1}{7} + 2\text{arc tan}\frac{1}{8}$.

This expansion is extremely useful for the computation of π by means of the series artan$x$ $= x - \frac{x^3}{3} + \frac{x^5}{5} - + \ldots$; in fact, if we substitute for x the values 1/5, 1/7 or 1/8, we obtain with just a few terms a high degree of accuracy, since the terms diminish rapidly. However, we can perform the calculation even more conveniently by basing it on the formula

$$\frac{\pi}{4} = \text{arc tan }\frac{120}{119} - \text{arc tan}\frac{1}{239} = 4 \text{ arc tan}\frac{1}{5} - \text{arc tan}\frac{1}{239}$$

obtained by similar arguments as those above.

Calculation of Logarithms: For the computation of logarithms, we transform the series

$$\frac{1}{2}\log\frac{1+x}{1-x} = x + \frac{x^3}{3} + \frac{x^5}{5} + \ldots(|x|<1);$$

where $0 < x < 1$, by the substitution

$$\frac{1+x}{1-x} = \frac{p^2}{p^2-1}, \qquad x = \frac{1}{1p^2-1}$$

into the series

$$\log p = \frac{1}{2}\log(p-1) + \frac{1}{2}\log(p+1) + \frac{1}{2p^2-1} + \frac{1}{3(2p^2-1)^3} + \ldots,$$

where $2p - 1 > 1$, that is, $p > 1$. If p is an integer and $p + 1$ can be resolved into smaller integral factors, this last series expresses the logarithm of p in terms of the logarithms of smaller integers plus a series, the terms of which diminish very rapidly and the sum of which can therefore be calculated accurately enough from only a few terms. Hence this series allows us to compute the logarithms of any prime number and hence that of any number, provided we have already calculated the value of log 2.

The accuracy of this determination of $\log p$ can be estimated more easily by means of the geometric series than from the general remainder formula. In fact, we have for the remainder R_n of the series, i.e., the sum of all the terms following the term $\frac{1}{n(2p^2-1)^n}$

$$R_n \frac{1}{(n+2)\left(2p^2-1\right)^{n+2}}\left(1+\frac{1}{\left(2p^2-1\right)^2}+\frac{1}{\left(2p^2-1\right)^4}+...\right)$$

$$=\frac{1}{(n+2)\left(2n^2-1\right)^n}\cdot\frac{1}{\left(2p^2-1\right)^2-1}$$

and this formula yields immediately the required error estimate.

For example, let us compute $\log_e 7$, using the first four terms of the series. We have

$$P = 7, \qquad 2P^2 - 1 = 97,$$

$$\log 7 = 2\log 2 + \frac{1}{2}\log 3 + \frac{1}{97} + \frac{1}{3.97^3} + ...,$$

$$\frac{1}{97} \approx 0.01030928, \frac{1}{3.97^3} \approx 0.00000037,$$

$$2\log 2 \approx 1.38629436, \frac{1}{2}\log 3 \approx 0.54930614,$$

whence $\log_e 7 \approx 1.945,910,15$.

Estimation of the error yields $R_n < \frac{1}{5.94^3} \times \frac{1}{97^3-1} < \frac{1}{36\times 10^3}$

However, we must note that each of the four numbers which we have added is only given to within an error, so that the last place in the above value of $\log_e 7$ might be wrong by 2. However, as a matter of fact, the last place is also correct.

Numerical Solution of Equations

Fig. Newton's method of approximation

In conclusion, we shall add some remarks about the numerical solution of the equation $f(x) = 0$, where $f(x)$ need not necessarily be a polynomial and we are, of course, only concerned with the determination of real roots. Every such numerical method is based on the scheme of starting

with some known approximation x_0 of one of the roots and then improving this approximation, whence this first approximation for the desired root of the equation is found; the quality of the approximation is not a concern. We may perhaps take a rough guess as a first approximation or, better still, obtain it from the graph of the function $y=f(x)$, the intersection of which with the x-axis yields the required root (of course with an error depending on the scale and the accuracy of the drawing).

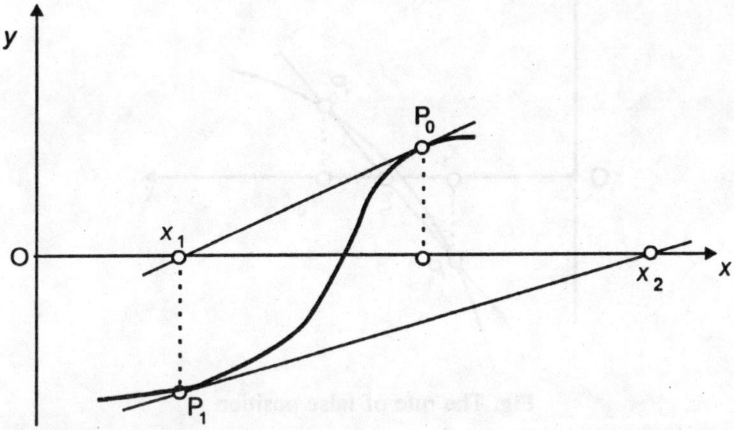

Newton's Method: The following procedure, due to Newton, is based on the fundamental principle of the differential calculus - the replacement of a curve by a straight line, the tangent - in the immediate neighbourhood of the point of contact. If we have an approximate value x_0 for a root of the equation $f(x) = 0$, we consider the point on the graph of the function $y = f(x)$ with the co-ordinates $x = x_0$, $y = f(x_0)$. We wish to find the intersection of the curve with the x-axis; we find an approximation at the point x_0, $y_0=f(x_0)$, where the tangent intersects the x-axis. The abscissa x_1 of this intersection of the tangent with the x-axis then represents a new, possibly better approximation to the required root of the equation than x_0.

By virtue of the geometrical meaning of the derivative, Fig. yields at once

$$\frac{f(x_0)}{x_0 - x_1} = f'(x_0),$$

whence we obtain the formula for the calculation of the new value x_1:

$$x_1 = x_0 - \frac{f(x_0)}{f'(x_0)}$$

If this procedure yields a better approximation than x_0, we repeat the process and find x_2, etc.; if the curve has the form shown in Fig., these approximations will approach more and more closely to the required solution.

The usefulness of this process depends essentially on the nature of the curve $y=f(x)$. In Fig., we see that the successive estimates converge with greater and greater accuracy to the required root. This is due to the fact that the curve has its convex side turned towards the x-axis. However, we see that if we choose the original value x_0 badly, our construction does not lead at all to the required root. We see from this that while using Newton's method we must examine each individual case in order to determine the degree of accuracy with which we have really solved the equation. We shall return to this topic later on.

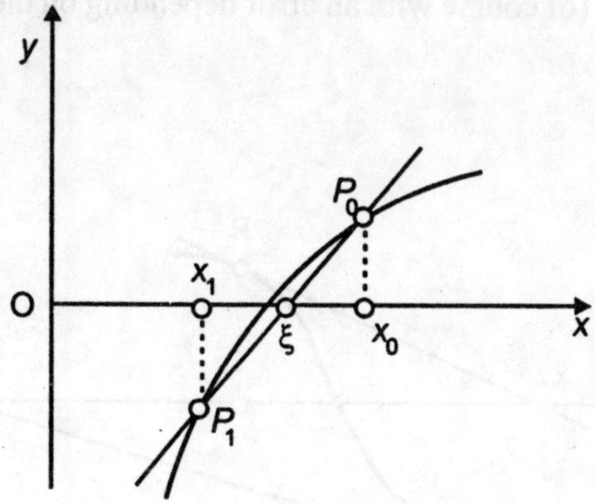

Fig. The rule of false position

The Rule of False Position: Newton's method, in which the tangent to the curve has a decisive role, is only the limiting case of an older method, known as the rule of false position, in which the secant takes the place of the tangent. Let us assume that we know two points (x_0, y_0) and (x_1, y_1) in the neighbourhood of the required intersection with the x-axis. If we replace the curve by the secant joining these two points, the intersection of this secant with the x-axis will under certain circumstances be an improved approximation to the required root of the equation. If the abscissa of this point is denoted by ξ, we have the equation

$$\frac{\xi - x_0}{f(x_0)} = \frac{\xi - x_1}{f(x_1)}$$

from which we compute ξ:

$$\xi = \frac{x_0 f(x_1) - x_1 f(x_0)}{f(x_1) - f(x_0)}$$

$$= \frac{x_0 f(x_1) - x_0 f(x_0) + x_0 f(x_0) - x_1 f(x_0)}{f(x_1) - f(x_0)}$$

or

$$\xi = x_0 - \frac{f(x_0)}{\{f(x_1) - f(x_0)\}/(x_1 - x_0)}$$

This formula, which determines the next approximation ξ from x_0 and x_1, is called the rule of false position. We can employ it with advantage if one value of the function is positive and the other negative, where $y_0 > 0$ and $y_1 < 0$. Repetition of this process will always lead us to the required result, if we use at each step a positive and a negative value of the function, between which the required root must necessarily lie.

The above formula of Newton results from the rule of false position as a limiting case, if we let x_1 tend to x_0. In fact, the denominator of the second term on the right hand side of the statement of the rule of false position tends to $f'(x_0)$ as x_1 tends to x_0.

The Method of Iteration: Another way of approximating roots of an equation $f(x) = 0$ is the method of iteration. i.e., set $\phi(x) = f(x) + x$ and write it in the form $x = \phi(x)$. We then assume that ξ is the true value of a solution of our equation and x_0 a first approximation. We obtain a second approximation x_1 by setting $x_1 = \phi(x_0)$, a third approximation x_2 by setting $x_2 = \phi(x_1)$, etc. In order to investigate the convergence of these approximations, we apply the mean value theorem; recalling that $x = \phi(\xi)$, we have

$$\xi - x_1 = \phi(\xi) - \phi(x_0) = (\xi - x_0)\phi'(\bar{\xi})$$

where $\bar{\xi}$ lies between ξ and x_0. This shows that, if for $|\xi - x| < |\xi - x_0|$ the absolute value of the derivative $f'(x)$ is less than $k < 1$, then the successive approximations converge, because

$$|\xi - x_1| < k|\xi - x_0|, |\xi - x_2| < k^2|\xi - x_0|, \ldots, |\xi - x_n| < k^n|\xi - x_0|,$$

whence the errors tend to zero, The smaller is the absolute value of the derivative $\phi'(x)$ in the neighbourhood of ξ, the more rapid is the convergence.

If $\phi'(x) > 1$ in the neighbourhood of ξ, the approximations do not any longer tend to ξ. We can then use the inverse function or else the following device: We choose a first approximation x_0, calculate $A = f'(x_0)$ and write

$$\phi(x) = -\frac{1}{A}f(x) + x$$

Then the equation $f(x) = 0$ can be written in the form $x = \phi(x)$, and now $\phi'(x) = -f'(x)/A + 1$, which has the value 0 at $x = x_0$, whence its absolute value will usually be less than a constant $k < 1$, provided $|\xi - x| < |\xi - x_0|$.

Returning to Newton's method, we can now investigate its suitability for application at any given point. The equation $f(x) = 0$ is equivalent to

$$x = \phi(x) = x - \frac{f(x)}{f'(x)}$$

provided that $f'(x) \neq 0$. Applying the method of iteration to this last equation, we obtain from a first approximation x_0 a second approximation

$$x_1 = x_0 - f(x_0)/f'(x_0);$$

in other words, the same second approximation as Newton's method gives when it is applied to the equation $f(x) = 0$. We thus see that the smaller is the value of

$$\phi'(x) = \frac{f(x)f''(x)}{(f'(x))^2},$$

the more rapidly converge the successive approximations. In words, Newton's formula converges rapidly for large values of $f'(x_0)$ and small values of $f(x_0)$ and curvature, as intuition would lead us to suspect.

We can also obtain an estimate of the accuracy of Newton's method, if we recall that, since $f(\xi) = 0$, the derivative $\phi'(\xi) = 0$. Applying Taylor's theorem, we have

$$\xi - x_1 = \phi(\xi) - \phi(x_0) = \frac{(\xi - x_0)^2}{2}\phi''(\bar{\xi}),$$

where $\bar{\xi}$ lies between ξ and x_0. Thus, if the error of the original estimate is small, the method converges much more rapidly than the method of iteration applied directly to $f(x) = 0$.

For example, if

$$\phi''(x) = \frac{\{f'(x)\}^2 f''(x) + f'(x)f(x)f'''(x) - 2f(x)\{f''(x)\}^2}{\{f'(x)\}^3} \quad (a)$$

is everywhere less than 10, then a first approximation, which has an error by less than 0.001, will yield a second approximation with an error of less than 0.000,005.

Stirling's Formula

In very many applications, especially in statistics and in the theory of probability, we find it necessary to have a simple approximation to $n!$ as an elementary function of n. Such an expression is given by the following theorem which bears the name of its discoverer James Stirling 1692 - 1770:

As $n \approx \dfrac{n!}{\sqrt{2\pi}n^{n+1/2}e^{-n}} \to 1$; in more detail $\sqrt{2\pi}n^{n+1/2}e^{-n} < n! < \sqrt{2\pi}n^{n+1/2}e^{-n}\left(1+\dfrac{1}{4n}\right)$.

In other words, the expressions $n!$ and $\sqrt{2\pi}n^{n+\frac{1}{2}}e^{-n}$ differ only by a small percentage for large n - as we say, the two expressions are asymptotically equal—and at the same time the factor $1 + 1/4n$ gives us an estimate of the degree of accuracy of the approximation.

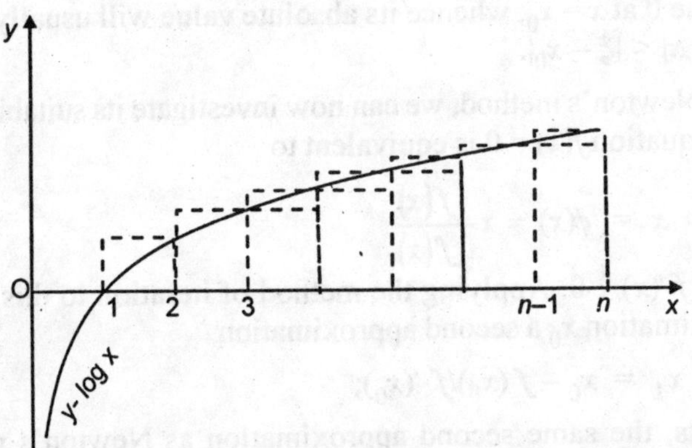

We are led to this remarkable formula, if we attempt to evaluate the area under the curve $y = \log x$. By integration(4.4.1), we find that A_n, the exact area under this curve between the ordinates $x = 1$ and $x = n$, is given by

$$\int_1^n \log x dx = x\log x - x\log x - x\Big|_1^n = n\log n - n + 1$$

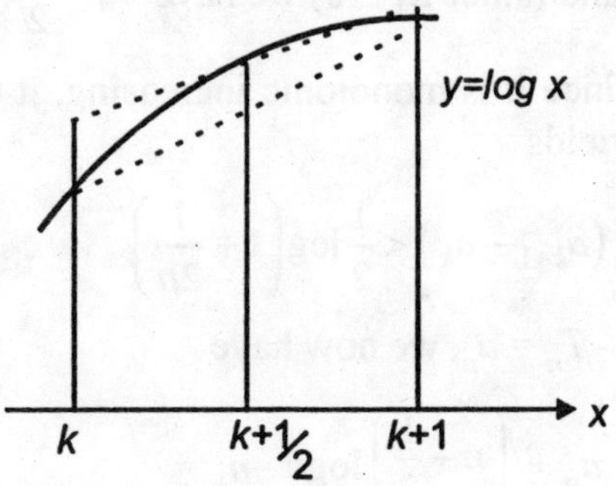

However, if we estimate the area by the trapezoidal rule, erecting ordinates at x=1, x=2, ... , $x = n$, we obtain an approximate value T_n for the area:

$$T_n = \log 2 + \log 3 + ... + \log(n-1) + \frac{1}{2}\log n$$

$$= \log n! - \log n$$

If we make the reasonable assumption that A_n and T_n are of the same order of magnitude, we find at once that $n!$ and $n^{n+\cdot\cdot}e^{-n}$ are of the same order of magnitude, which is essentially what Stirling's formula states.

In order to make this argument precise, we first show that $a_n = A_n - T_n$ is bounded, whence it follows immediately that $T_n = A_n\left(1-\frac{a_n}{A_n}\right)$ is of the same order of magnitude as A_n.

The difference a_{k+1} - a_k is the difference between the area under the curve and the area under the secant in the strip $k < x < k+1$. Since the curve is concave downward and lies above the secant, $a_{k+1}-a_k$ is positive, and $(a_n-a_{n-1}) + (a_{n-1}-a_{n-2}) + ... + (a_2-a_1) + a_1$ is monotonic increasing. Moreover, the difference a_{k+1}-a_k is clearly less than the difference between the area under the tangent at $x = k + 1/2$ and the area under the secant, whence we have the inequality

$$a_{k+1} - ak < \log\left(k+\frac{1}{2}\right) - \frac{1}{2}\log k - \frac{1}{2}\log(k+1)$$

$$= \frac{1}{2}\log\left(1+\frac{1}{2k}\right) - \frac{1}{2}\log\left(1+\frac{1}{\left(k+\frac{1}{2}\right)}\right)$$

$$< \frac{1}{2}\log\left(1+\frac{1}{2k}\right) - \frac{1}{2}\log\left(1+\frac{1}{2(k+1)}\right)$$

If we add these inequalities for $k = 1, 2, \ldots, n-1$, all the terms on the right hand side except two of them will cancel out and (since $a_1 = 0$) we have $a_n < \frac{1}{2}\log\frac{3}{2} - \frac{1}{2}\log\left(1+\frac{1}{2n}\right) < \frac{1}{2}\log\frac{3}{2}$

Hence a_n is bounded; since it is monotonic increasing, it tends to a limit a as n}}. Our inequality for $a_{k+1} - a_k$ now yields

$$a - a_n = \sum_{k=n}^{\infty}(a_{k+1} - a_k) < \frac{1}{2}\log\left(1+\frac{1}{2n}\right)$$

Since, by definition, $A_n - T_n = a_n$, we now have

$$\log n! = 1 - a_n + \left(n + \frac{1}{2}\right)\log n - n,$$

or, writing $a_n = e^{1-a}$, $n! = a_n n^{n+\frac{1}{2}} e^{-n}$

The sequence a_n is monotonic decreasing and tends to the limit $a = e^{1-a}$, whence

$$1 < \frac{a_n}{a} = e^{a-a_n} < e^{\frac{1}{2}\log(1+1/2n)} = \sqrt{\left(1+\frac{1}{2n}\right)} < 1 + \frac{1}{4n}$$

Hence we have $an^{n+\frac{1}{2}}e^{-n} < n! < an^{n+\frac{1}{2}}e^{-n}\left(1+\frac{1}{4n}\right)$.

There only remains for us to find the actual value of the limit a. Here we employ the formula proved in 4.4.4:

$$\sqrt{\pi} = \lim_{n\to\infty}\frac{(n!)^2 2^{2n}}{(2n)!\sqrt{n}}$$

Replacing $n!$ by $a_n n^{n+\ldots} e^{-n}$ and $(2n)!$ by $a_{2n} 2^{2n+ýÿ} n^{2n+ýÿ} e^{-2n}$, we immediately obtain

$$\sqrt{\pi} = \lim_{n\to\infty}\frac{a_n{}^2}{a_{2n}\sqrt{2}} = \frac{a^2}{a\sqrt{2}},$$

whence $a = \sqrt{2\pi}$ The proof of Stirling's formula is thus complete.

In addition to its theoretical interest, Stirling's formula is a very useful tool for the numerical calculation of $n!$ when n is large. Instead of forming the product of a large number of integers, we have merely to calculate Stirling's expression using logarithms, which involves far fewer operations. Thus, we obtain for n=10 the value 3,598,696 from Stirling's expression, while the exact value is 3628800. The percentage error is barely 5/6 %.

Chapter 8

Infinite Series and Other Limiting Processes

PRELIMINARY REMARKS

The geometric series, Taylor series and several special examples, which we have already met, suggest that we may well study from a rather more general point of view those limiting processes under the heading of summation of infinite series . By its nature, any limiting value $S = \lim_{n\to\infty} s_n$ can be written as an infinite series, because, if n takes the values 1, 2, 3, ..., we need only set $a_n = s_n - s_{n-1}$ for $n > 1$ and $a_1 = s_1$, in order to obtain $s_n = a_1 + a_2 + ... + a_n$, and the value S thus appears as the limit of s_n - the sum of n terms as n increases. We express this fact by saying that S is the sum of the infinite series $a_1 + a_2 + a_3 + ...$

Thus, an infinite series is simply a way of representing a limit, where each successive approximation is found from the preceding one by addition of one more term. In principle, the expression of a number as a decimal is merely the representation of a number a in the form of an infinite series $a = a_1+a_2+a_3+ ...$ where, if $0 < a < 1$, the term a_n is put equal to $a_n\, 10^{-n}$ and a_n is a whole number between 0 and 9, inclusively. Since every limiting value can be written in the form of an infinite series, it may seem that a special study of series is superfluous. However, in many cases, it happens that limiting values occur naturally in the form of infinite series, which often exhibit particularly simple laws of formation. Of course, it is not true that every series has an easily recognizable law of formation. For example, the number p can certainly be represented as a decimal, yet we know no simple law which enables us to state the value of an arbitrary digit, say its 7000-th decimal. However, if we set aside the representation of π by a decimal and consider instead Gregory's series, we have an expression with a perfectly clear general law of formation.

Infinite series, in which approximations to the limit are found by repeated addition of new terms, are analogous to infinite products, in which the approximations to the limit arise from repeated multiplication by new factors. However, we shall not go deeply into the theory of infinite products.

The Concepts of Convergence and Divergence

The Fundamental Ideas: We consider an infinite series the general term of which we denote * by a_n; the series then has the form $a_1+a_2+\ldots = \sum_{\nu=1}^{\infty} a_\nu$.

The symbol with the summation sign on the right hand side is merely an abbreviated way of writing the expression on the left hand side.

* For formal reasons, we include the possibility that certain of the numbers a_n may be zero. If all a_n vanish onwards from a number N (i.e., when $n > N$), we speak of a terminating series.

If, as n increases, the n-th partial sum $s_n = a_1 + a_2 + \ldots + a_n = \sum_{\nu=1}^{n} a_\nu$ approaches a limit

$S = \lim_{n\to\infty} s_n$

we say that the series converges, and otherwise that it diverges. In the first case, we call S the sum of the series. We have already met with many examples of convergent series; for instance, the $1 + q + q^2 + \ldots$, geometric series which converges to the sum $1/(1-q)$ for $q < 1$, Gregory's series, the series for log 2, the series for e, and others converge. In the language of infinite series, Cauchy's convergence test is expressed as follows:

A necessary and sufficient condition for the convergence of a series is that the number $|s_m - s_n| = |a_{n+1} + a_{n+2} + \ldots + a_m|$ becomes arbitrarily small if m and n are chosen sufficiently large ($m > n$).

In other words: A series converges if and only if the following condition is met: Given a positive number ε, however small, it is possible to choose an index $N=N(\varepsilon)$, which, in general, increases beyond all bounds as $\varepsilon \to 0$, in such a way that the $|s_m - s_n|$ is less than e, provided only that $m > N$ and $n > N$.

We can make the meaning of the convergence test clearer by considering the geometric series with $q = 1/2$. If we choose $\varepsilon = 1/10$, we need only take $N = 4$, because

$$|s_m - s_n| = \frac{1}{2^n} + \ldots + \frac{1}{2^{m-1}}$$

$$= \frac{1}{2^{n-1}}\left(\frac{1}{2} + \frac{1}{2^2} + \ldots + \frac{1}{2^{m-n}}\right) < \frac{1}{2^{n-1}}$$

and $\frac{1}{2^{n-1}} < \frac{1}{10}, if\ n > 4$.

If we choose ε equal to 1/100, it is sufficient to take 7 as the corresponding value of N, as is easily verified.

Obviously, it is a necessary condition for the convergence of a series that $\lim_{n\to\infty} a_n = 0$.

Otherwise, undoubtedly, the convergence criterion cannot be fulfilled. However, this necessary condition is by no means sufficient for convergence; on the contrary, it is easy to find infinite series the general term a_n of which approaches 0 as n increases, yet their sum does not exist, because the partial sum s_n increases without limit as n increases.

An example of this is the series $1+\frac{1}{\sqrt{2}}+\frac{1}{\sqrt{3}}+...+\frac{1}{\sqrt{n}}+...$ with the general term $\frac{1}{\sqrt{n}}$. We immediately see that

$$s_n > \frac{1}{\sqrt{n}}+...+\frac{1}{\sqrt{n}} = \frac{n}{\sqrt{n}} = \sqrt{n}$$

The n-th partial sum increases beyond all bounds as n increases, whence the the series diverges

`The same is true for the classical example of the harmonic series $1+\frac{1}{2}+\frac{1}{3}+\frac{1}{4}+...$, for which

$$a_{n+1}+...+a_{2n} = \frac{1}{n+1}+...+\frac{1}{2n} > \frac{1}{2n}+...+\frac{1}{2n} = \frac{1}{2}$$

Since n and $m = 2n$ can be taken as large as we please, the series diverges, because Cauchy's test is not fulfilled; in fact, the n-th partial sum tends obviously to infinity, since all the terms are positive. On the other hand, the series, formed from the same numbers with alternating signs, $1-\frac{1}{2}+\frac{1}{3}-\frac{1}{4}+\frac{1}{5}-+..+\frac{(-1)^{n-1}}{n}+...$ converges and has the sum log 2.

It is by no means true that in every divergent series s_n tends to +1 or –1. Thus, in the case of the series 1 – 1 + 1 – 1 + 1 + – ..., we see that the partial sum s_n has alternately the values 1 and 0, and, on account of this backwards / forwards oscillation neither approaches a definite limit nor increases numerically beyond all bounds.

As the convergence and divergence of an infinite series is concerned, the following fact which, though self-evident, is very important should be noted: The convergence or divergence of a series is not changed by insertion or removal of finite numbers of terms. As far as convergence or divergence is concerned, it does not matter in the least whether we begin the series at the term a_0, or a_1, or a_5, or at any other arbitrarily chosen term.

Absolute Convergence and Conditional Convergence:. The series $1+\frac{1}{2}+\frac{1}{3}+\frac{1}{4}...$ diverges; but if we change the sign of every second term, the resulting series converges. On the other hand, the geometric series 1-*q*+*qýÿ*-*qýÿÿýÿýÿ* converges and has the sum l/(l + *q*), provided that $0 < q < 1$; on converting all the signs to +, we obtain the series $1 + q + q^2 + q^3 + ...$, which also converges and has the sum l/(l - *q*).

There appears here a distinction which we must examine more closely. With a series all terms of which are positive, there are only two possible cases; either it converges or the partial sum increases beyond all bounds with n. In fact, the partial sums, being a monotonic increasing sequence, must converge, if they remain bounded. Convergence occurs, if the terms approach zero sufficiently rapidly as n increases; on the other hand, divergence occurs, if the terms do not approach zero at all or if they approach zero too slowly. However, in series, where some terms are positive and others negative, the changes of sign may bring about convergence, since a too great increase in the partial sums, due to the positive terms, is compensated by negative terms, as a final result of which a definite limit is approached.

In order to grasp this fact better with a series $\sum_{\nu=1}^{\infty} a_\nu$, which has positive and negative terms, we compare it with the series which has the same terms all with positive signs, that is,

$$|a_1|+|a_2|+... = \sum_{\nu=1}^{\infty}|a_\nu|$$

If this series converges, then for sufficiently large values of n and $m > n$, the expression $|a_1|+|a_{n+2}|+...+|a_m|$ will certainly be as small as we please; on account of the relation $|a_{n+1}+...+a_m| \leq |a_{n+1}|+...+|a_m|$, the expression on the left hand side is also arbitrarily small, whence the original series $\sum_{\nu=1}^{\infty} a_\nu$ converges. In this case, the original series is said to be absolutely convergent. Its convergence is due to the numerical smallness of its terms and does not depend on the change of the signs.

On the other hand, if the series with all the terms taken positively diverges and the original series still converges, we say that the original series is conditionally convergent. Conditional convergence results from the terms of opposite signs compensating one another.

For conditional convergence, Leibnitz's convergence test is frequently of use: If the terms of a series have alternating sign and, moreover, their absolute values $|a_n|$ tend monotonically to 0 (so that $|a_{n+1}| < |a_n|$), the series $\sum_{\nu=1}^{\infty} a_\nu$ converges.

In the proof, we assume that $a_1 > 0$, which does not essentially limit the generality of the argument, and write our series in the form $b_1 - b_2 + b_3 - + ...$, where all the terms b_n are now positive, b_n tends to 0 and the condition $b_{n+1} < b_n$ is satisfied. If we bracket the terms together in the two ways $b_1 - (b_2 - b_3) - (b_4 + b_5) - ...$ and $(b_1 - b_2) + (b_3 + b_4) + (b_5 + b_6) + ...$

we see at once that the partial sums satisfy the two relations:

$$s_1 > s_2 > s_3 > ... > s_{2m+1} > ...$$

$$s_2 < s_4 < s_6 < ... < s_{2m} < ...$$

On the other hand, $s_{2n} < s_{2n+1} < s_1$ and $s_{2n+1} > s_{2n} > s_2$, whence the odd partial sums s_1, s_3, ... form a monotonic decreasing sequence, which in no case drops below the value s_2 and this sequence has a limit L. The even partial sums s_2, s_4, ... likewise form a monotonic increasing sequence the terms of which in no case exceed the fixed number s_1, whence this sequence must have a limiting value L'. Since the numbers s_{2n} and s_{2n+1} differ only by the number b_{2n+1}, which approaches 0 as n increases, the limiting values L and L' are equal to each other. In other words, the even and the odd partial sums

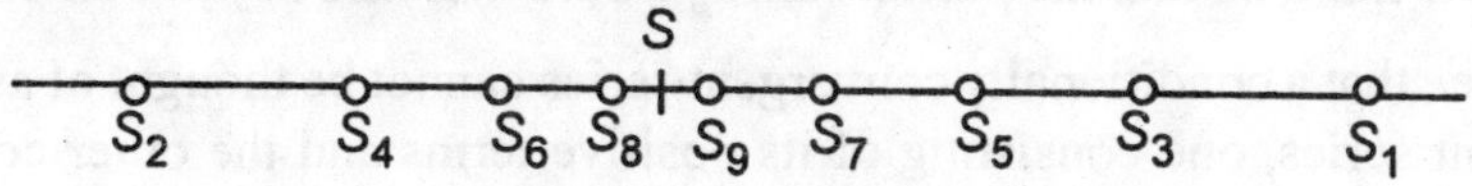

Fig. Convergence of An Alternating Series

approach the same limit, which we will now denote by S. However, this implies that our series is convergent, as has been asserted; its sum is S.

In conclusion, we make another general remark about the fundamental difference between absolute convergence and conditional convergence. We consider a convergent series $\sum_{\nu=1}^{\infty} a_\nu$ and denote the positive terms of the series by $p_1, p_2, p_3, \ldots$ and the negative terms by $-q_1, -q_2, -q_3, \ldots$ If we form the n-th partial sum $s_n = \sum_{\nu=1}^{n} a_\nu$ of the given series, a certain nmnber, say, n' positive terms and a certain number, say n'' negative terms must appear, where $n' + n'' = n$. Moreover, if the number of positive terms as well as that of negative terms in the series is infinite, then the two numbers n'' and n'' will increase beyond all bounds with n. We see immediately that the partial sum s_n is simply equal to the partial sum $\sum_{\nu=1}^{n'} p_\nu$ of the positive terms of the series plus the partial sum $-\sum_{\nu=1}^{n''} q_\nu$ of the negative terms. If the given series converges absolutely, then both the series of positive terms $\sum_{\nu=1}^{\infty} p_\nu$ and the series of the absolute values of the negative terms $\sum_{\nu=1}^{\infty} q_\nu$ certainly converge. In fact, as m increases, the partial sums $\sum_{\nu=1}^{m} p_\nu$ *and* $\sum_{\nu=1}^{m} q_\nu$ are monotonic non-decreasing with the upper bound $\sum_{\nu=1}^{\infty} |a_\nu|$.

The sum of an absolutely convergent series is then simply equal to the sum of the series consisting only of the positive terms plus the sum of the series consisting of the negative terms, or, in other words, it is equal to the difference of the two series with positive terms.

However, $\sum_{\nu=1}^{n} a_\nu = \sum_{\nu=1}^{n'} p_\nu - \sum_{\nu=1}^{n''} q_\nu$; as n increases, n' and n'' must also increase beyond all bounds, and the limit of the left hand side must therefore be equal to the difference of the two sums on the right hand side. If the series contains only a finite number of terms of one particular sign, the facts are correspondingly simplified. On the other hand, if the series does not converge absolutely, but does converge conditionally, then both the series $\sum_{\nu=1}^{\infty} p_\nu$ and $\sum_{\nu=1}^{\infty} q_\nu$ must be divergent, because, if both were convergent, the series would converge absolutely, contrary to our hypothesis. If only one diverges, say $\sum_{\nu=1}^{\infty} p_\nu$, and the other converges, then separation into positive and negative parts, $s_n = \sum_{\nu=1}^{n'} p_\nu - \sum_{\nu=1}^{n''} q_\nu$, shows that the series could not converge; in fact, as *n* increases, *n'* and $\sum_{\nu=1}^{n'} p_\nu$ would increase beyond all bounds, while the term $\sum_{\nu=1}^{n''} q_\nu$ would approach a definite limit, so that the partial sum s_n would increase beyond all bounds.

Hence, we see that a conditionally convergent series cannot be thought of as the difference of two convergent series, one consisting of its positive terms and the other consisting of the absolute values of its negative terms.

Closely connected with this fact, there arises another difference between absolutely and conditionally convergent series which we shall now mention briefly.

Rearrangement of Terms: It is a property of finite sums that we may change the order of the terms or, as we say, rearrange the terms at will without changing their values. There arises the question: What is the exact meaning of a change of the order of terms in an infinite series and does such a rearrangement leave the value of the sum unchanged? While in the case of finite sums there arises no difficulty, for example, in adding the terms in reverse order, in the case of infinite series such a possibility does not exist; there is no last term with which to begin. Now a change of the order of terms in an infinite series can only mean this: A series $a_1 + a_2 + a_3 + \ldots$ is rearranged into a series $b_1 + b_2 + b_3 + \ldots$, provided that every term a_n of the first series occurs exactly once in the second series and conversely. For example, the amount by which a_n is displaced may increase beyond all bounds as *n* does; the only point is that it must appear somewhere in the new series. If some of the terms are moved to later positions in the series, other terms must, of course, be moved to earlier positions. For example, the series $1 + q + q^2 + q^4 + q^3 + q^8 + q^7 + q^6 + q^5 + q^{16} + \ldots$ is a rearrangement of the geometric series $1 + q + q^2 + \ldots$

As regards the change of order, there is a fundamental distinction between absolutely convergent and conditionally convergent series. In absolutely convergent series, rearrangement of terms does not affect the convergence and the value of the sum of the series is not changed, exactly as in the case of finite sums.

On the other hand, in conditionally convergent series, the value of the sum of the series can be changed at will by suitable rearrangement of the series, and the series can even be made to diverge, if so desired.

The first of these facts, referring to absolutely convergent series, is easily established. Let us assume, to begin with, that our series has only positive terms and consider the *n*-th partial sum $s_n = \sum_{\nu=1}^{n} a_\nu$. All the terms of this partial sum occur in the *m*-th partial sum $t_m = \sum_{\nu=1}^{m} b_\nu$ of the rearranged series, provided only that *m* is chosen large enough. Hence $t_m \geq s_n$. On the other hand, we can determine an index *n'* so large that the partial sum $s_{n'} = \sum_{\nu=1}^{n'} a_\nu$ of the first series contains all the terms b_1, b_2, b_m. It then follows that $t_m \leq s_{n'} \leq A$, where *A* is the sum of the first series. Thus, for all sufficiently large values of *m*, we have $s_n \leq t_m \leq A$; since s_n can be made to differ from *A* by an arbitrarily small amount, it follows that the rearranged series also converges, in fact, to the same limit *A* as the original series.

If the absolutely convergent series has both positive and negative terms, we may regard it as the difference of two series each of which has only positive terms. Since in the rearrangement of the original series each of these two series merely undergoes rearrangement and therefore converges to the same value as before, the same is true of the original series when rearranged. In fact, by the case just considered, the new series is absolutely convergent and is therefore the difference of the two rearranged series of positive terms.

The fact just proved may seem to the beginner to be a triviality. The fact that it really does require a proof and that in this proof absolute convergence is essential can be demonstrated by an example of the opposite behaviour of conditionally convergent series. Consider the familiar series $1-\frac{1}{2}+\frac{1}{3}-\frac{1}{4}+\frac{1}{5}-\frac{1}{6}+\frac{1}{7}-\frac{1}{8}+-\ldots=\log 2$.

Write below it the result of multiplication by the factor 1/2 $\frac{1}{2}-\frac{1}{4}+\frac{1}{6}-\frac{1}{8}+-\ldots=\frac{1}{2}\log 2$

and add the two series, combining the terms in vertical columns. We thus obtain $1-\frac{1}{3}-\frac{1}{2}+\frac{1}{5}+\frac{1}{7}-\frac{1}{4}+\frac{1}{9}+\frac{1}{11}-++\ldots=\frac{3}{2}\log 2$.

This last series can obviously be obtained by rearranging the original series and yet the value of the sum of the series has been multiplied by the factor 3/2. It is easy to imagine the effect of the discovery of this apparent paradox on the mathematicians of the Eighteenth Century, who were accustomed to operate with infinite series without regard to their convergence.

We shall give the proof of the theorem stated above concerning the change in the sum of a conditionally convergent series which arises from a change of order of its terms, although we shall have no occasion to make use of the result.

Let $p_1, p_2, \ldots$ be the positive terms and $-q_1, -q_2, \ldots$ the negative terms of the series. Since the absolute value $|a_n|$ tends to 0 as n increases, the numbers p_n and q_n must also tend to 0 with n. Moreover, as we have already seen, the sum $\sum_1^\infty p_\nu$ must diverge and the same is true for $\sum_1^\infty q_\nu$.

This abbreviated notation for $\sum_1^\infty p_\nu$ and analogous expressions for other series will be used often in what follows.

We can now find easily a rearrangement of the original series which has an arbitrary number a as its limit. In order to be specific, let a be positive. We then add the first n_1 positive terms, just enough to ensure that the sum $\sum_1^{n_1} p_\nu$ is less than a. Since the sum $\sum_1^{n_1} p_\nu$ increases with n_1 beyond all bounds, it is always possible, by using enough terms, to make the partial sum larger than a. The sum will then differ from the exact value a by at most p_n. We now add just enough negative terms $-\sum_1^{m_1} p_\nu$ in order to ensure that the sum $-\sum_1^{m_1} p_\nu -\sum_1^{m_1} q_\nu$ is less than a. This is also possible, as follows from the divergence of the series from the divergence of the series $\sum_1^{\infty} q_v$. The difference between this sum and a is now at most q_{m1}. We now add just enough other positive terms $\sum_{n_1+1}^{n_2} p_v$ in order to make the partial sum again larger than a, as is again possible since the series of positive terms diverges. The difference between the partial sum and a is now at most p_{ms}. We again add just just enough negative terms $\sum_{m_1+1}^{m_2} q_{v'}$, which start next after the last one previously used, to make the sum once more less than a and continue in the same way. The values of the sums thus obtained will oscillate about the number a and, when the process is carried out far enough, the oscillation will only take place between arbitrarily narrow bounds; in fact, since the terms p_n and q_n themselves tend to 0 when n is sufficiently large, the length of the interval in which the oscillation occurs will also tend to 0, and the theorem is proved.

In the same manner, we can rearrange s series in such a way as to make it diverge; we merely have to choose such large numbers of positive terms as compared with the negative ones that there occurs no longer compensation.

Operations with Infinite Series: It is clear that two convergent infinite series $a_1 + a_2 + ... = S$ and $b_1 + b_2 + ... = T$ can be added term by term, i.e., that the series formed from the terms $c_n = a_n + b_n$ converges and has the sum $S + T$, because

$$\sum_{v=1}^{n} c_v = \sum_{v=1}^{n} a_v + \sum_{v=1}^{n} b_v \to S+T$$

This theorem is really nothing more than another statement of the fact (1.6.4) that the limit of the sum of two terms is the sum of their limits.

It is also clear that, if we multiply each term of a convergent infinite series by the same factor, the series remains convergent and its sum is multiplied by the same factor.

In the cases just mentioned, it is immaterial whether the convergence is absolute or conditional. On the other hand, a further study, which is not necessary for us here, shows that, if two infinite series are multiplied by each other by the method used in multiplying finite sums, the product series will not usually converge or have the product of the two sums as its sum unless at least one of the two series is absolutely convergent.

Tests for Convergence and Divergence

We have already encountered a test of a general nature for the convergence of series, which applies to series with terms of alternating signs and decreasing absolute values and which asserts that such series are at least conditionally convergent. In what follows, we shall only consider criteria referring to absolute convergence.

The Comparison Test: All such considerations of convergence depend on the comparison of the series in question with a second series; this second series is chosen in such a way that its convergence can readily be tested. The general comparison test may be stated as follows:

If all the numbers b_1, b_2 ... are positive and the series $\sum_{\nu=1}^{\infty} b_\nu$ converges and if $|a_n| \leqq b_n$ for all values of n, then the series $\sum_{\nu=1}^{\infty} a_n$ is absolutely convergent.

If we apply Cauchy's test, the proof becomes almost trivial, because, if m n, we have $|a_n + ... + a_m| \leqq |a_n| + ... + |a_m| \leqq b_n + ... + b_m$.

Since the series $\sum_{\nu=1}^{\infty} b_\nu$ converges, the right hand side is arbitrarily small, provided that n and m are sufficiently large, whence for such values of n and m the left hand side is also arbitrarily small, so that, by Cauchy's test, the given series converges. The convergence is absolute, since our argument applies equally well to the convergence of series of absolute values $|a_n|$.

The analogous proof for the following fact will be left to the reader:

If $|a_n| \geqq b_n > 0$ and the series $\sum_{\nu=1}^{\infty} b_\nu$ diverges, then the series $\sum_{n=1}^{\infty} a_\nu$ is certainly not absolutely convergent.

Comparison with the Geometric Series: In applications of the test, the most frequently used comparison series is the geometric series. We obtain immdediately the theorem:

The series $\sum_{n=1}^{\infty} a_n$ is absolutely convergent, if from a certain term onwards there holds a relation of the form

$$|a_n| < cq^n,$$

where c is a positive number independent of n and q is any fixed positive number less than 1.

This test is usually expressed in one of the following, weaker forms:

The series $\sum_{n=1}^{\infty} a_n$ converges absolutely, if there holds from a certain term onwards a relation of the form

$$\left|\frac{a_{n+1}}{a_n}\right| < q,$$

where q is again a positive number less than 1 and independent of n, or, if from a certain term onward there holds a relation of the form

$$\sqrt[n]{|a_n|} < q,$$

where q is again a positive number less than 1. In particular, the conditions of these tests are satisfied if there applies a relation of the form

$$\lim_{n\to\infty}\left|\frac{a_{n+1}}{a_n}\right| = k < 1$$

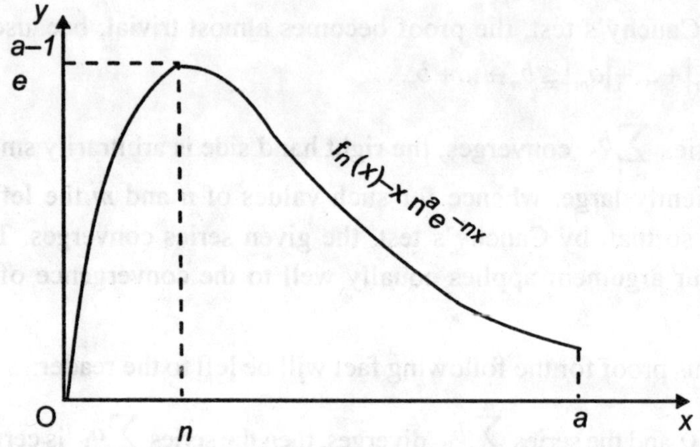

These statements are easily established in the following manner:

Let the criterion II*a*, the ratio test, be satisfied from the suffix n_0 onward, i.e., when $n > n_0$. For the sake of brevity, let $a_{n_0+m+1} = b_m$ and find that $|b_1| < q|b_0|, |b_2| < q|b_1| < q^2|b_0|, |b_3| < q$ $|b_2| < q^3|b_0|$, etc.; hence $|b_m| < q^m|b_0|$,

which establishes our statement. For the Criterion II*b* - the root test -we have at once $|a_n|$ $< q_n$, and our statement follows immediately.

Finally, in order to prove III, we consider an arbitrary number q such that $k < q < 1$. Then, from a certain n_0 onwards, i.e.,, when $n > n_0$, it is certain that $\left|\frac{a_{n+1}}{a_n}\right| < q$ or $\sqrt[n]{|a_n|} < q$, as the case may be, since from a certain term onwards the values of $\left|\frac{a_{n+1}}{a_n}\right| < q$ *of* $\sqrt[n]{|a_n|}$ differ from k by less than $(g - k)$. The statement is then established by reference to the results already proved.

We stress the point that the four tests derived from the original criterion $|a_n| < cq^n$ are not equivalent to each other or to the original one, i.e., that they cannot be derived from one another in both directions. We shall soon see from examples that, if a series satisfies one of these conditions, it need not by any means satisfy all the other ones.

* More exactly: If III*a* is fulfilled, then II*a* ie fulfilled: if III*b*, then II*b*; if III*a*, then III*b*; if II*a*, then II*b*, and if any of the four is satisfied, then so is I. None of these statements can he reversed.

For the sake of completeness, it may be pointed out that a series certainly diverges if from a certain term onwards $|a_n| > c$ for a properly chosen positive number c or if from a certain term onwards $\sqrt[n]{|a_n|} > 1$, or if

$$\lim_{n\to\infty}\left|\frac{a_{n+1}}{a_n}\right| = k, \qquad \lim_{n\to\infty}\left|\sqrt[n]{a_n}\right| = k,$$

where $k > 1$. In fact, as we immediately recognize, in such a series the terms cannot tend to zero as n increases, whence the series must diverge. (Under these conditions, the series cannot even be conditionally convergent.)

Our tests furnish sufficient conditions for the absolute convergence of a series, i.e., if they are satisfied, we can conclude that the series converges absolutely. However, they are definitely not necessary conditions, i.e., absolutely convergent series can be found which do not satisfy these conditions.

For example, the knowledge that

$$\lim_{n\to\infty}\left|\frac{a_{n+1}}{a_n}\right| = 1, \qquad \lim_{n\to\infty}\left|\sqrt[n]{a_n}\right| = 1,$$

does not entitle us to make any statement about the convergence of the series. Such a series may converge or diverge. For example, the series $\sum_{n=1}^{\infty}\frac{1}{n}$, for which $\lim_{n\to\infty}\sqrt[n]{|a_n|} = 1$ *and* $\lim_{n\to\infty}\sqrt[n]{\left|\frac{a_{n+1}}{a_n}\right|} = 1$ is divergent, as we have seen before. On the other hand, we shall soon see that the series $\sum_{n=1}^{\infty}\frac{1}{n^2}$, which satisfies the same relations, converges.

As an example of the application of our tests, we first consider the series $q + 2q^2 + 3q^3 + \dots + nq^n + \dots$

For this series,

$$\lim_{n\to\infty}\sqrt[n]{|a_n|} = |q|\cdot\lim_{n\to\infty}\sqrt[n]{n} = |q|,$$

$$\lim_{n\to\infty}\left|\frac{a_{n+1}}{a_n}\right| = |q|\cdot\lim_{n\to\infty}\frac{n+1}{n} = |q|.$$

It follows from the ratio test as well as from the root test even in the weaker form III that the series converges, if $|q| < 1$.

On the other hand, if we consider the series $1 + 2q + q^2 + 2q^3 + \dots + q^{2n} + 2q^{2n+1} + \dots$, we can no longer prove convergence by the ratio test when ýÿ ⟩ $|q| < 1$, because then $|2q^{2n+1}/q^{2n}| = 2|q|$ ⟩ 1. But the root test yields immediately $\lim\sqrt[n]{|a_n|} = |q|$ and shows that the series converges provided $|q| < 1$, which, of course, we could have observed directly.

Comparison with an Integral

We now proceed to a discussion of convergence which is independent of the preceding considerations. We shall carry it out for the particularly simple and important case of the series

$$\sum_{n-1}^{\infty} \frac{1}{n^{\alpha}} = 1 + \frac{1}{2^{\alpha}} + \frac{1}{3^{\alpha}} + \ldots,$$

where the general term a_n is $1/n^{\alpha}$, $\alpha > 0$. In order to investigate the properties of this series, we consider the graph of the function $y = 1/x^{\alpha}$ and mark off on the x-axis the integral values $x = 1, x = 2$. We first construct the rectangle of height $1/n^{\alpha}$ over the interval n-1 ýÿ x ýÿ n of the x-axis ($n > 1$) and compare it with the area of the region bounded by the same interval of the x-axis, the ordinates at the ends and the curve $y = 1/x^{\alpha}$. Secondly, we construct the rectangle of height $1/n^{\alpha}$ lying above the interval n {x} n+1 and similarly compare it with the area of the region lying above the same interval and below the curve.

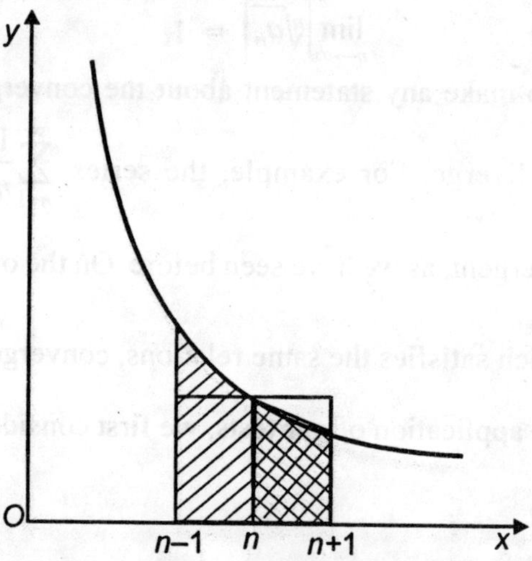

Fig. Comparison of a series with an integral

In the first case, the area under the curve is obviously larger than the area of the rectangle; in the second case, it is less than the area of the rectangle. In other words, $\int_{n}^{n+1} \frac{dx}{x^{\alpha}} < \frac{1}{n^{\alpha}} < \int_{n-1}^{n} \frac{dx}{x^{\alpha}}$, as we may also prove directly from the integral itself. Writing down this inequality for $n = 2, n = 3, \ldots, n = m$ and summing, we obtain the estimate * for the m-th partial sum $s_m = \sum_{n=1}^{m} \frac{1}{n^{\alpha}}$

$$1 + \int_{2}^{m+1} \frac{dx}{x^{\alpha}} < s_m < 1 + \int_{1}^{m} \frac{dx}{x^{\alpha}}$$

* From this relation follows at once that the sequence of numbers $C_n = 1 + \frac{1}{2} + \frac{1}{3} + \ldots + \frac{1}{n} - \log n$ is bounded below. Since we see from the inequality $\frac{1}{n+1} < \int_n^{n+1} \frac{dx}{x} = \log(n+1)$that the sequence is monotonic decreasing, it must approach a limit

$$\lim_{n\to\infty} C_n = \lim_{n\to\infty} \left(1+\frac{1}{2}+\frac{1}{3}+\ldots+\frac{1}{n}-\log n\right)$$

The number C with the value 0.577,2 is called *Euler's constant*. In contrast to the other important special numbers of analysis, such as *p* and *e*, no other expression with a simple law of formation has been found for Euler's constant.

Now, as m increases, the integral $\int_1^m \frac{1}{x^\alpha} dx$ tends to a finite limit or increases without limit according to whether $\alpha > 1$ or $\alpha > 1$. Consequently, the monotonic sequence of the numbers s_m is bounded or increases beyond all bounds according to whether $\alpha >$ or $\alpha > 1$, whence follows the theorem:

The series $\sum_{n=1}^{\infty} \frac{1}{n^\alpha} = \frac{1}{1^\alpha} + \frac{1}{2^\alpha} + \cdots$ converges - and, of course, absolutely - if and only if, $\alpha > 1$.

The divergence of the harmonic series, which we have proved previously in a different way, is an immediate consequence of this. In particular, we see that the series $\frac{1}{1^2}+\frac{1}{2^2}+\frac{1}{3^2}+\ldots$, $\frac{1}{1^3}+\frac{1}{2^3}+\frac{1}{3^3}+\ldots$, ..., converge.

The series $\sum_{\nu=1}^{\infty} \frac{1}{\nu^\alpha}$, the convergence of which we have just studied, frequently serve as comparison series in investigations of convergence. For example, we see at once that for $\alpha > 1$ the series $\sum_{\nu=1}^{\infty} \frac{c_\nu}{\nu^\alpha}$ converges absolutely if the absolute values $|c_n|$ of the coefficients remain less than a fixed bound independent of *n*.

Sequences and Series of Functions

General Remarks: The terms of the infinite series hitherto considered have been constants, whence these series (when convergent) always represented definite numbers. But both in theory and in applications, the series of outstanding importance are those in which the terms are functions of a variable, so that the sum of the series is also a function of the variable, as in the case of Taylor series.

We shall therefore consider a series $g_1(x) + g_2(x) + g_s(x) + ...$, in which the functions $g_n(x)$ are defined in an interval $a > x > b$. We will denote the n-th partial sum $g_1(x) + g_2(x) + ... + g_n(x)$, by $f_n(x)$. Then the sum $f(x)$ of our series, where it exists, is simply the limit $\lim_{n\to\infty} f_n(x)$.

We may therefore regard the sum of an infinite series of functions as the limit of a sequence of functions $f_1(x), f_2(x), f_n(x), ...$ Conversely, for any such sequence of functions $f_1(x), f_2(x), ...,$ we can form an equivalent series by setting $g_1(x) = f_1(x)$ ... and $g_n(x) = f_n(x) - f_{n-1}(x)$ for $n > 1$. Hence, when it is convenient, we can pass from the consideration of series to that of sequences and conversely.

Limiting Processes with Functions and Curves: We shall now state exactly what we mean by saying that a function $f(x)$ is the limit of a sequence $f_1(x), f_2(x), \dots, f_n(x)$, in an interval $a \} x \} b$. The definition is: The sequence $f_1(x), f_2(x), ...$ converges in that interval to the limit function $f(x)$, if at each point x of the interval the values $f_n(x)$ converge in the usual sense to the value $f(x)$. . In this case, we shall write $\lim_{n\to\infty} f_n(x) = f(x)$

According to Cauchy's test , we can express the convergence of the sequence without necessarily knowing or stating the limit function $f(x)$. In fact, our sequence of functions will converge to a limit function if, and only if, at each point x in our interval and for every positive number e the quantity $|f_n(x) - f_m(x)|$ is less than e, provided that the numbers n and m are chosen large enough, i.e., larger than a certain number $N = N(\varepsilon)$. This number $N(\varepsilon)$ usually depends on e and x and increases beyond all bounds as ε tends to zero.

We have frequently encountered cases of limits of sequences of functions. We mention only the definition of the power x^α for irrational values of α by the equation

$$x^\alpha = \lim_{n\to\infty} x^{r_n}$$

where $r_1, r_2, ..., r_n, ...$ is a sequence of rational numbers tending to α or the equation

$$e^\alpha = \lim_{n\to\infty}\left(1+\frac{x}{n}\right)^n,$$

where the functions $f_n(x)$ on the right hand side are polynomials of degree n.

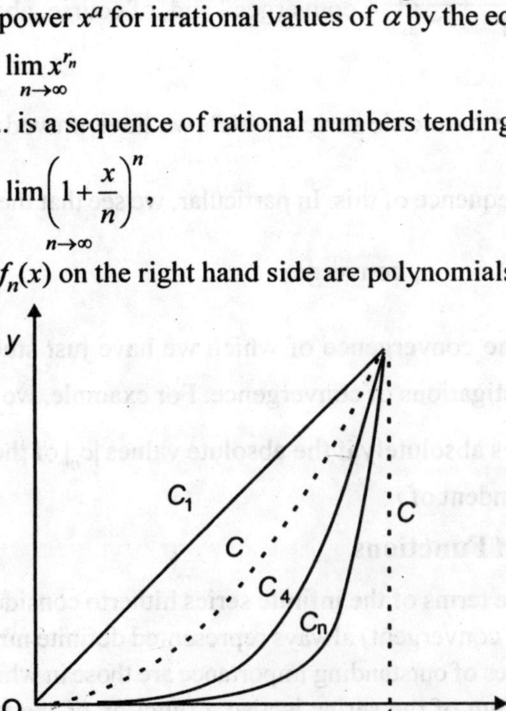

Fig. Limit curve and limit function

The graphical representation of functions by means of curves suggests that we can also speak of limits of sequences of curves, saying, for example, that the graphs of the above limit functions x^a and e^x are to be regarded as the limit curves of the graphs of the functions x^{r_n} *and* $\left(1+\frac{x}{n}\right)^n$ respectively. However, there is a fine distinction between the passages to the limit with functions and with curves. Until the middle of the Nineteenth Century, this distinction was not sufficiently recognized and only by having a clear idea of it can we avoid apparent paradoxes. We shall illustrate this point by an example.

Consider the functions

$$f_n(x) = x^n \ (n = 1, 2, \ldots)$$

in the interval $0 \leq x \leq 1$. All these functions are continuous and the limit function

$$\lim_{n\to\infty} f_n(x) = f(x)$$

exists. However, this limit function is not continuous. On the contrary, since for all values of n the value of the function $f_n(1) = 1$, the limit $f(1) = 1$,

while, on the other hand, for $0 \leq x < I$, the limit $f(x) \lim_{n\to\infty} f_n(x) = 0$, as we have seen in 1.5.6

The function $f(x)$ is therefore a discontinuous function which at $x=1$ has the value 1 and for all other values of x in the interval has the value 0.

This discontinuity becomes clear, if we consider the graphs C_n of the functions $y = f_n(x)$. These are continuous curves, all of which pass through the origin and the point $x = 1, y = 1$ and which draw in closer and closer to the x-axis as n increases. The curves possess a limit curve C which is not at all discontinuous, but consists of the portion of the x-axis between $x=0$ and $x=1$ and the portion of the line $x=1$ between $y=0$ and $y=1$. The curves therefore converge to a continuous limit curve with a vertical portion, while the functions converge to a discontinuous limit function. We thus recognize that this discontinuity of the limit function expresses itself by the occurrence in the limit curve of a portion perpendicular to the x-axis. Such a portion must involve a discontinuity in the limit function and, in fact, such a portion is always present when the limit function is discontinuous. This limit curve is not the graph of the limit function nor can any curve with a vertical portion be the graph of any single-valued function $y = f(x)$; in fact, corresponding to the value of x at which the vertical portion occurs, the curve gives an infinite number of values of y and the function only one. Hence the limit of the graphs of the functions $f_n(x)$ is not the same as the graph of the limit $f(x)$ of these functions.

Naturally, corresponding statements also apply to infinite series.

Uniform and Non-uniform Convergence

General Remarks and Examples: The distinction between the concepts of convergence of functions and of convergence of curves introduces a phenomenon of which it is essential that the student should clearly recognize it. This is the so-called non-uniform convergence of sequences or infinite series of functions. Since it is well known that, as a rule, beginners encounter difficulties here, we should discuss the matter in some detail.

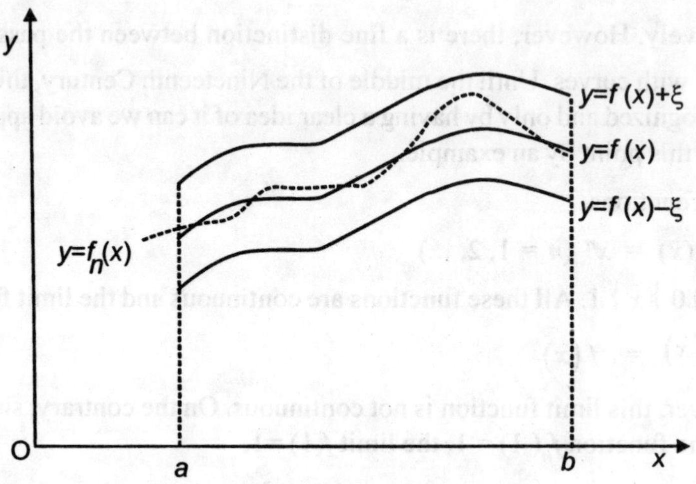

Fig. To illustrate uniform convergence

By definition, a function $f(x)$ is the limit of a sequence $f_1(x), f_2(x), \ldots$ in an interval $a \,\}\, x \,\}\, b$,only means that the usual limit relationship $f(x) = \lim_{n\to\infty} f(x)$ holds at each point x of the interval. From a naive point of view, one might expect that it would follow automatically from this concept of convergence that if we assign an arbitrary degree of accuracy, say ε=1/1000 or ε=1/100, then from a certain index N onwards all the functions $f_n(x)$ would lie between $f(x)$ + e and $f(x) - \varepsilon$ for all values of x, so that their graphs $y = f_n(x)$ will lie entirely in the strip indicated. Thus, for every positive ε , there is a corresponding number $N = N(\varepsilon)$, which, of course, will ordinarily increase beyond all bounds as $\varepsilon < 0$, such that, for $n > N$, the difference $|f(x) - f_n(x)| < \varepsilon$, no matter where x is chosen in the interval.

If the accuracy of the approximation can be made at least equal to a pre-assigned number ε everywhere in the interval at the same time, i.e., by everywhere choosing the same number $N(\varepsilon)$ independent of x, we may say that the approximation is uniform. One is at first astonished to find that the naive assumption that convergence is necessarily uniform is altogether wrong; in other words, that convergence may very well be non-uniform.

A Test of Uniform Convergence: The preceding considerations show that the uniform convergence of a sequence or series is a special property which is not possessed by all sequences and series. We shall now reformulate the concept of uniform convergence. The convergent series $g_1(x) + g_2(x) + \ldots$ is said to be uniformly convergent in an interval, if the sum $f(x)$ can be approximated to within ε (where ε is an arbitrarily small positive number) by taking a number of terms which is sufficiently large and which is the same throughout the interval.

First of all, we assume that the series $g_1(x) + g_2(x) + ...$ converges at every point of a certain interval to a limit function $f(x)$; we denote by $f_n(x)$ the n-th partial sum of the series $f_n(x) = g_1(x) + g_n(x)$ and by $R_n(x)$ the remainder of the series after n terms:

$$R_n(x) = f(x) - f_n(x)$$

The series $g_1(x) + g_2(x) + ...$ is said to be uniformly convergent in the interval if there corresponds to every positive number ε a number N, which only depends on e and not on x, such that for $n > N$ one has the inequality $|R(x)|=|f(x)-f_n(x)|<e$ for all values of x *in* the interval.

Expressed more figuratively, the partial sum $f_n(x)$ represents the sum $f(x)$ to within an error of less than e everywhere in the interval at the same time, provided only that n is chosen large enough. By Cauchy's test, we readily see that the series converges uniformly, if, and only if the difference $|f_n(x) - f_m(x)|$ can be made less than an arbitrary quantity e everywhere in the interval by choosing n and m larger than a number N independent of x. In fact, firstly, if the convergence is uniform, we can make both $|f_n(x) - f(x)|$ and $|f_m(x) - f(x)|$ less than $e/2$ by choosing n and m larger than a number N independent of x, whence it follows that $|f_n(x) - f_m(x)| < e$; and, secondly, if $|f_n(x) - f_m(x)| < e$, for all values of x whenever n and m are larger than N, then, on choosing any fixed value of $n > N$ and letting m increase beyond all bounds, we have

$$|f_n(x) - f(x)| = \lim_{m\to\infty} |f_n(x) - f_m(x)| \leq \varepsilon,$$

for every value of x, whence the convergence is uniform.

If we wish to speak of the uniform convergence of a sequence of functions, we have only to make minute changes in the above definition; the sequence $f_1(x), f_2(x), ...$ converges uniformly to $f(x)$ in an interval, if the difference $|f(x)-f_n(x)|$ can be made less than ε everywhere in the interval by choosing n larger than a number N independent of x. As above, a necessary and sufficient condition for the uniform convergence of the sequence is that $|f_n(x) - f_m(x)|<\varepsilon$ for all values of x when both n and m are larger than a certain number N, which depends on ε but not on x.

We shall soon see that it is just this condition of uniform convergence which makes infinite series and other limiting processes with functions into convenient and useful tools of analysis. Fortunately, in the limiting processes, usually encountered in the Calculus and its applications, non-uniform convergence is a sort of exceptional phenomenon which will scarcely trouble us in our present applications of analysis.

In most cases, the uniformity of the convergence of a series is established by means of the criterion:

If the terms of the series $\sum_{\nu=1}^{\infty} g_\nu(X)$ satisfy the condition $|g_\nu(X)| \leq a_\nu$, where the numbers a_ϖ are constants which form a convergent series $\sum_{\nu=1}^{\infty} a_\nu$ then the series $\sum_{\nu=1}^{\infty} g_\nu(X)$ converges uniformly (and, we incidentally remark, absolutely).

In fact, we have then $\left|\sum_{\nu=n}^{m} g_\nu(x)\right| \le \sum_{\nu=n}^{m} \left|g_\nu(x)\right| \le \sum_{\nu=n}^{m} a_\nu$, and since, by Cauchy's test, the sum $\sum_{\nu=n}^{m} a_\nu$, can be made arbitrarily small by choosing n and $m > n$ large enough, this expresses exactly the necessary and sufficient condition for uniform convergence.

A first example is offered by the geometric series $1 + x + x + \ldots$, where x is restricted to the interval $|x| \rangle q$, a positive number less than 1. The terms of the series are then numerically less than or equal to the terms of the convergent geometric series $\sum q^\nu$

A further example is the trigonometric series $\frac{c_1 \sin(x-\delta_1)}{1^2} + \frac{c_2 \sin(x-\delta_2)}{2^2} + \frac{c_1 \sin(x-\delta_3)}{3^3} + \ldots$, provided that $|c_n| < c$, a positive constant independent of n. Because we have then

$$g_n(x) = \frac{c_n \sin(x-\delta_n)}{n^2}, \text{ so that } \left|g_n(x)\right| < \frac{c}{n^2}$$

Hence the uniform and absolute convergence of the trigonometric series follows from the convergence of the series

Continuity of the Sum of a Uniformly Convergent Series of Continuous Functions: As we have already hinted, the significance of the uniform convergence of an infinite series lies in the fact that a uniformly convergent series in many respects behaves exactly like the sum of a finite number of functions. For example, the sum of a finite number of continuous functions is itself continuous and, correspondingly, we have the theorem:

If a series of continuous terms converges uniformly in an interval, its sum is also a continuous function.

The proof is quite simple. We subdivide the series $f(x) = g_1(x) + g_2(x) + \ldots$ into the m-th partial sum $f_n(x)$ plus the remainder $R_n(x)$. As usual, $f_n(x) = g_1(x) + \ldots g_n(x)$.

If now any positive number ε is assigned, we can,by virtue of the uniform convergence, choose the number n so large that the remainder is less than $\varepsilon/4$ throughout the entire interval, whence $\left|R_n(x+h) - R_n(x)\right| < \frac{\varepsilon}{2}$ for every pair of numbers x and $x + h$ in the interval. The partial sum $f_n(x)$ consists of the sum of a finite number of continuous functions, whence it is continuous; thus, for each point x in the interval, we can choose a positive δ so small that $\left|f_n(x+h) - f_n(x)\right| < \frac{\varepsilon}{2}$, provided $|h| < \delta$ and the points x and $m + h$ lie in the interval. It then follows that

$$\begin{aligned} \left|f(x+h) - f(x)\right| &= \left|f_n(x+h) - f_n(x) + R_n(x+h) - R_n(x)\right| \\ &\le \left|f_n(x+h) - f_n(x)\right| + \left|R_n(x+h) - R_n(x)\right| < \varepsilon, \end{aligned}$$

which expresses the continuity of our function.

The significance of this theorem becomes clear when we recall that the sums of non-uniformly convergent series of continuous functions are not necessarily continuous, as our previous examples show. We may conclude from the preceding theorem that, if the sum of a convergent series of continuous functions has a point of discontinuity, then in every neighbourhood of this point the convergence is non-uniform, whence every representation of discontinuous functions by series of continuous functions is based on the use of non-uniformly convergent limiting processes.

Integration of Uniformly Convergent Series: A sum of a finite number of continuous functions can be integrated term by term, i.e., the integral of the sum can be found by integrating each term separately and adding the integrals. In the case of a convergent infinite series, the same procedure is possible, provided that the series converges uniformly in the interval of integration.

A series $\sum_{\nu=1}^{\infty} g_\nu(X) = f(X)$ which converges uniformly in an interval can be integrated term by term in that interval, or, more precisely, if a and x are two numbers in the interval of uniform convergence, the series $\sum_{\nu=1}^{\infty} \int_a^X g_\nu(t)\,dt$ converges and, in fact, converges uniformly with respect to x for each fixed value of a, its sum being equal to $\int_a^X f(t)\,dt$. .

In order to prove this we write, as before,

$$f(x) = \sum_{\nu=1}^{\infty} g_\nu(x) = f_n(x) + R_n(x)$$

We have assumed that the separate terms of the series are continuous, whence, by 8.4.3, the sum is also continuous and therefore integrable. Now, if ε is any positive number, we can find a number N so large that, for every $n > N$, one has $|R_n(x)| < \varepsilon$ for every value of x in the interval. By the first mean value theorem of the integral calculus, we have $\left|\int_a^x \{f(t) - f_n(t)\}\,dt\right| \leq \varepsilon\, l$,

where l is the length of the interval of integration. Since the integration of the finite sum $f_n(x)$ can be performed term by term, this yields $\left|\int_a^x f(t)\,dt - \sum_1^n \int_a^x g_\nu(t)\,dt\right| \leq \varepsilon\, l$

However, since $\varepsilon\, l$ can be made as small as we please, this yields $\sum_{\nu=1}^{\infty} \int_a^x g_\nu(t)\,dt = \lim_{n\to\infty} \sum_{\nu=1}^{n} \int_a^x g_\nu(t)\,dt = \int_a^x f(t)\,dt$, as was to be proved.

If we wish to conside instead of infinite series sequences of functions, our result can be expressed as follows:

If in an interval the sequence of functions $f_1(x)$, $f_2(x)$, ... tends uniformly to the limit function $f(x)$, then $\int_a^b f(x)\,dx = \lim_{n\to\infty} \int_a^b f_n(x)\,dx$ for every pair of numbers a and b in the interval; in other words, we can then interchange the order of the operations of integration and going to the limit.

This fact is far from being a triviality. It is true that, from a naive point of view such as prevailed in the Eighteenth Century, the interchangeabi]ity of the two processes is hardly to be doubted; however, a glance at the examples in 8.4.1, shows that in the case of non-uniform convergence the above equation might not hold. We need only consider Example 2, in which the integral of the limit function is 0, while the integral of the function $f_n(x)$ over the interval $0 \rangle x \rangle 1$, i.e., the area of the triangle has the value $\int_a^b f_n(x)dx = n^{a-2}$, and, when $\alpha \rangle 2$, this does not tend to zero. We see here immediately from figure that the reason for the difference between $\int_0^1 f_n(x)dx = n^{\alpha-2}$, and $\lim_{n\to\infty} \int_0^1 f_n(x)dx$ lies in the non-uniformity of the convergence.

On the other hand, by considering values of α in the interval $1 \rangle \alpha \rangle 2$, we see that the equation $\lim_{n\to\infty} \int_0^1 f_n(x)dx = \int_0^1 f(x)dx$ can be valid, although the convergence is non-uniform. As a further example, we can integrate term by term the series $\sum_0^\infty g_n(x)$, where $g_n(x) = x^n - x^{n-1}$ for $n \rangle 1$ and $g_0(x) = 1$, between 0 and 1, even though it does not converge uniformly. Thus, while uniformity of convergence is a sufficient condition for term-by-term integrability, it is by no means a necessary condition. Neglect of this aspect may easily lead to a misunderstanding.

Differentiation of Infinite Series: The behaviour of uniformly converging series or sequences with respect to differentiation is quite different from that with respect to integration. For example, the sequence of functions $f_n(x) = \dfrac{\sin n^2\pi}{n}$ certainly converges uniformly to the limit function $f(x) = 0$, but the derivative $f_n'(x) = n \cos n^2 x$ does not converge everywhere to the derivative of the limit function $f'(0) = 0$, as we see by considering $x = 0$. In spite of the uniformity of the convergence, we cannot therefore change the order of the process of differentiation and pass on to the limit.

Naturally, corresponding statements hold for infinite series. For example, the series $\sin x + \dfrac{\sin 2^4 x}{2^2} + \dfrac{\sin 3^4 x}{3^2} + \ldots$ is absolutely and uniformly convergent, because its terms are numerically not greater than the terms of the convergent series $\dfrac{1}{1^2} + \dfrac{1}{2^2} + \dfrac{1}{3^2} + \cdots$.

However, if we differentiate the series term by term, we obtain the series $\cos x + 2^2 \cos 2^4 x + 3^2 \cos 3^4 x + \ldots$, which obviously does not converge everywhere, since, for example, it diverges at $x = 0$.

The only useful criterion which assures us in special cases that term-by-term differentiation is permissible is given by the theorem:

If, on differentiating a convergent infinite series $\sum_{\nu=0}^{\infty} G_\nu(X) = F(X)$ term by term, we obtain a uniformly convergent series of continuous terms $\sum_{\nu=0}^{\infty} g_\nu(X) = f(X)$, then the sum of this last series is equal to the derivative of the sum of the first series.

Hence, this theorem expressly requires that, after differentiating the series term by term, we must still investigate whether the result of the differentiation is a uniformly convergent series or not.

The proof of the theorem is almost trivial, because, by the theorem in 8.4.4, we can integrate term by term the series obtained by differentiation. Recalling that $g_n(t) = G_n'(t)$, we obtain

$$\int_a^x f(t)\,dt = \int_a^x \left(\sum_{\nu=0}^{\infty} g_\nu(t)\right) dt = \sum_{\nu=0}^{\infty} \int_a^x g_\nu(t)\,dt = \sum_{\nu=0}^{\infty} \left(G_\nu(x) - G_\nu(a)\right)$$
$$= F(x) - F(a)$$

Since this is true for every value of x in the interval of uniform convergence, it follows that $f(x) = f(a)$ as was to be proved.

8.5 Power Series

Among infinite series, power series occupy the main position. We mean by a power series a series of the type

$$P(x) = c_0 + c_1 x + c_2 x^2 + \ldots = \sum_{\nu=0}^{\infty} c_\nu x^\nu$$

(a power series in x) or, more generally,

$$P(x) = c_0 + c_1 x(x - x_0) + c_2(x - x_0)^2 + \ldots = \sum_{\nu=0}^{\infty} c_\nu (x - x_0)^\nu$$

(a power series in $(x - x_0)$), where x_0 is a fixed number. If we introduce into the last series $\xi = x - x_0$ as a new variable, it becomes a power series $\sum_{\nu=0}^{\infty} c_\nu \xi^\nu$ in the new variable ξ, whence, without any loss of generality, we can confine our attention to power series of the more special form $\sum_{\nu=0}^{\infty} c_\nu x^\nu$

In 6.2, we have considered the approximate representation of functions by polynomials and were thus led to the expansion of functions in Taylor series, which, in fact, are power series. We shall now study power series in somewhat greater detail and obtain expansions of the most important functions in series in simpler and more convenient ways than before.

Convergence Properties of Power Series: There are power series which converge for no value of x, except, of course, for $x = 0$, e.g., the series $x + 2^2x^2 + 3^3x^3 + \ldots + n^n x^n + \ldots$

In fact, if $x < 0$, we can find an integer N such that $|x| > 1/N$. Then the absolute values of all the terms $n^n x^n$, for which $n > N$, will be larger than 1 and, as n increases, $nn^n x^n$ will increase beyond all bounds, so that the series fails to converge.

On the other hand, there are series which converge for every value of x; for example, the power series for the exponential function

$$e^x = 1 + x + \frac{x^2}{2!} + \frac{x^3}{3!} + \ldots,$$

the convergence of which for every value of x follows at once from the ratio test in 8.2.2. The $(n+1)$-th term, divided by the n-th term, yields x/n and, whatever number x is chosen, this ratio tends to zero as n increases. The behaviour of power series with regard to convergence is expressed in the fundamental theorem:

If a power series in x converges for a value $x = \xi$, it converges absolutely for every value x such that $|x| < |\xi|$ and the convergence is uniform in every interval $|x| \} \eta$, where η is any positive number less than $|\xi|$. Hence h may lie as close to $|\xi|$ as we please.

The proof is simple. If the series $\sum_{\nu=0}^{\infty} c_\nu \xi^\nu$ converges, its terms tend to 0 as n increases. From this follows the weaker statement that all the terms lie below a bound M independent of n, i.e., $|c_n \xi^n| < M$. If now q is any number such that $0<q<1$ and we restrict x to the interval $|x| \}$ $q|\xi|$, then $|c_n x^n| < |c_n x^n| q^\nu < Mq^\nu$. Hence, in this interval, the terms of the series are smaller in absolute value than the terms of the convergent geometric series $\Sigma M q^\nu$, whence follows, by the theorem, the absolute and uniform convergence of the series in the interval $-q|\xi| < x < q|x|$.

If a power series does not converge everywhere, i.e., if there is a value $x = x$ for which it diverges, it must diverge for every value of x such that $|x| > |\xi|$. In fact, if it were convergent for such a value of x, by the theorem above, it would have to converge for any numerically smaller value of ξ.

Hence, we recognize that a power series, which converges for at least one value of x other than 0 and which diverges for at least one value of x, has an interval of convergence, i.e., there exists a definite positive number r such that the series diverges for $|x| > \rho$ and converges for $|x| < \rho$. No general statement can be made for $|x| = \rho$. The limiting cases, in which the series converges only for $x=0$ and that which converges everywhere, are expressed symbolically by writing $\rho = 0$ and $\rho = \}$, respectively.

It is possible to find this interval of convergence directly from the coefficients c_ν of the series. If the limit $\lim_{n\to\infty} \sqrt[n]{|c_n|}$ exists, then

$$\rho = \frac{1}{\lim_{n\to\infty} \sqrt[n]{|c_n|}}$$

In general, ρ is given by the formula

$$\rho = \frac{1}{\overline{\lim_{n\to\infty}} \sqrt[n]{|c_n|}}$$

where $\overline{\lim}$ is the symbol for the upper limit, as it has been defined in A1.5.

For example, $\rho = 1$ for the geometric series $1 + x + x + \ldots$; at the end-points of the interval of convergence, the series diverges. Similarly, for the series for the inverse tangent

$$\text{arc tan } x = x - x^3/3 + x^5/5 - + \ldots,$$

we have $\rho = 1$ and at both of the interval's end-points $x = 1$ the series converges, as we recognize at once from Leibnitz's test.

We deduce from the uniform convergence the important fact that within its interval of convergence (if such an interval exists) a power series represents a continuous function.

Integration and Differentiation of Power Series: On account of the uniformity of convergence, it is always permissible to integrate a power series

$$f(x) = \sum_{\nu=0}^{\infty} c_\nu x^\nu$$

term by term over any closed interval lying entirely within the interval of convergence. We thus obtain the function

$$F(x) = c + \sum_{\nu=0}^{\infty} \frac{c_\nu}{\nu+1} x^{\nu+1},$$

for which

$$F'(x) = f(x)$$

Moreover, since $|c_\nu/(n+1)| \} |c_\nu|$ for all values of ν, the series obtained by integration converges more rapidly than the original series.

We can also differentiate a power series term by term within its interval of convergence, thus obtaining the equation

$$f'(x) = \sum_{\nu=1}^{\infty} \nu c_\nu x^{\nu-1}$$

In order to prove this statement, we need only show that the series on the right hand side converges uniformly, if x is restricted to an interval lying entirely within the interval of convergence.

Now, let x be a number, lying as close to ρ as we please, for which $\sum_{\nu=1}^{\infty} c_\nu \xi^\nu$ converges; then, as we have seen before, all the numbers $|c_n \xi_n|$ lie below a bound M independently of ν, so that $\left|c_\nu \xi^{\nu-1}\right| < \frac{\text{M}}{|\xi|} = N$ Now let q be any number such that $0 < q < 1$; if we restrict x to the interval $|x| \} q|\xi|$, the terms of the series under discussion are not larger than those of the series $\sum_{\nu=1}^{\infty} \left|\nu c_\nu q^{\nu-1} \xi^{\nu-1}\right|$, and therefore less than those of the series $\sum_{\nu=1} N\nu q^{\nu-1}$ .However, in this last series, the ratio of the $(n+1)$-th term to the n-th term is $\{(n+1)/n\}q$, which tends to q as n

increases. Since $0<q<1$, it follows by Criterion III*a* that this series converges. Hence the series, obtained by differentiation, converges uniformly and by the theorem at the end of last section represents the derivative $f'_n(x)$ of the function $f(x)$, which proves our statement.

If we apply this result again to the power series

$$f'(x) = \sum_{\nu=1}^{\infty} \nu c_\nu x^{\nu-1},$$

we find on differentiating term by term that

$$f''(x) = \sum_{\nu=2}^{\infty} \nu(\nu-1)c_\nu x^{\nu-2}$$

and, by continuing the process, we arrive at the theorem:

Every function represented by a power series can be differentiated as often as we please within the interval of convergence and the differentiation can be performed term by term.

As an explicit expression for the *k*-th derivative, we obtain

$$f^{(k)}(x) = \sum_{\nu=k}^{\infty} \nu(\nu-1)\ldots(\nu-k+1)c_\nu x^{\nu-k},$$

or in a slightly different form

$$\frac{f^{(k)}(x)}{k!} = \sum_{\nu=k}^{\infty}\binom{\nu}{k}c_\nu x^{\nu-k} = \sum_{\nu=0}^{\infty}\binom{k+\nu}{k}c_{k+\nu}x^{\nu}$$

These two formulae are frequently useful.

Operations with Power Series: The preceding theorems on the behaviour of power series are our justification for operating in the same way with power series as with polynomials. Obviously, two power series can be added or subtracted by adding or subtracting the corresponding coefficients (8.1.4). It is also clear that a power series, like any other convergent series, can be multiplied by a constant factor by multiplying each term by that factor. On the other hand, the multiplication and division of two power series requires a somewhat more detailed study, for which we refer the reader to the Appendix. We merely mention here without a proof that two power series $f(x) = \sum_{\nu=0}^{\infty} a_\nu x^\nu$ and $g(x) = \sum_{\nu=0}^{\infty} b_\nu x^\nu$ can be multiplied by each other like polynomials. In order to be specific, we have the theorem:

Throughout the common part of the interval of convergence of two power series, their product is given by the convergent power series $\sum_{\nu=0}^{\infty} c_\nu x^\nu$ with the coefficients

$$c_0 = a_0 b_0,$$
$$c_1 = a_0 b_1 + a_1 b_0,$$
$$c_2 = a_0 b_2 + a_1 b_1 + a_2 b_0,$$

....

$$c_n = a_0 b_n + a_1 b_{n-1} + \dots + a_n b_0,$$

....

Theorem of Uniqueness of Power Series: In the theory of power series, the following fact is important:

If both power series $\sum_{\nu=0}^{\infty} a_\nu x^\nu$ and $\sum_{\nu=0}^{\infty} b_\nu x^\nu$ converge in an interval which contains the point $x = 0$ inside and if in that interval the two series represent the same function $f(x)$, then they are identical, that is, the equation $a_n = b_n$ is true for every value of n.

In other words:

If at all, a function $f(x)$ can be represented by a power series in x in only one way,

Briefly speaking, the representation of a function by a power series is unique.

In order to prove this, we need only note that the difference of the two power series, i.e., the power series $\phi(x) = \sum_{\nu=0}^{\infty} c_\nu x^\nu$ with the coefficients $c_\nu = a_\nu - b_\nu$ represents the function $\phi(x) = f(x) - f(x) = 0$ in the interval, i.e., this last power series converges to the limit 0 everywhere in the interval. In particular, for $x = 0$, the sum of the series must be 0, i.e., $c_0 = 0$, so that $a_0 = b_0$. We now differentiate the series in the interior of the interval and obtain $\phi'(x) = \sum_{\nu=1}^{\infty} \nu c_\nu x^{\nu-1}$ However, $\phi'(x)$ is also 0 throughout the interval, whence, in particular, we have for $x = 0$: $c_1 = 0$ or $a_1 = b_1$. Continuing this process of differentiation and then setting $x = 0$, we find successively that all the coefficients c_n are equal to zero, which proves the theorem.

Moreover, we see that we can draw from the above discussion the conclusion:

If we take the ν-th derivative of a series $f(x) = \sum a_\nu x^\nu$ and then set $x = 0$, we at once obtain

$$a_\nu = \frac{1}{\nu!} f^{(\nu)}(0)$$

i.e., Every power series which converges for points other than $x = 0$ is the Taylor series of the function which it represents.

The uniqueness of the expansion is here expressed by the fact that the coefficients are uniquely determined by the function itself.

Expansion of Given Functions in Power Series

Within its interval of convergence, every power series represents a continuous function with continuous derivatives of all orders. We shall now discuss the converse problem of the expansion of a given function in a power series. In theory, we can always do this by means of of the n-th derivative and with the estimation of the remainder. But we can often reach our goal more simply by means of the device:

We first write down the statement $f(x) = \sum_{\nu=0}^{\infty} c_\nu x^\nu$, where, to begin with, the coefficients c_ν are unknown. Then we determine by some known property of the function $f(x)$ the coefficients and prove subsequently the convergence of the series. The series represents a function and there remains only to prove that this function is identical to $f(x)$. Due to the uniqueness of the expansion in power series, we know that no other series than the one just found can be the required expansion.

We shall now consider some examples of the use of this method. In fact, we have already obtained the series for artan x and log $(1+x)$ by a method, which forms part of the range of ideas. We simply integrated term by term the series for the derivatives of these functions, which we knew to be geometric series.

The Exponential Function: Our problem is to find a function for which $f'(x) = f(x)$ and $f(0) = 1$. If we write down the series with undetermined coefficients $f(x) = c_0 + c_1x + c_2x^2 + \ldots$, and differentiate it, we obtain $f'(x) = c_1 + 2c_2x + 3c_3x^2 + \ldots$.

Since, by assumption, these two power series must be identical, we have the equation $nc_n = c_{n-1}$, which is true for all values of $n \geq 1$. If we observe that, due to the relation $f(0)=1$, the coefficient c_0 must have the value 1, we can compute all the coefficients successively and obtain the power series $f(x) = 1 + \frac{x}{1!} + \frac{x^2}{2!} + \frac{x^3}{3!} + \ldots$

As is easily shown by the ratio test, this series converges for all values of x and therefore represents a function for which the relations $f'(x) = f(x), f(0) = 1$ are actually fulfilled.

Now, the function e^x certainly possesses these properties; we readily deduce that the function $\phi(x)$ is identical to e^x. In fact, if we form the quotient $f(x)=f(x)/e^x$ and differentiate it, we find

$$\varphi(\xi) = \frac{e^\alpha f'(x) - e^\alpha f'(x)}{e^{2x}} = 0$$

Hence, the function $\phi(x)$ is a constant and since it has the value 1 for $x = 0$, it must be identically equal to 1, thus proving that our power series and the exponential function are identical.

The Binomial Series: We can now return to the binomial series (6.3.3), this time using the method of undetermined coefficients. We wish to expand the function $f(x) = (1+x)^\alpha$ in a power series, whence we write $f(x) = (1 + x)^\alpha = c_0 + c_1x + c_2x^2 + \ldots$,

where the coefficients c_n are undetermined. We now note that obviously our function satisfies the relation $(1 + x) = f'(x) = \alpha f(x) = \sum_{\nu=0}^{\infty} \alpha c_\nu x^\nu$

On the other hand, if we differentiate the series for $f(x)$ term by term and multiply by $(1+x)$, we obtain $(1 + x) f'(x) = c_1 + (2c_2 + c_1)x + (3c_3 + 2c_2)x^2 + \ldots$; since these two power series for $f(x)$ must be identical, we have $\alpha c_0 = c_1$, $\alpha c_1 = 2c_2 + c_1$, $\alpha c_2 = 3c_3 + 2c_2$, ...

It is now certain that $c_0 = 1$, since our series must have the value 1 for $x = 0$; we thus find in succession the expressions

$$c_1 = \alpha,\ c_1 = \frac{(\alpha-1)\alpha}{2},\ c_1 = \frac{(\alpha-2)(\alpha-1)\alpha}{3\cdot 2},\ldots,$$

for the coefficients, and, in general, as is readily established,

$$c_\nu = \frac{(\alpha-\nu-1)(\alpha-\nu+2)\ldots(\alpha-1)\alpha}{\nu(\nu-1)\ldots 2.1} = \binom{\alpha}{\nu}$$

Substituting these values for the coefficients, we have the series $\sum_{\nu=0}^{\infty}\binom{\alpha}{\nu}x^\nu$; we must yet investigate the convergence of this series and show that it actually represents $(1 + x)^n$.

By the ratio test, we find that, when α is not a positive integer, the series converges if $|x| < 1$ and diverges if $|x| > 1$; in fact, then the ratio of the $(n + 1)$-th term to the n-th term is $\{(\alpha - n + 1)/n\}x$ and the absolute value of this expression tends to $|x|$ as n increases beyond all bounds.

We state here without proof the exact conditions under which this series converges. If the index a is an integer, the series terminates and is therefore valid for all values of x (as it becomes the ordinary binomial theorem). For all other values of α, the series is absolutely convergent for $|x|<1$ and diverges for $|x|>1$. For $x = +1$, the series converges absolutely if $\alpha > 0$, conditionally if $-1 < \alpha < 0$, and it diverges if $\alpha = -1$. Finally, at $x = -1$, the series is absolutely convergent if $a > 0$, divergent if $a < 0$.

Hence, if $|x| < 1$, our series represents a function $f(x)$ which satisfies the condition $(1 + x)f'(x) = af(x)$, as follows from the method of forming the coefficients. Moreover, $f(0) = 1$. But these two conditions ensure that the function $f(x)$ is identical to $(1 + x)^\alpha$. In fact, setting

$$\phi(x) = f(x)/(1 + x)^\alpha,$$

we find that

$$\phi(x) = \frac{(1+x)^\alpha f'(x) - \alpha(1+x)^{\alpha-1} f(x)}{(1+x)^{2\alpha}}$$

hence $f(x)$ is a constant and is always equal to 1, since $\phi(0) = 1$. We have thus proved that for $|x| < 1$

$$(1 + x)^\alpha = \sum_{\nu=0}^{\infty}\binom{\alpha}{\nu}x^\nu$$

which is the binomial series.

We present here the following special cases of the binomial series:

The geometric series: $\frac{1}{1+x} = (1+x)^{-1} = 1 - x + x^2 - x^3 + x^4 - + \ldots = \sum_{\nu=0}^{\infty}(-1)^\nu x^\nu$; the series

$\frac{1}{(1+x)^2} = (1+x)^{-2} = 1 - 2x + 3x^2 - 4x^3 + - ... = \sum_{\nu=0}^{\infty}(-1)^{\nu}(\nu+1)x^{\nu}$, which may otherwise be obtained from the geometric series by differentiation and the

$$\sqrt{(1+x)} = (1+x)^{\frac{1}{2}} = 1+\frac{1}{2}x-\frac{1}{2.4}x^2+\frac{1.3}{2.4.6}x^3-\frac{1.3.5}{2.4.6.8}x^4+-...,$$

$$\frac{1}{\sqrt{(1+x)}} = (1+x)^{-\frac{1}{2}} = 1+\frac{1}{2}x+\frac{1.3}{2.4}x^2-\frac{1.3.5}{2.4.6}x^3\frac{1.3.5.7}{2.4.6.8}x^4-+...,$$

series the first two or three terms of which are useful expressions.

The Series for arsin x: This series can be established readily by expansion of the expression $1/\sqrt{(1-t^2)}$ as a binomial series

$$\left(1-t^2\right)^{-\frac{1}{2}} = 1+\frac{1}{2}t^2+\frac{1.3}{2.4}t^4+...$$

This series converges if $|t| \rangle 1$, whence it converges uniformly for $|t| < q < 1$. On integrating term by term between 0 and x, we obtain

$$\text{arc sin}x = x+\frac{1}{2}\frac{x^3}{3}+\frac{1.3}{2.4}\frac{x^5}{5}+...;$$

by the ratio test, we find that this series converges if $|x| < 1$ and diverges if $|x| > 1$.

The derivation of this series by Taylor's theorem would be decidedly less convenient owing to the difficulty of estimating the remainder.

The Series for arsinh $x = \log\{x + (1+x)\}$: We obtain this expansion by a similar method. By the binomial theorem, we write down the series for the derivative of arsinh x:

$$\frac{1}{\sqrt{1+x^2}} = 1-\frac{1}{2}x^2+\frac{1.3}{2.4}x^4-\frac{1.3.5}{2.4.6}x^6+-...,$$

and then integrate term by term to obtain the expansion

$$\text{arc sinh}x = x-\frac{1}{2}\frac{x^3}{3}+\frac{1.3}{2.4}\frac{x^5}{5}-+...;$$

with the interval of convergence $-1 < x < 1$.

Example of Multiplication of Series: The expansion of the function $\frac{\log(1+x)}{1+x}$ is a simple example of the application of the rule for the multiplication of power series. We have only to multiply the logarithmic series $\log(1+x) = x - \frac{x^2}{2} + \frac{x^3}{3} + \frac{x^4}{4} + - ...$ by the geometric series $\frac{1}{1+x} = 1 - x + x^2 - x^3 + x^4 - + ...$; the reader should verify that this yields the remarkable expansion

$$\frac{\log(1+x)}{1+x} = x - \left(1+\frac{1}{2}\right)x^2 + \left(1+\frac{1}{2}+\frac{1}{3}\right)x^3 - \left(1+\frac{1}{2}+\frac{1}{3}+\frac{1}{4}\right)x^4 + - \ldots \text{ for } |x| < 1.$$

Example of Term-by-Term Integration (Elliptic Integral):. In previous applications, we have encountered the elliptic integral

$$K = \int_0^{\pi/2} \frac{d\varphi}{\sqrt{(1-k^2\sin^2\varphi)}} (k^2 < 1)$$

(the period of oscillation of a pendulum . In order to evaluate the integral, we can first expand the integrand by the binomial theorem:

$$\frac{1}{\sqrt{(1-k^2\sin^2\varphi)}} = 1 + \frac{1}{2}k^2\sin^2\varphi + \frac{1.3}{2.4}k^4\sin^4\varphi + \frac{1.3.5}{2.4.6}k^6\sin^6\varphi + \ldots,$$

Since k } sin } φ is never larger than k} , this series converges uniformly for all values of φ and we may integrate term by term:

$$K = \int_0^{\pi/2} \frac{d\varphi}{\sqrt{1-k^2\sin^2\varphi}} = \int_0^{\pi/2} d\varphi + \frac{1}{2}k^2 \int_0^{\pi/2} \sin^2\varphi\, d\varphi + \frac{1.3}{2.4}k^4 \int_0^{\pi/2} \sin^4\varphi\, d\varphi + \ldots$$

The integrals here have already been calculated in 4.4.4. Substitution of their values yields

$$K = \int_0^{\pi/2} \frac{d\varphi}{\sqrt{1-k^2\sin^2\varphi}} = \frac{\pi}{2}\left\{1 + \left(\frac{1}{2}\right)^2 k^2 + \left(\frac{1.3}{2.4}\right)^2 k^4 + \left(\frac{1.3.5}{2.4.6}\right)k^6 + \ldots\right\}$$

More examples on the theory of series are given in A8.1.

Power Series with Complex Terms

Introduction of Complex Terms into Power Series: The similarity between certain power series representing functions which are apparently unrelated led Euler to set up a purely formal link between them, found by giving complex values, in particular, pure imaginary values to the variable x. We shall first do this formally, unhindered by questions of rigour and investigate the results of the process.

The first striking relation of this sort is obtained, if we replace the quantity x in the series for e^x by a pure imaginary $i\phi$, where ϕ is a real number. If we recall the fundamental equation for the imaginary unit i, that is, $i = -1$, from which it follows that $i^3 = -i$, $i^4 = 1$, $i^5 = i$, ..., then, on separating the real and imaginary terms of the series, we obtain

$$e^{1\phi} = \left(1 - \frac{\phi^2}{2!} + \frac{\phi^4}{4!} + \frac{\phi^6}{6!} + - \ldots\right) + i\left(\phi - \frac{\phi^3}{3!} + \frac{\phi^5}{5!} - \frac{\phi^7}{7!} + - \ldots\right)$$

or, in another form, $e^{1\phi} = \cos\phi + i\sin\phi$

This is the well-known and important Euler formula. which is as yet purely formal. It is consistent with De Moivre's theorem which is expressed by the equation $(\cos\phi + i\sin\phi)(\cos\psi + i\sin\psi) = (\cos\phi+\psi) + i\sin(\phi+\psi)$.

By virtue of Euler's formula, this equation merely states that the relation $e^x . e^v = e^{x+v}$ continues to hold for pure imaginary values $x = i\phi, y = i\psi$.

If we replace the variable x in the power series for cos x by the pure imaginary ix, we obtain at once the series for cosh x; this relation can be expressed by the equation

$$\cosh x = \cos ix.$$

In the same way, we obtain

$$\sinh x = \frac{1}{i}\sinh ix.$$

Since Euler's formula also yields

$$e^{-if} = \cos\phi - i\sin\phi,$$

we arrive at the exponential expressions for the trigonometric functions

$$\sin x = \frac{e^{ix} - e^{-ix}}{2i}, \qquad \cos x = \frac{e^{ix} + e^{-ix}}{2}$$

These are exactly analogous to the exponential expressions for the hyperbolic functions and are, in fact, transformed into them by the relations

$$\cosh x = \cos ix, \qquad \sinh x = \frac{1}{i}\sin ix$$

Naturally, corresponding formal relations can also be obtained for the functions tan x, tanh x, cot x and coth x, which are interrelated by the equations

$$\tanh x = \frac{1}{i}\tan ix \qquad \coth x = i\cot ix$$

Finally, similar relations can also be found for the inverse trigonometric and hyperbolic functions. For example, we immediately find from

$$y = \tan x = \frac{e^{ix} - e^{-ix}}{i\left(e^{ix} + e^{-ix}\right)} = \frac{e^{2ix} - 1}{i\left(e^{2ix} + 1\right)}$$

that

$$e^{2ix} = \frac{1+iy}{1-iy}$$

It we take the logarithms of both sides of this equation and then write x instead of y and artan x instead of x, we obtain the equation

$$\operatorname{artan} x = \frac{1}{2i}\log\frac{1+ix}{1-ix},$$

which expresses a remarkable connection between the inverse tangent and the logarithm. If we replace in the known power series for $\log(1+x)/(1-x)$ (6.1.1) x by ix, we actually obtain the power series for artan x:

$$\text{artan}x = \frac{1}{i}\left(ix+\frac{(ix)^3}{3}+\frac{(ix)^5}{5}+\ldots\right)=x-\frac{x^3}{3}+\frac{x^5}{5}-+\ldots$$

The above relations have as yet a purely formal character and naturally demand a more exact statement as to the meaning which they are intended to convey. In the next sub-section, we shall indicate how this can be given with the help of function theory.

However, we shall later on only require Euler's formula $e^{i\phi}=\cos\phi+i\sin\phi$ This being so, we can avoid here a thorough analysis. We need only regard the symbol $e^{i\phi}$ as a formal abbreviation $\cos\phi+i\sin\phi$, for in which case De Moivre's formula $e^{i\phi}\cdot e^{i\psi}=e^{i(\phi+\psi)}$ appears merely as a consequence of the elementary addition theorem of trigonometry. From this formal point of view, in order to make the relation $e^x\cdot e^v=e^{x+y}$ e*. e* remain valid for any complex arguments, we set up the further definition

$$e^x = e^{\xi}(\cos\eta+i\,\text{xin}\eta)$$

where $x=\xi+i\eta$ (ξ,η real).

A Glance at the General Theory of functions of a Complex Variable: Although the purely formal point of view, indicated above, is by itself free from objection, it is still desirable to recognize in the above formula something more than a mere formal connection. In order to pursue this aim, we are led to the general theory of functions, as (for the sake of brevity) we call the general theory of the so-called analytic functions of a complex variable. In this, we may use as our starting point a general discussion of the theory of power series with complex variables and complex coefficients. The construction of such a theory of power series presents no difficulties once we have defined the concept of limit in the domain of complex numbers; in fact, it follows almost exactly the theory of real power series . But as we shall not make any use of these matters in what follows, we shall content ourselves here by stating certain facts and omit the proofs. It is found that the following generalization of the theorem of 8.5.1 holds for complex power series:

If a power series converges for any complex value $x=\xi$ whatever, then it converges absolutely for every value x for which $|x|<|\xi|$; if it diverges for a value $x=x$, then it diverges for every value x for which $|x|>|\xi|$. A power series, which does not converge everywhere, but does converge for some other point in addition to $x=0$, has a circle of convergence, that is, there exists a number $\rho>0$ such that the series converges absolutely for $|x|<\rho$and diverges for $|x|>\rho$.

Having established the concept of functions of a complex variable, represented by power series, and having developed the rules for operating with such functions, we can think of the functions e^x, $\sin x$, $\cos x$, artan x, etc., of the complex variable x as simply defined by the power series which represent them for real values of x. Then, all the above formal relationships reduce to trivialities.

We shall merely indicate by two examples, how this introduction of complex variables helps us to understand the elementary functions. The geometric series for l/(l + xýÿ) ceases to converge when x leaves the interval -1 } x } l and so does the series for artan x, although there are no peculiarities in the behaviour of these functions at the ends of the interval of convergence; in fact, they and all their derivatives are continuous for all real values of x. On the other hand, we can readily understand that the series for l/(l – x) and log (1 – x) cease to converge as x passes through the value 1, since they become infinite there. But this divergence of the series for the inverse tangent and the series $\sum_{\nu=0}^{\infty}(-1)^{\nu} x^{2\nu}$ $for |x|>1$ becomes clear immediately, if we also consider complex values of x. In fact, we find that when $x = i$, the sum functions become infinite and so cannot be represented by a convergent series. Hence, by our theorem about the circle of convergence, the series must diverge for all values of x such that $|x| > |i|$; in particular, for real values of x, the series diverge outside the interval -1 } x { 1.

Another example is given by the function $f(x) = e^{-1/x2}$ for x ψ0, which, in spite of its apparently regular behaviour, cannot be expanded in a Taylor series. As a matter of fact, this function ceases to be continuous, if we take pure imaginary values of $x = i\xi$ into account. The function then takes the form e^{1/ξ^2} and increases beyond all bounds as ξ } 0. It is therefore clear that no power series in x can represent this function for all complex values of x in a neighbourhood of the origin, no matter how small a neighbourhood we choose.

These remarks on the theory of functions and power series of a complex variable must be sufficient for us here.

Chapter 9

Fourier Series

Besides the power series, there is another class of infinite series which has a particularly important role both in pure and applied mathematics. These are the Fourier series, in which the individual terms are trigonometric functions and the sum is a periodic function.

Periodic Functions

General Remarks: Periodic functions of the time, i.e., functions which repeat their course after a definite interval of time, are encountered in many .applications. In most machines, a periodic process occurs in rhythm with the rotation of a flywheel, e.g., the alternating current developed by a dynamo. Periodic functions are also associated with all vibration phenomena.

A periodic function with period $2l$ is represented by the equation $f(x + 2l) = f(x)$, true for all values of x. We call especially attention to the fact that $2l$ is called the period.

In representing periodic functions, it is often convenient to denote the independent variable x by a point on the circumference of a circle instead of the usual point on a straight line. If a function $f(x)$ has the period $2p$, i.e., if the equation $f(x + 2\pi) = f(x)$ holds for all values of x and we denote by x the angle at the centre of a circle of unit radius, which is included between an arbitrary initial radius and the radius to a variable point on the circumference, then the periodicity of the function $f(x)$ is expressed simply by the fact that there corresponds to each point on the circumference just one value of the function. For example, in the case of a machine, periodicity may be expressed in terms of the position of a point on the flywheel.

It is worth noting that in addition to the period $2l$, the function $f\{x)$ necessarily has also the period $4l$, since $f(x+ 4l) =f(x+ 2l) = f(x)$; similarly, $f(x)$ has the periods $6l$, $8l$, ...; and it is also possible (though not necessarily true) that $f(x)$ may have shorter periods such as l or $l/5$. Graphically speaking, the graph of the function has in any two consecutive intervals of length $2l$ exactly the same shape. In order to have available a second interpretation, which some readers may prefer, we may think of the variable x as the time and write, accordingly, sometimes t instead of x; the function $f(x)$ then represents a periodic process or, as we shall also say, a vibration (or oscillation). The period $2l = T$ is then called the period of vibration (or period of oscillation).

If any arbitrary function $f(x)$ is given in a definite interval, say. $-l$ ýÿ x ýÿ l, it can always be extended as a periodic function; we have only to define $f(x)$ outside the interval by the equation $f(x + 2nl) = f(x)$, where n is an arbitrary positive or negative integer. We must point out here that, if $f(x)$ is continuous in the interval $-l$ } x } l, but $f(-l)$ } $f(l)$, our extended periodic function will be discontinuous at the points l, $3l$ ..., where $l = p$. Moreover, in this case, the extension fails to yield a single-valued function $f(x)$ at the points $x = l$, $3l$, since, for example, we have defined $f(3l)$ as $f(l+2l)$, which gives $f(3l) = f(l)$, and we also have defined it as /(-l+ 4l), which yields $f(3l) = f(-l)$. We avoid this difficulty by not extending the function as defined for $-l$ ýÿ x ýÿ l, but as defined either for $-l < x < l$ or $-l < x < l$, i.e., we either discard the original value $f(-l)$ or the original value $f(+l)$.

We should point out here a general fact,, which relates to periodic functions and is expressed by the equation

$$\int_{-l-a}^{l-a} f(x)dx = \int_{-l}^{l} f(x)dx$$

or, in words: The integral of a periodic function over an interval, the length of which is one period $T == 2l$, has always the same value, no matter where the interval lies.

In order to prove this, we need only note that, by virtue of the equation $f(\xi - 2l) = f(\xi)$, the substitution $x = \xi - 2l$ yields

$$\int_{\zeta}^{\beta} f(x)\,dx = \int_{\alpha+2l}^{\beta+2l} f(\xi)\,d\xi = \int_{\alpha+2l}^{\beta+2l} f(x)\,dx$$

In particular, for $\alpha = -l - \alpha$ and $\beta = -l$, it follows that

$$\int_{-l-a}^{-l} f(x)\,dx = \int_{l-a}^{-l} f(x)\,dx,$$

whence

$$\int_{-l-a}^{-l} f(x)\,dx = \int_{l-a}^{-l} f(x)\,dx + \int_{-l}^{l-a} f(x)\,dx$$

$$= \int_{l-a}^{-l} f(x)\,dx + \int_{-l}^{l-a} f(x)\,dx = \int_{-l}^{l} f(x)dx,$$

which proves our statement. If we recall the geometrical meaning of the integral, the statement is made obvious by figure.

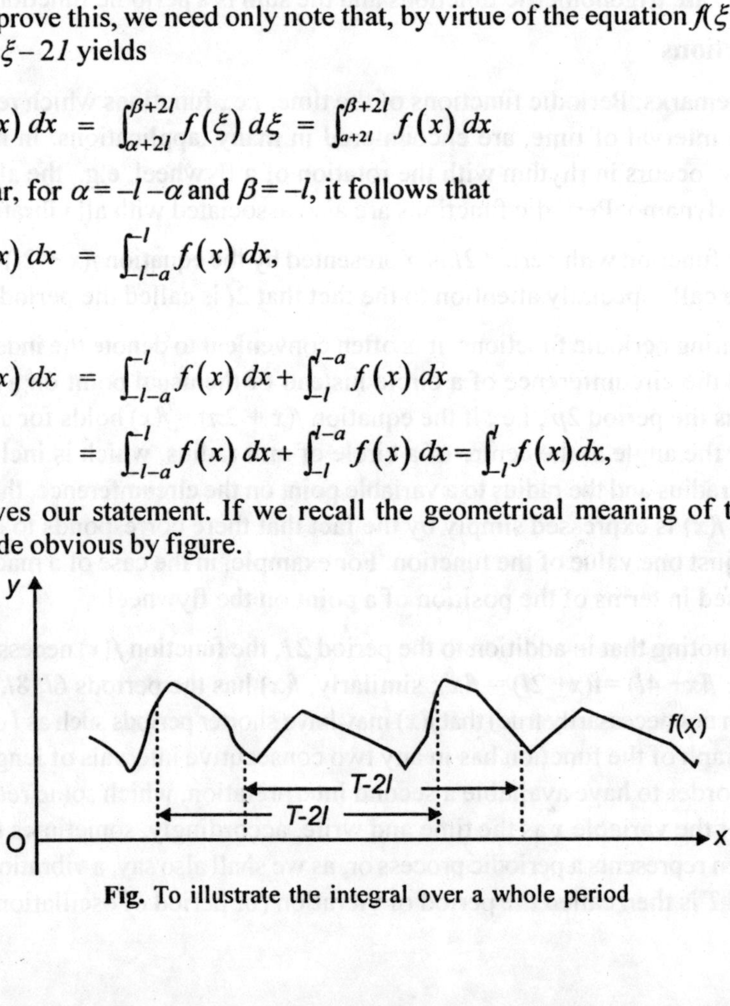

Fig. To illustrate the integral over a whole period

The simplest periodic functions, from which we shall later compose the most general periodic functions, are a sinωx and a cosωx, or, more generally, $a\sin\omega(x-\xi)$ and $a\cos w(x-\xi)$, where a(0), ω(0) and x are constants. We call the processes represented by these functions* sinusoidal vibrations or simple harmonic vibrations (or oscillations). The period of vibration is $T=2\pi/\omega$. The number ω is called the circular frequency of the vibration**; since $1/T$ is the number of vibrations in unit time or frequency, ω is the number of vibrations in time 2π. The number a is called the amplitude of the vibration; it represents the maximum value of the function $a\sin\omega(x-\xi)$ or $a\cos w(x-\xi)$, since both sine and cosine have the maximum value 1. The number $\omega(x-\xi)$ is called the phase and the number wx the phase displacement.

We obtain these functions graphically by stretching the sine curve in the ratios 1 : ω along the x-axis and a : 1 along the y-axis and then translating the curve by a distance ξ along the x-axis in the positive direction.

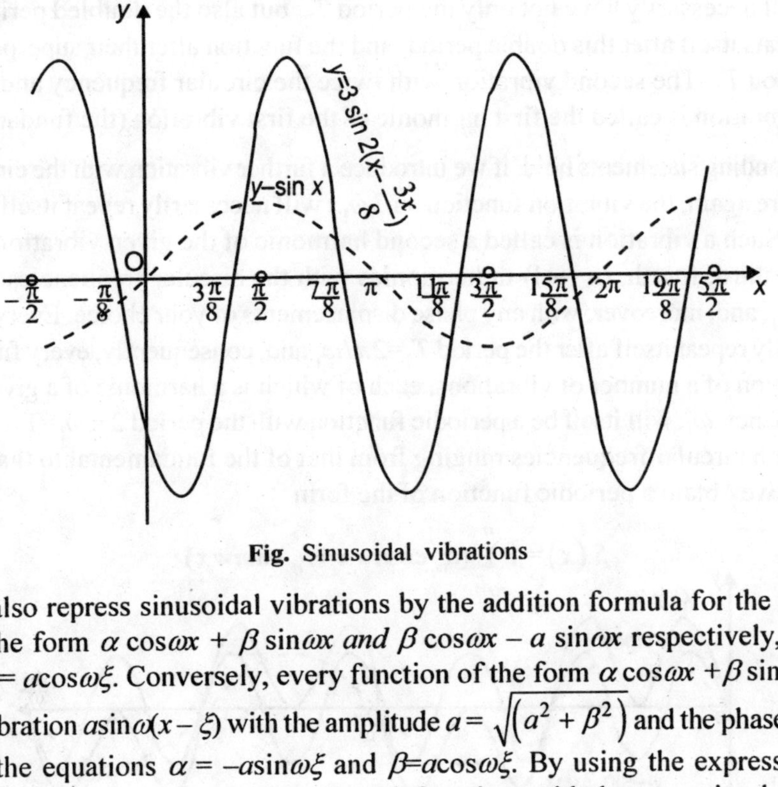

Fig. Sinusoidal vibrations

We can also repress sinusoidal vibrations by the addition formula for the trigonometric functions in the form $\alpha\cos\omega x+\beta\sin\omega x$ *and* $\beta\cos\omega x-a\sin\omega x$ respectively, where $\alpha=-a\sin\omega\xi$ and $\beta=a\cos\omega\xi$. Conversely, every function of the form $\alpha\cos\omega x+\beta\sin\omega x$ represents a sinusoidal vibration $a\sin\omega(x-\xi)$ with the amplitude $a=\sqrt{(a^2+\beta^2)}$ and the phase displacement $\omega\xi$ given by the equations $\alpha=-a\sin\omega\xi$ and $\beta=a\cos\omega\xi$. By using the expression $a\sin wx+b\cos wx$, we see that the sum of two or more such functions with the same circular frequency w always represent another sinusoidal vibration with the circular frequency w.

Superposition of Sinusoidal Vibrations. Harmonics. Beats:. Although many vibrations are found to be sinusoidal (5.4.4), it is nevertheless true that most periodic motions have a more complicated character, being obtained by superposition of sinusoidal vibrations. Mathematically speaking, this simply means that the motion, e.g., the distance of a point from its initial position as a function of the time, is given by a function which is the sum of a number of pure periodic functions of the above type. The sine waves of the function are then piled up on top of one another (that is, their ordinates are added), or, as we say, they are superposed. In this superposition, we assume that all the circular frequencies (and, of course, the periods) of the superposed vibrations differ, because the superposition of two sinusoidal vibrations with the same circular frequency gives us another sinusoidal vibration with the same circular frequency (but with a different amplitude and phase displacement), as shown above.

If we consider the simplest case, the superposition of two sinusoidal vibrations with the circular frequencies ω_1 and ω_2, we find that there are two fundamentally different cases, depending on whether the two circular frequencies have a rational ratio or not, or, as we say, whether they are commensurable or incommensurable. We begin with the first case and, by way of an example, take the second circular frequency to be twice the first: $\omega_2=2\omega_1$. The period of the second vibration will then be half the period of the first vibration, $2\pi/2\omega_1 = T_2 = T_1/2$, whence it will necessarily have not only the period T_2, but also the doubled period T_1, since the function repeats itself after this double period; and the function after their superposition will also have the period T_1. The second vibration with twice the circular frequency and half the period of the first vibration is called the first harmonic of the first vibration (the fundamental).

Corresponding statements hold, if we introduce a further vibration with the circular frequency $\omega_3 = 3\omega_1$. Here again, the vibration function $\sin 3\omega_1 x$ will necessarily repeat itself with the period $2\pi/\omega_1 = T_1$. Such a vibration is called a second harmonic of the given vibration. Similarly, we can consider third, fourth, (n — l)-th harmonics with the circular frequencies $\omega_4 = 4\omega_1$, $\omega_5 = 5\omega_1$, $\omega_n = nw_1$, and, moreover, with any phase displacements of your choice. Every such harmonic will necessarily repeat itself after the period $T_1=2\pi/\omega_1$ and, consequently, every function obtained by superposition of a number of vibrations, each of which is a harmonic of a given fundamental circular frequency ω_1, will itself be a periodic function with the period $2\pi/\omega_1=T_1$. By superposing vibrations with circular frequencies ranging from that of the fundamental to that of the (n – 1)-th harmonic, we obtain a periodic function of the form

$$S(x)= a\sum_{n=1}^{n} (a_n \cos nwx+b_n \sin nwx)$$

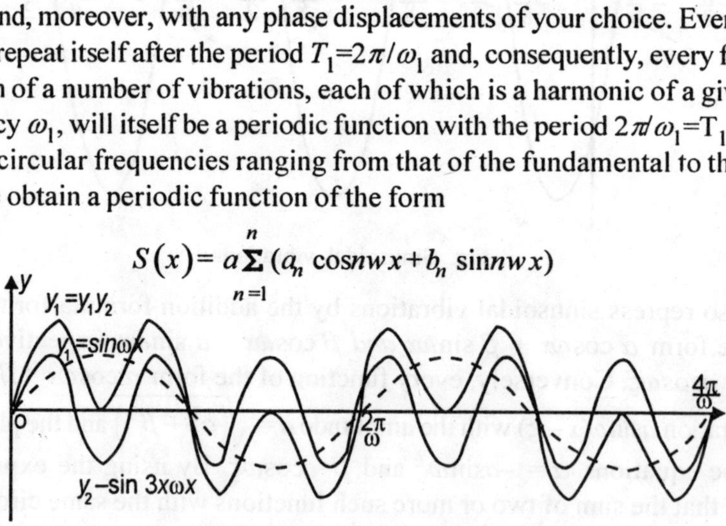

Fig. Combination of vibrations

The proportions of figure correspond to the assumption $w=1$.

Since this function contains $2n + 1$ constants, which we can choose arbitrarily, we are thus able to generate very complicated curves which are not at all like the original curves. The term harmonic originated in acoustics, where we find that there corresponds to a fundamental vibration with circular frequency a note of a certain pitch, then the first, second, third, etc., harmonics correspond to the sequence of harmonics of the fundamental, i.e., to the octave, octave + fifth, double octave, etc. (In acoustics, also the terms overtone, (upper) partial are used.

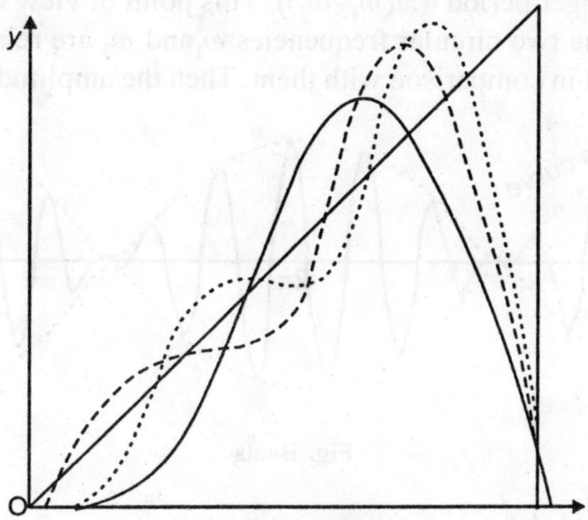

———— $\sin x - \frac{\sin 2x}{2}$

- - - - - - - $\sin x - \frac{\sin 2x}{2} + \frac{\sin 3x}{3}$

··········· $\sin x - \frac{\sin 2x}{2} + \frac{\sin 3x}{3} - \frac{\sin 4x}{4}$

Fig. Combination of vibrations

In general, in the case of superposition of vibrations in which the circular frequencies have rational ratios, these circular frequencies can all be represented as integral multiples of a common fundamental circular frequency. However, the superposition of two vibrations with incommensurable circular frequencies ω_1 and ω_2 represents an intrinsically different type of phenomenon. In that case, the process resulting from the superposition of sinusoidal vibrations will no longer be periodic. We cannot go here into the mathematical discussions that arise from this, but will merely remark that such functions always have an approximately periodic character or, as we say, are almost periodic. In 1935, such functions had just been studied in great detail.

A final remark on the superposition of sinusoidai vibrations concerns the phenomenon of beats. If we superpose two vibrations, both with unit amplitude but different circular frequencies ω_1 and ω_2, and, for the sake of simplicity, take the same value of ξ for both (the generalization to arbitrary phase can be left to the reader), then we are merely concerned with the behaviour of the function

$$y = \sin\omega_1 x + \sin\omega_2 x \ (\omega_2 > \omega_1 > 0)$$

A well-known trigonometrical formula yields

$$y = 2\cos\frac{1}{2}(\omega_1 - \omega_2)x \sin\frac{1}{2}(\omega_1 + \omega_2)x$$

This equation represents a phenomenon which we may think of as follows: We have a vibration with the circular frequency $(\omega_1 + \omega_2)$ and the period $4\pi/(\omega_1+\omega_2)$. However, this vibration does not have a constant amplitude; on the contrary, its amplitude is given by $2\cos(\omega_1 + \omega_2)x$, which varies with the longer period $4\pi/(\omega_1-\omega_2)$). This point of view is particularly useful and easy to interpret when the two circular frequencies $\omega_1$and ω_2 are relatively large, while their difference $\omega_1-\omega_2$ is small in comparison with them. Then the amplitude

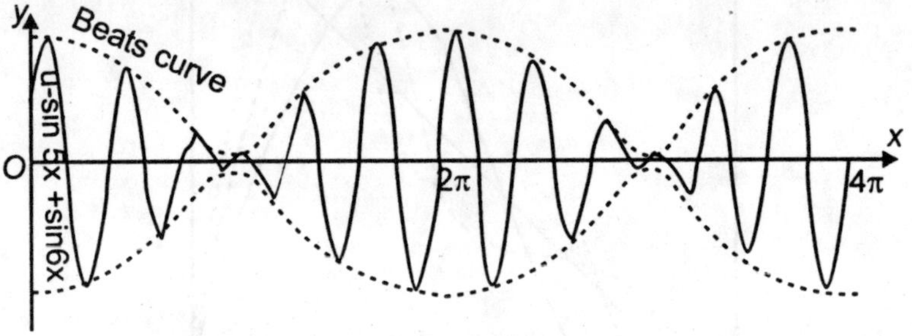

Fig. Beats

of the vibration $2\cos\frac{1}{2}(\omega_1 - \omega_2)x$ with period $4\pi/(\omega_1 - \omega_2)$ will vary only slowly compared with the period of vibration and this change of amplitude will repeat itself periodically with the long period $4\pi/(\omega_1 - \omega_2)$. These rhythmic changes of amplitude are called beats. Ev'eryone is acquainted with this phenomenon in acoustics, and perhaps also in wireless telegraphy, in which the circular frequencies $\omega_1$and ω_2 are, as a rule, far above those which the ear can detect, while the difference $\omega_1- \omega_2$ falls in the range of audible notes. The beats can then be heard, while the original vibrations remain imperceptible to the ear.

Use of complex notation

General Remarks: An investigation of vibration phenomena and periodic functions gains in formal simplicity if we employ complex numbers, combining each pair of trigonometric functions $\cos w\, x$ and $\sin w\, x$ to form an expression of the type $\cos w\, x + i\sin \omega\, x = e^{i\omega x}$(8.7.1). We must here bear in mind that one equation between complex quantities is equivalent to two equations between real quantities and that our results must always be interpreted and made intelligible in the real domain.

If we replace everywhere the trigonometric functions by exponential functions in accordance with the formulae

$$2\cos\theta = e^{i\theta} + e^{i\theta},$$

$$2i\sin\theta = e^{i\theta} - e^{i\theta}$$

we express sinusoidal vibrations in terms of the complex quantities $e^{i\omega x}$, $e^{-i\omega x}$ or $ae^{i\omega(x-\xi)}$, $ae^{-i\omega(x-\xi)}$, respectively, where a, ω and $\omega\xi$ are the real quantities: Amplitude, circular frequency and phase displacement. The real vibrations are obtained from this complex expression simply by taking real and imaginary parts.

The convenience of this mode of representation for many purposes is due to the fact that the derivatives of the real vibrations with respect to the time are obtained by differentiating the complex exponential function just as if it were a real constant, as is expressed by the formula

$$\frac{d}{dx}a\{\cos\omega(x-\xi)+i\sin\omega(x-\xi)\}$$

$$= a\omega\{-\sin\omega(x-\xi)+i\cos\omega(x-\xi)\}$$

$$= ia\omega\{\cos\omega(x-\xi)+i\sin\omega(x-\xi)\},$$

or

$$\frac{d}{dx}ae^{i\omega(x-\xi)} = ia\omega e^{i\omega(x-\xi)}$$

Application to the Study of Alternating Currents: We shall now illustrate these matters by means of an important example and denote here the independent variable, the time, by t instead of x.

We consider an electric circuit with resistance R and inductance L, on which an external electromotive force (voltage) E is impressed. In the case of a direct current, E is constant and the current I is given by Ohm's law $E = RI$.

However, if we are dealing with an alternating current, E is a function of the time t, and consequently so is I; Ohm's law then becomes

$$E - L\frac{dI}{dt} = RI$$

In the simplest case, to which we will restrict ourselves here, the external electromotive force E is sinusoidal with circular frequency ω. Now, instead of taking this oscillation in the form $a\cos t\, t$ or $a\sin w\, t$, we combine both possibilities formally in the complex form

$$E = \varepsilon e^{i\omega t} = \varepsilon\cos\omega t + i\varepsilon\sin\omega t,$$

where $\varepsilon\,(>0)$ represents the amplitude. We shall operate with this complex voltage as if it were a real parameter and thus obtain a complex current I. Then, the significance of the relation thus found between the complex qnantities E and I is that the current corresponding to an electromotive force $\varepsilon\cos\omega\, t$ is the real part of I, while the current corresponding to an electromotive force $\varepsilon\sin\omega t$ is the imaginary part of I. The complex current can be calculated immediately, if we write down for I an expression of the form $I = \alpha e^{i\omega t} = \alpha\,(\cos\omega t + i\sin\omega t)$, i.e., if we assume that I is also sinusoidal with circular frequency ω. The derivative of I is then given formally by the expression

$$\frac{dI}{dt} = i\alpha\omega e^{i\omega t} = \alpha\omega(-\sin\omega t + i\cos\omega t)$$

On substitution of these quantities into the generalized form of Ohm's law and dividing by the factor $e^{i\omega t}$, we obtain the equation $\varepsilon - \alpha Li c\omega = R\alpha$, or

$$\alpha = \frac{\varepsilon}{R + i\omega L},$$

so that $E = (R + i\omega L)I = WI$.

We may regard this last equation as Ohm's law for alternating currents in complex form, if we call the quantity $W = R + i\omega L$ the complex resistance of the circuit. Ohm's law is then the same as for direct current: The current is equal to the voltage divided by the resistance.

If we write the complex resistance in the form

$$W = \omega e^{i\delta} = \omega \cos\delta + i\omega \sin\delta,$$

where

$$\omega = \sqrt{(R^2 + L^2\omega^2)}, \quad \tan\delta = \frac{\omega L}{R}$$

we obtain $I = \dfrac{\varepsilon}{\omega} e^{i(\omega t - \delta)}$,

According to this formula, the current has the same period (and circular frequency) as the voltage; the amplitude a of the current is connected with the amplitude e of the electromotive force by the equation

$$a = \frac{\varepsilon}{\omega},$$

and, in addition, there is a phase difference between the current and the voltage. The current does not reach its maximum at the same time as the voltage, but at a time δ/ω later and the same is of course true for the minimum. In electrical engineering, the quantity $\omega = \sqrt{(R^2 + L^2\omega^2)}$ is frequently called the impedance or the alternating current resistance of the circuit for the circular frequency ω; the phase displacement, usually stated in degrees, is called the lag.

Complex Representation of the Superposition of Sinusoidal Vibrations:. So far, the complex notation has been used to denote the combination of two sinusoidal vibrations. But a single vibration or a compound vibration of the type

$$S(x) = a + \sum_{\nu=1}^{n} (a_\nu \cos\nu x + b_\nu \sin\nu x)$$

(for the sake of simplicity, we have taken $\omega = 1$) can also be reduced to complex form by the substitution

$$\cos\nu x = \frac{1}{2}\left(e^{i\nu x} + e^{-i\nu x}\right) \quad \sin\nu x = \frac{1}{2}\left(e^{i\nu x} - e^{-i\nu x}\right)$$

The above expression then assumes the form

$$S(x) = \sum_{\nu=-\eta}^{n} a_\nu e^{i\nu x},$$

where the complex numbers a_ν are linked to the real numbers a, a_ν and b_ν by the equations

$$a_\nu = a_\nu + a_{-\nu}, \; a = a_0, \; b_\nu = i(a_\nu - a_{-\nu})$$

In order that the equation $a_\nu = a_\nu + a_{-\nu}$ shall formally include the case $n = 0$, we often put $a = a_0 = a_0/2$.

Conversely, we may regard any arbitrary expression of the form $\sum_{\nu=-n}^{n} a_\nu e^{i\nu n}$ as a function which represents the superposition of vibrations written in complex form. In order that the result of this superposition may be real, it is only necessary that $a_\nu + a_{-\nu}$ should be real and $a_\nu - a_{-\nu}$ pure imaginary, i.e., that a_ν and $a_{-\nu}$ are conjugate complex numbers.

Deduction of a trigonometric Formula: By using complex notation, we obtain a very simple proof of a formula which we shall require below. This is the trigonometric summation formula

$$\alpha_n(\alpha) = \frac{1}{2} + \cos\alpha + \cos 2\alpha + \ldots + \cos 2n\alpha = \frac{\sin\left(n + \frac{1}{2}\right)\alpha}{2\sin\frac{1}{2}\alpha}$$

which holds for all values of a except the values 0, $2p$, $4p$, ... In order to prove this, we replace the cosine function by its exponential expression and thus write the sum $\sigma_n(\alpha)$ in the form

$$\sigma_n(\alpha) = \frac{1}{2}\sum_{\nu=-n}^{n} e^{i\nu a}$$

On the right hand side, we have a geometric progression with the common ratio $q = e^{i\alpha} \neq 1$. Using the ordinary formula for the sum, we have

$$\sigma_n(\alpha) = \frac{1}{2} e^{-in\alpha} \frac{1 - q^{2n+1}}{1 - q} = \frac{1}{2} \frac{e^{-ina} - e^{(n+1)i\alpha}}{1 - e^{i\alpha}}$$

On multiplying the numerator and denominator by $e^{-ia/2}$, we obtain

$$\sigma_n(\alpha) = \frac{\sin\left(n + \frac{1}{2}\right)\alpha}{2\sin\frac{1}{2}\alpha}$$

as has been stated above.

Fourier Series

The function

$$S(x) = a + \sum_{\nu=1}^{n} (a_\nu \cos\nu x + b_\nu \sin\nu x),$$

resulting from the superposition of sinusoidal vibrations, contains $2n + 1$ arbitrary constants a, a_ν, b_ν. There arises now the question whether these constants can be chosen such that in the interval $-\pi < x < +p$ the sum $S(x)$ shall approximate to a given function $f(x)$, and if so, how they are to be found. More precisely, we ask whether the given function $f(x\}$ can be expanded in an infinite series

$$f(x) = a+\sum_{\nu=1}^{\infty}(a_\nu \cos\nu x + b_\nu \sin\nu x),$$

If we assume for the moment that this expansion of the function $f(x)$ is actually possible and that the series converges uniformly in the interval $-\pi < x < +\pi$, we readily obtain a simple relation between the function $f(x)$ and the coefficients $a = a_0$, a_ν and b_ν. We multiply the above hypothetical expansion by $\cos nx$ and integrate term by term, which is permissible on account of its uniform convergence. By virtue of the orthogonality relations

$$\int_{-\pi}^{+\pi} \sin mx \sin nx\, dx = \begin{cases} 0, & \text{if } m \neq n, \\ \pi, & \text{if } m \neq n \neq 0, \end{cases}$$

$$\int_{-\pi}^{+\pi} \sin mx \cos nx dx = 0,$$

$$\int_{-\pi}^{+\pi} \cos mx \cos nx dx = \begin{cases} 0, & \text{if } m \neq n, \\ \pi, & \text{if } m \neq n \neq 0, \end{cases}$$

proved at the end of 4.3, we obtain at once for the coefficients the formula

$$a_\nu = \frac{1}{\pi}\int_{-\pi}^{+\pi} f(x)\cos\nu x\, dx$$

Similarly, multiplying the series by $\sin \nu x$ and integrating, we find

$$b_\nu = \frac{1}{\pi}\int_{-\pi}^{+\pi} f(x)\sin\nu x dx$$

These formulae assign a definite sequence of coefficients a_ν and b_ν, usually called Fourier coefficients, to every function $f(x)$ which is defined and continuous in the interval $-\pi < x < +\pi$, or has only a finite number of jump discontinuities there. If the function $f(x)$ is given, we can employ these quantities a_ν, b_ν to form the Fourier partial sum

$$S_n(x) = \frac{1}{2}a_0 + \sum_{\nu=1}^{n}(a_\nu \cos\nu x + b_\nu \sin\nu x),$$

and we may also write down formally the corresponding infinite Fourier series. Our problem is now to distinguish simple classes of functions $f(x)$ for which these Fourier series actually converge and represent them.

In order to formulate the result which we wish to prove, we introduce the definition:

A function $f(x)$ is said to be sectionally continuous in an interval, if it is itself sectionally continuous (i.e., is continuous in the interval except for a finite number of jump discontinuities) and if, in addition, its first derivative $f(x)$ is sectionally continuous.

We shall imagine that the function $f(x)$, originally defined in the interval $-\pi \}x\} \pi$, is periodically extended. At each point at which the function $f(x)$ has a jump discontinuity, we shall, if necessary, alter the function and assign to it the value which is the arithmetic mean of the left hand and right-hand limits of $f(x)$, i.e., we shall write

$$f(x) = \frac{1}{2}(f(x-0)+f(x+0)),$$

where $f(x - 0)$ and $f(x + 0)$ are simply the limits of $f(x)$ as x approaches from the left and from the right hand side, respectively. This equation is obviously true for every point x at which $f(x)$ is continuous.

Our goal is now the theorem: If the function $f(x)$ if sectionally smooth and satisfies the above equation, then its Fourier series converges at any point x and represents the function.

Note that this theorem can be proved for more general classes of functions. However, the result formulated here is sufficient for all applications.

Moreover, we shall prove the theorem: In every closed interval in which the function $f(x)$ (imagined to be periodically extended) is continuous as well as sectionally smooth, the Fourier series converges uniformly.

Finally: If the function $f(x)$ is sectionally smooth and has no discontinuities, the Fourier series converges absolutely.

The proofs of these theorems will be postponed. We merely wish to emphasize here that the functions, which can be expanded according to these theorems, have a very high degree of arbitrariness; it is by no means necessary that the function should be given by a single analytical expression.

In the next section, we shall display the extraordinary fertility of Fourier expansions by discussing a number of examples.

Examples of Fourier Series

Preliminary Remarks: We shall assume that our functions $f(x)$ have the period 2π and are defined in the interval $-\pi < x < +\pi$, . Beyond this interval, to the left and right hand sides, they are to be extended periodically as in 8.1.1.

Before going into details, we note that if $f(x)$ is an even function, then clearly $f(a)$ sin nx is odd and $f(x)$ cos νx is even, so that

$$b_\nu = \frac{1}{\pi}\int_{-\pi}^{+\pi} f(x)\sin \nu x dx = 0;$$

$$a_\nu = \frac{2}{\pi}\int_{-\pi}^{\pi} f(x)\cos \nu x dx.$$

We thus obtain a cosine series. On the other hand, if $f(x)$ is an odd function, then

$$a_\nu = \frac{1}{\pi}\int_{-\pi}^{+\pi} f(x)\cos \nu x dx = 0;$$

$$b_\nu = \frac{2}{\pi}\int_{0}^{\pi} f(x)\sin \nu x dx,$$

and we obtain a sine series.

Consequently, if the function $f(x)$ is initially given only in the interval $0 < x < \pi$, we can extend it in the interval $-\pi < x < 0$ either as an odd function or as an even function, and, correspondingly, expand it in the interval $0 < x < \pi$ in a sine series or in a cosine series.

Expansion of the Functions $\psi(x) = x$ and $\psi(x) = x$: For the odd function x, we have

$$b_\nu = \frac{2}{\pi}\int_0^\pi f(x)\sin \nu x dx,$$

and, on integration by parts,

$$\frac{\pi}{2}b_\nu = \left.\frac{-x\cos\nu x}{\nu}\right|_0^\pi + \frac{1}{\nu}\int_0^\pi \cos\nu x dx = (-1)^{\nu+1}\frac{\pi}{\nu}$$

Hence, we obtain for the periodic function $\psi(x)$, which is equal to x in the interval $-\pi < x < \pi$

$$\psi(x) = 2\left(\frac{\sin x}{1} - \frac{\sin 2x}{2} + \frac{\sin 3x}{3} - + \ldots\right)$$

If we set $x = \pi/2$, we obtain Gregory's series

$$\frac{\pi}{4} = 1 - \frac{1}{3} + \frac{1}{5} - + \ldots,$$

with which we are already familiar. The function $\psi(x)$, represented by this series, is not a continuous function; on the contrary, it jumps by 2π at the points $x = k\pi$, $k = 1, 3. 5, \ldots$. At these points of discontinuity, that is, at the points $x = k\pi$, $k = 1, 3, 5, \ldots$, each term of the series is zero, whence the function itself is zero. Hence, at the points of discontinuity, the series represents the arithmetic mean of the left hand and right hand limits.

If x is any fixed number between $-\pi$ and π and we replace x in the above series by $(x - \xi)$, we obtain the series

$$\psi(x - \xi) = 2\left(\frac{\sin(x-\xi)}{1} - \frac{\sin 2(x-\xi)}{2} + \frac{\sin 3(x-\xi)}{3} - + \ldots\right)$$

$$= -\frac{2}{1}\sin\xi\cos x + \frac{2}{1}\cos\xi\sin x + \frac{2}{2}\sin 2\xi\cos 2x$$

$$-\frac{2}{2}\cos 2\xi\sin 2x - \frac{2}{3}\sin 3\xi\cos 3x + \frac{2}{3}\cos 3\xi\sin 3x + \ldots$$

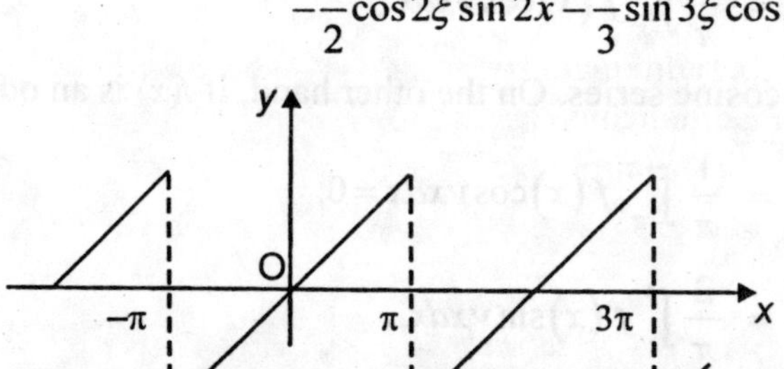

This series may also be written in the form of a Fourier series with the coefficients

$$a_0 = 0,\ a_n = 2\frac{(-1)^n}{n}\sin n\xi,\ b_n = 2\frac{(-1)^{n-1}}{n}\cos n\xi$$

which tend to zero as n increases; this series represents a function with the discontinuities, described above, at the points $x = \xi\ \pi,\ \xi\ 3\pi, \ldots$

For the even function $\varphi(x) = x$, on integrating twice by parts, we find

$$a_\nu = \frac{2}{\pi}x^2\cos\nu x dx = (-1)^\nu\frac{4}{\nu^2}\ (\nu > 0),\ a_0 = \frac{2\pi^2}{3},$$

Differentiating this series term by term and dividing by 2, we recover formally the series for $\psi(x) = x$.

Expansion of the Function $x\cos x$:. For this odd function, we find

$$a_\nu = 0,\ b_\nu = \frac{2}{\pi}\int_0^\pi x\cos x\sin\nu x dx$$

Using the formula

$$\int_0^\pi x\sin\mu x\,dx = (-1)^{\mu-1}\frac{\pi}{\mu}\quad(\mu = 1, 2, 3, \ldots)$$

above, we find

$$b_\nu = \frac{2}{\pi}\int_0^\pi x\cos x\sin\nu x\,dx = \frac{1}{\pi}\int_0^\pi x\big(\sin(\nu+1)x + \sin(\nu-1)x\big)dx$$

$$= \frac{(-1)^{\nu+2}}{\nu+1} + \frac{(-1)^\nu}{\nu-1} = (-1)^\nu\frac{2\nu}{\nu^2-1}\quad(\nu = 2, 3, \ldots),$$

$$b_1 = -\frac{1}{2}$$

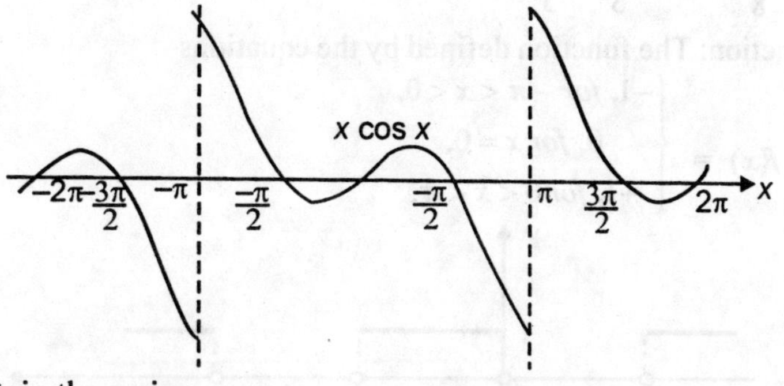

Hence we obtain the series

$$x\cos x = -\frac{1}{2}\sin x + 2\sum_{\nu=2}^{\infty}\frac{(-1)^\nu\nu}{\nu^2-1}\sin\nu x,$$

and, if we add the series in 9.4.2, the series

$$x\cos x = -\frac{1}{2}\sin x + 2\sum_{\nu=2}^{\infty}\frac{(-1)^{\nu}\,\nu}{\nu^2-1}\sin\nu x,$$

When the function, which is equal to $x\cos x$ in the interval $-\pi < x < \pi$, is extended periodically beyond this interval, the same discontinuities appear as are exhibited by the function $\psi(x)$. On the other head, if the function $x(1+\cos x)$ is periodically extended, it remains continuous at the end-points of the interval and, in fact, its derivative also remains continuous, since the discontinuities are eliminated by the factor $1+\cos x$ which vanishes together with its derivative at the end-points.

The Function $f(x) = |x|$: This function is even, whence $b_\nu = 0$ and

$$a_\nu = \frac{2}{\pi}\int_0^{\pi} x\cos\nu x\,dx,$$

and, on integration by parts, we obtain

$$\int_0^{\pi} x\cos\nu x\,dx = \frac{1}{\nu}x\sin\nu x\Big|_0^{\pi} - \frac{1}{\nu}\int_0^{\pi}\sin\nu x\,dx$$

$$= \begin{cases} 0, \textit{if } \nu \textit{ is even and } \neq 0, \\ -\dfrac{2}{\nu^2}, \textit{if } \nu \textit{ is odd}, \end{cases}$$

whence

$$f(x) = \frac{\pi}{2} - \frac{4}{\pi}\left(\cos x + \frac{\cos 3x}{3^2} + \frac{\cos 5x}{5^2} + \ldots\right)$$

Setting $x = 0$, we obtain the remarkable formula

$$\frac{\pi^2}{8} = 1 + \frac{1}{3^2} + \frac{1}{5^2} + \ldots$$

The Step Function: The function defined by the equations

$$f(x) = \begin{cases} -1, \textit{for } -\pi < x < 0, \\ 0, \textit{for } x = 0, \\ +1, \textit{for } 0 < x < \pi, \end{cases}$$

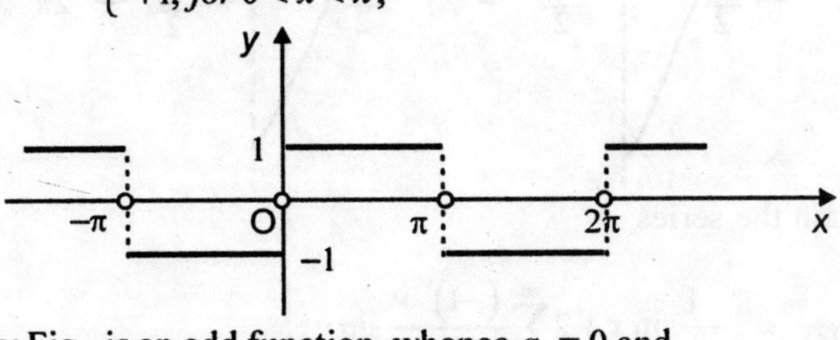

as, indicated by Fig., is an odd function, whence $a_\nu = 0$ and

$$b_\nu = \frac{2}{\pi}\int_0^\pi \sin\nu x\,dx = \begin{cases} 0 \text{ if } \nu \text{ is even,} \\ \dfrac{4}{\pi\nu} \text{ if } \nu \text{ is odd,} \end{cases}$$

so that its Fourier series is

$$f(x) = \frac{4}{\pi}\left(\frac{\sin x}{1} + \frac{\sin 3x}{3} + \ldots\right)$$

For $x = \pi/2$, in particular, this again yields Gregory's series.

Note that this series can be derived *formally* from that for $|x|$ by term by term differentiation.

The Function $f(x) = |\sin x|$: The even function $f(x) = |\sin x|$ can be expanded in a sine series with the coefficients a_n given by

$$\frac{\pi}{2}a_\nu = \int_0^\pi \sin x\cos\nu x\,dx = \frac{1}{2}\int_0^\pi \{\sin(\nu+1)x - \sin(\nu-1)x\}\,dx$$

$$= \begin{cases} 0 \text{ if } \nu \text{ is odd,} \\ \dfrac{-2}{\nu^2-1} \text{ if } \nu \text{ is even,} \end{cases}$$

We thus obtain

$$f(x) = |\sin x| = \frac{2}{\pi} - \frac{4}{\pi}\sum_{\mu=1}^{\infty}\frac{\cos 2\mu x}{4\mu^2-1}$$

Expansion of the Function cos μx. Resolution of cotan into Partial Fractions. The Infinite Product for sin x:. Let $f(x) = \cos\mu x$ for $-\pi < x < \pi$, where m is not an integer. Since $f(x)$ is even, we obtain $b_\nu = 0$, while

$$\frac{\pi}{2}a_\nu = \int_0^\pi \cos\mu x\cos\nu x\,dx = \frac{1}{2}\int_0^\pi \{\cos(\mu+\nu)x + \cos(\mu-\nu)x\}\,dx$$

$$= \frac{1}{2}\left\{\frac{\sin(\mu+\nu)\pi}{\mu+\nu} + \frac{\sin(\mu-\nu)\pi}{\mu-\nu}\right\}$$

$$= \frac{\mu(-1)^\nu}{\mu^2-\nu^2}\sin\mu\pi$$

We thus find

$$\cos\mu x = \frac{2\mu\sin\mu\pi}{\pi}\left(\frac{1}{2\mu^2} - \frac{\cos x}{\mu^2-1^2} + \frac{\cos 2x}{\mu^2-2^2} - + \ldots\right)$$

This function remains continuous at the points $x = \} \pi$. If we set $x = \pi$, divide both sides of the equation by sin μx and then write x instead of μ, we obtain

$$\cos\pi x = \frac{2x}{\pi}\left(\frac{1}{2x^2} + \frac{1}{x^2-1^2} + \frac{1}{x^2-2^2} + \ldots\right)$$

This is the so-called resolution of cotan x into partial fractions, a very important formula which is frequently discussed in analysis. We now rewrite this series in the form

$$\cos \pi x - \frac{1}{\pi x} = \frac{2x}{\pi}\left\{\frac{1}{1^2-x^2}+\frac{1}{2^2-x^2}+\ldots\right\}$$

If x lies in an interval $0 \rbrace x \rbrace q < 1$, the absolute value of the n-th term on the right hand side is less than $\frac{2}{\pi}\frac{1}{n^2-q^2}$ whence the series converges uniformly in this interval and can be integrated term by term. We thus obtain

$$\pi\int_0^x\left(\cot \pi t-\frac{1}{\pi t}\right)dt = \lim_{a\to 0} \log\frac{\sin \pi a}{\pi a} = \log\frac{\sin \pi x}{\pi x}$$

on the left hand side and

$$\log\left(1-\frac{x^2}{1^2}\right)+\log\left(1-\frac{x^2}{2^2}\right)+\ldots = \lim_{n\to\infty}\sum_{\nu=1}^{n}\log\left(1-\frac{x^2}{\nu^2}\right)$$

on the right hand side after multiplying both sides by π. If we step over from the logarithm to the exponential function, we find

$$\frac{\sin \pi x}{\pi x} = e^{\lim\limits_{n\to\infty}\sum\limits_{\nu=1}^{n}\log\left(1-x^2/\nu^2\right)} = \lim_{n\to\infty} e^{\lim\limits_{n\to\infty}\sum\limits_{\nu=1}^{n}\log\left(1-x^2/\nu^2\right)}$$

$$= \lim_{n\to\infty}\prod_{\nu=1}^{n}\left(1-\frac{x^2}{\nu^2}\right)$$

Hence

$$\sin \pi x = \pi x\left(1-\frac{x^2}{1^2}\right)\left(1-\frac{x^2}{2^2}\right)\left(1-\frac{x^2}{3^3}\right)\ldots$$

Thus, we have obtained the famous expression for sin x as an infinite product. Setting x = ýÿ, we obtain Wallis' product

$$\frac{\pi}{2} = \prod_{\nu=1}^{\infty}\frac{2\nu}{2\nu-1}\cdot\frac{2\nu}{2\nu+1} = \frac{2}{1}\cdot\frac{2}{3}\cdot\frac{4}{3}\cdot\frac{4}{5}\ldots,$$

The formula for sin πx is particularly interesting, because it shows directly that the function sin πx vanishes at the points $x = 0, 1, 2, \ldots$. In this respect, it corresponds to the factorization of a polynomial when its zeroes are known.

The Convergence of Fourier Series

We now proceed to establish rigorously the theorems which were stated in 9.3 and illustrated in 9.4.

The Convergence of the Fourier Series of a Sectionally smooth Function:. We first recall that, if $f(x)$ is any function which is defined and sectionally continuous (i.e., continuous, except at most at a finite number of jump discontinuities) in the interval $-\pi < x < \pi$, we can form its Fourier coefficients according to the formulae

$$a_\nu = \frac{1}{\pi}\int_{-\pi}^{+\pi} f(t)\cos\nu t\,dt,$$

$$b_\nu = \frac{1}{\pi}\int_{-\pi}^{+\pi} f(t)\sin\nu t\,dt,$$

and write down formally the series $\frac{1}{2}a_0 + \sum_{\nu=1}^{\infty}(a_\nu \cos\nu x + b_\nu \sin\nu x)$

This series is called the Fourier series corresponding to $f(x)$, irrespectively of whether it converges or not. We will now find the conditions, which must be imposed on $f(x)$, in order to ensure that the Fourier series corresponding to $f(x)$ converges and represents $f(x)$. We will assume that $f(x)$ is extended periodically beyond the interval $-\pi < x < \pi$.

We shall now prove the theorem: If the function $f(x)$ is sectionally smooth, i.e. $f(x)$ and its derivative $f'(x)$ are sectionally continuous, and satisfies at each point of discontinuity (s) the condition $f(x)=\{f(x-0)+f(x+0)\}$, then the Fourier series corresponding to $f(x)$ converges at every point and represents the function $f(x)$.

In order to prove this theorem, we consider the partial sums

$$S_n(x) = \frac{1}{2}a_0 + \sum_{\nu=1}^{n}(a_\nu \cos\nu x + b_\nu \sin\nu x)$$

If we substitute for the coefficients the above integral expressions and then interchange the order of integration and summation, we obtain

$$S_n(x) = \frac{1}{\pi}\int_{-\pi}^{+\pi}\sum_{\nu=1}^{n}(\cos\nu t\cos\nu x + \sin\nu t\sin\nu x)\,dt,$$

or, by the addition theorem for the cosine,

$$S_n(x) = \frac{1}{\pi}\int_{-\pi}^{+\pi} f(t)\left\{\frac{1}{2} + \sum_{\nu=1}^{n}\cos\nu(t-x)\right\}dt$$

If we now apply the summation formula obtained above, this becomes

$$S_n(x) = \frac{1}{2\pi}\int_{-\pi}^{+\pi} f(t)\frac{\sin\left(n+\frac{1}{2}\right)(t-x)}{\sin\frac{1}{2}(t-x)}dt$$

Finally, applying the transformation $t=(\tau-x)$ and noting the periodicity of the integrand, we obtain

$$S_n(x) = \frac{1}{2\pi}\int_{-\pi}^{+\pi} f(x+\tau)\frac{\sin\left(n+\frac{1}{2}\right)\tau}{\sin\frac{1}{2}\tau}d\tau$$

Starting with this form of the partial sum $S_n(x)$, we can prove that it tends to $f(x)$, by means of the lemma:

If the function $s(x)$ is sectionally continuous in the interval $a \rangle x \rangle b$, then the integral

$$I = \int_a^b s(t)\sin \lambda t \, dt$$

tends to 0 as λ increases.

In the proof, we may assume that $s(x)$ is continuous in the entire interval, since otherwise we merely need carry out the argument for each sub-interval in which $s(x)$ is continuous

As in the similar argument in A8.2, we note that, if λ is positive, the function $\sin\lambda t$ is alternately positive and negative in successive intervals of length π/λ. For large values of λ, the contributions to the integral from adjacent intervals almost cancel one another, since, on account of the continuity, the values of $s(x)$ in two such adjacent intervals differ only slightly from each another. We make use of this circumstance by transforming the integral I by the substitution $t = \tau + h$, where $h = \pi/\lambda$; then, since $\sin \lambda t = -\sin \lambda\tau$, we obtain

$$I = -\int_{a-h}^{b-h} s(\tau + h)\sin \lambda\tau \, d\tau$$

If we again replace the letter τ by t and then add the two expressions for I, we find

$$2I = -\int_{a-h}^{a} s(t+h)\sin \lambda t \, dt + \int_a^{b-h}\{s(t) - s(t+h)\}\sin \lambda t \, dt + \int_{b-h}^{b} s(t)\sin \lambda t \, dt$$

If M is an upper bound for the absolute value of $s(x)$, i.e., if for all values of x in the interval under consideration $|s(x)| \rangle M$, then there follows at once from this expression for I the inequality

$$2|I| \leq 2Mh + \int_a^{b-h}|s(t) - s(t+h)|\,dt.$$

Now, let ε be any positive number; if we choose λ so large that in the entire interval $a < t < b - h$ the expression $|s(t) - s(t+h)|$ remains less than $\varepsilon/(b-a)$ and also $Mh = M\pi/\lambda < \varepsilon/2$, then $|I]< \varepsilon$; consequently, since ε can be chosen as small as we please, $\lim\limits_{\lambda\to\infty} I = 0$

If we assume that $s(x)$, besides being continuous, has a sectionally continuous derivative $s'(x)$, the proof of this lemma follows simply on integration by parts. In fact,

$$\int_a^b s(t)\sin \lambda t \, dt = \frac{1}{\lambda}\left\{s(a)\cos \lambda a - s(b)\cos \lambda b + \int_a^b \theta'(t)\cos \lambda t \, dt\right\}$$

We see here at once that, as λ increases, the right hand side tends to zero.

Besides this lemma, we need the integration formula

$$\int_0^\pi \frac{\sin\left(n + \frac{1}{2}\right)t}{2\sin\frac{1}{2}t}\,dt = \frac{\pi}{2},$$

which is true for every positive integer n. We establish this readily by using our summation formula for the cosine, since

$$\int_0^{\pi} \frac{\sin\left(n+\frac{1}{2}\right)t}{2\sin\frac{1}{2}t}\,dt = \int_0^{\pi}\left(\frac{1}{2}+\sum_1^{n}\cos\nu t\right)dt = \frac{\pi}{2}$$

Proof of the Main Theorem: By means of the lemma, the main theorem is readily proved, i.e., the formula

$$\lim_{n\to\infty} S_n(x) = \lim_{n\to\infty}\frac{1}{2\pi}\int_{-\pi}^{+\pi} f(x+t)\frac{\sin\left(n+\frac{1}{2}\right)t}{2\sin\frac{1}{2}t}\,dt = f(x)$$

We begin by subdividing the interval of integration at the origin. For fixed values of x, the function

$$s(t) = \frac{f(x+t)-f(x+0)}{2\sin\frac{1}{2}t}$$

is sectionally continuous in the interval $0 \rangle t \rangle \pi$. In fact, this is obvious when $0 \rangle t \rangle \pi$, while the continuity at $t = 0$ follows from the assumed existence of the right hand derivative

$$\lim_{t\to 0, t>0}\frac{f(x+t)-f(x+0)}{t} = \lim_{t\to+0}\frac{f(x+t)-f(x+0)}{2\sin\frac{1}{2}t}\cdot\frac{2som\frac{1}{2}t}{t}$$

$$= \lim_{t\to+0}\frac{f(x+t)-f(x+0)}{2\sin\frac{1}{2}t}$$

Hence, as $\lambda = n +$ increases, the integral

$$\frac{1}{\pi}\int_0^{\pi} s(t)\sin\lambda t\,dt = \frac{1}{2\pi}\int_0^{\pi} f(x+t)\frac{\sin\lambda t}{\sin\frac{1}{2}t}\,dt - \frac{1}{2\pi}\int_0^{\pi} f(x+0)\frac{\sin\lambda t}{\sin\frac{1}{2}t}\,dt$$

tends to zero.

However, since the factor $f(x+0)$ can be taken out of the second integral on the right hand side and, for $\lambda = n +$, the integral $\int_0^{\pi}\frac{\sin\lambda t}{2\sin\frac{1}{2}t}$ is equal to $\pi/2$, we obtain immediately

$$\lim_{\lambda\to\infty}\frac{1}{2\pi}\int_0^{\pi} f(x+t)\frac{\sin\lambda t}{\sin\frac{1}{2}t}\,dt = \frac{1}{2}f(x+0)$$

By setting in this equation $x = 0, f(t) = (\sin t)/t$ and then replacing t by u/λ, we obtain the important relation $\lim\limits_{\lambda\to\infty}\int_0^{\lambda\pi}\frac{\sin u}{u}du = \frac{\pi}{2}$.

In the same way, we obtain for the interval $-\pi < t < 0$

$$\lim_{\lambda\to\infty}\frac{1}{2\pi}\int_{-\pi}^{0} f(x+t)\frac{\sin\lambda t}{\sin\frac{1}{2}t}dt = \frac{1}{2}f(x-0)$$

and by addition

$$\lim_{\lambda\to\infty}\frac{1}{2\pi}\int_{-\pi}^{+\pi} f(x+t)\frac{\sin\lambda t}{\sin\frac{1}{2}t}dt = f(x)$$

Further Investigation of Convergence: In the neighbourhood of those points, where the function $f(x)$ is discontinuous, the Fourier series does not converge uniformly; in fact, by 8.4.3, a uniformly convergent series of continuous functions possesses a continuous sum. Nevertheless, we have the important theorem:

If a sectionally smooth periodic function has no discontinuities, its Fourier series converges absolutely and uniformly. The convergence of the Fourier series for any sectionally smooth function whatsoever is uniform in every closed interval which contains no point of discontinuity of the function.

In order to prove this theorem, we start from a fundamental inequality satisfied by the Fourier coefficients of any function $f(x)$ which is sectionally continuous. This so-called Bessel inequality states that for all values of n $\frac{1}{2}a_0^2 + \sum\limits_{\nu=1}^{n}\left(a_\nu^2 + b_\nu^2\right) \le \frac{1}{\pi}\int_{-\pi}^{+\pi}\{f(x)\}^2 dx$.

The proof follows from the fact that the expression

$$\int_{-\pi}^{+\pi}\left\{f(x) - \frac{1}{2}a_0 - \sum_{\nu-1}^{n}\left(a_\nu\cos\nu x + b_\nu\sin\nu x\right)\right\}^2 dx$$

is always positive or zero. If we evaluate the integral by expanding the bracket under the integral sign and recall the orthogonality relations and definitions of the Fourier coefficients, we obtain at once Bessel's inequality in the form $\int_{-\pi}^{+\pi}\{f(x)\}^2 dx - \pi\left\{\frac{1}{2}a_0^2 + \sum\limits_{\nu=1}^{n}\left(a_\nu^2 + b_\nu^2\right)\right\} \ge 0$

In addition to Bessel's inequality, we employ Schwarz's inequality : If $u_1, u_2, \ldots, u_n$ and $v_1, v_2, \ldots, v_n$ are arbitrary real numbers, it is always true that $\left(\sum\limits_{\nu=1}^{n}u_\nu v_\nu\right)^2 \le \sum\limits_{\nu=1}^{n}u_\nu^2 \cdot \sum\limits_{\nu=1}^{n}v_\nu^2$ the equality sign occurring only when the sequence u is proportional to the sequence v.

We now assume that the periodic function $f(x)$ is sectionally smooth as well as continuous. The derivative $g(x) = f'(x)$ is sectionally continuous and we easily show that c_ν and d_ν, the Fourier coefficients of $g(x)$, satisfy the relations

$$c_0 = 0,$$
$$c_\nu = \nu b_\nu \ (\nu \geq 1);$$
$$d_\nu = -\nu a_\nu \ (\nu \geq 1);$$

in fact, on integration by parts, we have

$$c_\nu = \frac{1}{\pi}\int_{-\pi}^{+\pi} g(x)\cos \nu x\, dx$$
$$= \frac{1}{\pi} f(x)\cos \nu x \Big|_{-\pi}^{+\pi} + \frac{\nu}{\pi}\int_{-\pi}^{+\pi} f(x)\sin \nu x\, dx = \nu b_\nu$$

similar proofs holding for the other statements.

Hence, Bessel's inequality applied to the function $g(x)$ yields

$$\sum_{\nu=1}^{n} \nu^2\left(a_\nu^2 + b_\nu^2\right) = \sum_{\nu=1}^{n}\left(c_\nu^2 + d_\nu^2\right) \leq \frac{1}{\pi}\int_{-\pi}^{+\pi}\{g(x)\}^2 dx.$$

If, for the sake of brevity, we denote the right hand side of this inequality by M and apply Schwarz's inequality, we find that for $m > n$

$$\sum_{\nu=n+1}^{m} \left|a_\nu \cos \nu x + b_\nu \sin \nu x\right| \leq \sum_{\nu=n+1}^{m} \sqrt{\left(a_\nu^2 + b_\nu^2\right)} = \sum_{\nu=n+1}^{m}\left\{\frac{1}{\nu}\left(\nu\sqrt{\left(a_\nu^2 + b_\nu^2\right)}\right)\right\} \leq M\sqrt{\left(\sum_{\nu=n+1}^{m}\frac{1}{\nu^2}\right)},$$

since $\sqrt{\left(a_\nu^2 + b_\nu^2\right)}$ is the amplitude of the periodic function $a_\nu \cos \nu x + b_\nu \sin \nu x$

However, owing to the $\sum_{\nu=1}^{\infty}\frac{1}{\nu^2}$, convergence of , the right hand side, which is independent of x, can be made as small as we please by choosing n and m large enough, which proves the absolute and uniform convergence of the series.

Incidentally, the same considerations show that for periodic functions with continuous derivatives of the $(h-1)$-th order and derivatives of the $(h-1)$-th order, which are at least sectionally continuous, the sum $\sum^{n} \nu^{2h}\left(a_\nu^2 + b_\nu^2\right)$ remains below a fixed bound. This gives us a definite statement about the order to which the Fourier coefficients vanish. For such a function, the Fourier series of the derivatives up to the order $(h-1)$ converge absolutely and uniformly.

In order to prove the above theorem for sectionally smooth functions which are discontinuous, we first consider a special function $\psi(x)$ of this type.

In the interval $-\pi < x < \pi$, we define $\psi(x)$ as equal to x, outside this interval, $\psi(x)$ is extended periodically. According to equation , its Fourier series is $2\left(\frac{\sin x}{1} - \frac{\sin 2x}{2} + \frac{\sin 3x}{3} - + \ldots\right)$

This series cannot be uniformly convergent, because its sum is the discontinuous function $y(x)$. However, we shall show that the convergence is uniform in every interval $-l \,\}\, x \,\}\, l$ for which $0 < l < \pi$.

The proof is based on a special artifice*. We observe that in the interval $-l \}x\} l$ the function $\cos x/2$ is never less than the positive quantity $\cos l/2 = \kappa$. If we multiply the absolute value of the difference between the m-th and n-th partial sums of the above series ($m > n$), i.e., the expression

$$|S_m(x) - S_n(x)| = 2\left|\frac{\sin(n+1)x}{n+1} - \frac{\sin(n+2)x}{n+2} + -\dots \frac{\sin mx}{m}\right|,$$

by the function $\cos x/2$, then, in accordance with the well-known trigonometric formula

$2\sin u \cos v = \sin(u+v) + \sin(u-v)$, we obtain the absolute value of the expression

If we combine the terms on the right hand side with the same numerators, we obtain

$$\frac{\sin\left(n+\frac{1}{2}\right)x}{n+1} \pm \frac{\sin\left(m+\frac{1}{2}\right)x}{m} + \frac{\sin\left(n+\frac{2}{2}\right)x}{(n+1)(n+2)} - \frac{\sin\left(m+\frac{5}{2}\right)x}{(n+2)(m+3)} + -\dots \mp \frac{\sin\left(m-\frac{1}{2}\right)x}{(m-1)m},$$

and, since $\cos x/2 < \kappa$ and $|\sin u| < 1$, the estimate

$$|S_m(x) - S_n(x)| \le \frac{1}{k}\left[\frac{1}{n+1} + \frac{1}{m} + \frac{1}{(n+1)(n+2)} + \dots + \frac{1}{(m-1)m}\right]$$

But the expression on the right hand side does not depend on x and, by virtue of the convergence of the series $\sum_{\nu=1}^{\infty} \frac{1}{\nu(\nu+1)}$, it can be made as small as we please by choosing n and m large enough. This implies the uniform convergence of the Fourier series, as we have asserted.

Now that we have obtained the expression for a particular discontinuous function, we can transfer the discontinuity to any arbitrary point in the interval by translation of the curve or of the co-ordinate system. In fact, the function

$$\psi(x-\xi) = 2\left(\frac{\sin(x-\xi)}{1} - \frac{\sin 2(x-\xi)}{2} + \frac{\sin 3(x-\xi)}{3} - + \dots\right)$$

is continuous except at the points $(2k+1)\pi + \xi$, where k is an integer. However, on passing these points, the function jumps by an amount -2π from the value π to the value $-\pi$, while at these points themselves the value of the function is zero.

If now $f(x)$ is any sectionally smooth function, which in the interval $-\pi \,\}\, x \,\}\, \pi$ is discontinuous only at the points $\xi_1, \xi_2, \dots, \xi_m$ and if on passing these points from the left to the right hand side the function jumps by the amounts $\delta_1, \delta_2, \dots, \delta_m$, respectively, then the function

$$f(x) + \frac{\delta_1}{2\pi}\psi(x+\pi-\xi_1) + \frac{\delta_2}{2\pi}\psi(x+\pi-\xi_2) + \dots + \frac{\delta_m}{2\pi}\psi(x+\pi-\xi_m)$$

will be continuous and sectionally smooth, whence, by the previous proof, it can be expanded in a uniformly convergent Fourier series. We now obtain the Fourier series of the function $f(x)$ by adding term by term the finite number of Fourier series corresponding to the functions

$$-\frac{\delta_1}{2\pi}\psi(x+\pi-\xi_1),\ldots,-\frac{\delta_m}{2\pi}\psi(x+\pi-\xi_m)$$

Hence the theorem is proved.

This result is quite adequate for most mathematical investigations and applications. However, we point out that the investigation of Fourier series has been advanced much further. The conditions for an expansion in Fourier series, which we have found here to be sufficient, are by no means necessary. Functions with far fewer continuity properties than those discussed here can be represented by Fourier series. There is an extensive literature devoted to these questions and to the general problem of the expandability of a function in a Fourier series. As a remarkable result of such investigations, we mention thee fact that there are continuous functions, the Fourier series of which do not converge in any interval, no matter how small. Such a result does not in any impugnway reduce the usefulness of Fourier series; on the contrary, it must be regarded as evidence that the concept of a continuous function involves fairly complicated possibilities, as has already been shown by the example of continuous functions which nowhere have a derivative.

A Sketch of the Theory of Functions of Several Variables

Up to this point, we have been concerned exclusively with functions of a single independent variable. We must now go on to consider functions of several independent variables. Even the applications of the calculus force us to take this step. In almost all the relationBhips which occur in nature, in fact, the functions in question do not depend on a single independent variable; on the contrary, the dependent variable is usually determined by two, three, or more independent variables. Thus, for example, the volume of an ideal gas is a function of a single variable, the pressure, if we keep the temperature constant, but not otherwise. As a rule, the temperature also varies and the volume depends upon a pair of values, namely, the value of the pressure and that of the temperature, whence it is a function of two independent variables.

Also from the point of view of pure mathematics, the need for a detailed study of functions of several independent variables is urgent. Here we shall be able to take advantage of what we have learned previously, so that in many cases we have only to make simple extensions of our arguments.

It is usually sufficient to consider the case of only two independent variables x and y, as long as no essentially new considerations are required for an extension to functions of three or more variables. Hence, in order to keep our statements and notation simple, we shall, as a rule, consider only two independent variables.

Chapter 10

The Concept of Function in the Case of Several Variables

Functions and their Ranges of Definition: Equations of the form

$$u = x^2 + y^2, \qquad u = x - y,$$
$$u = xy, \text{ or} \qquad u = \sqrt{(1 - x^2 - y^2)}$$

assign a functional value u to each pair of values (x, y). In the first three of our example, this correspondence holds for every system of values (x, y), while in the last case the correspondence has a meaning only for those pairs of values (x, y) for which the inequality $x + y < 1$ is true.

In these cases we say that u is a function of the independent variables x and y. In general, we use this expression whenever some law assigns a value of u as dependent variable, corresponding to each pair of values (x,y) belonging to a certain specified set. The relation between x, y and u may be stated in terms of a functional equation, as above, or by means of a verbal description such as u is the area of the rectangle with sides x and y, or it may follow from physical observations as for instance in the case of the magnetic declination at different latitudes and longitudes. The essential thing is that there exists a correspondence. Similarly, u is said to be a function of the three independent variables $x, y, z,$ if for each triad of values (x, y, u) of a certain set there exists a corresponding value of u given by some definite law; it is similar for the general case of functions of n independent variables $x_1, x_2, \ldots, x_n$.

The set of values which the pair (x, y) can assume is called the range of definition of the function $u = f(x, y)$. We shall restrict our attention to the simplest types of range of definition. We shall consider that (x, y) is limited either to a so-called rectangular region (domain) $a \leqq x \leqq b,\ c \leqq y \leqq d,$ or else to a circle, determined by an inequality of the form $(x - a)^2 + (y - b)^2 \leqq r^2$

In the case of functions of three variables x, y, z, we shall again consider only rectangular regions $a \leqq x \leqq b,\ c \leqq y \leqq d,\ e \leqq z \leqq f,$ and spherical regions $(x - a)^2 + (y - b)^2\ (z - c)^2 \leqq r^2$

In dealing with more than three independent variables, geometrical intuition fails us, but it is often convenient to also extend geometrical terminology to this case. Thus, for functions of n variables $x_1, x_2, ..., x_n$, we shall consider regions $a_1 \leqq x_1 \leqq b_1$, $a_2 \leqq x_2 \leqq b_2$, ..., $a_n \leqq x_n \leqq b_n$, and also regions $(x_1 - a_1)^2 + (x_2 - a_2)^2 + ... + (x_n - a_n)^2 \leqq r^2$, which we will call rectangular and spherical regions, respectively.

The Simplest Types of Functions. Just as in the case of functions of one variable, the simplest functions are the rational integral functions or polynomials. The most general polynomial of the first degree (linear function) has the form $u = ax + by + c$, where a, b and c are constants. The general polynomial of the second degree has the form $u = ax^2 + bxy + cy^2 + dx + ey + f$.

The general polynomial is a sum of terms of the form $a_{mn}x^m y^n$, where the quantities a_{mn} are arbitrary constants.

Rational fractional functions are quotients of polynomials; for example, there belongs to this class the linear fractional function

$$u = \frac{ax + by + c}{a'x + b'y + c'}$$

By extraction of roots, we pass on from the rational functions to certain algebraic functions, for example,

$$u = \sqrt{\left(\frac{x - y}{x + y}\right)} + \sqrt[3]{\left\{\frac{(x + y)^2}{x^3 + xy}\right\}}$$

An accurate definition of the term algebraic function is given in at the end of 10.5.1.

In the construction of more complicated functions of several variables, we almost always fall back on the well-known functions of a single variable, for example (10.4.1),

$$u = \sin(xy) \text{ or } \qquad u = \log(y^2 + \cos\frac{1}{2}x)$$

Geometrical Representation of Functions: Just as we represent functions of one variable by means of curves, we seek to represent functions of two variables geometrically by means of surfaces; we shall consider hereafter only those functions which can actually be represented in this way. We achieve this representation very simply by considering a rectangular co-ordinate system in space with co-ordinates x, y and u, and marking off above each point (x, y) of the range (R) of definition of the function the point P with the third co-ordinate $u = f(x, y)$. As the point (x, y) ranges over the region R, the point P describes a surface in space. We take this surface as the geometrical representation of the function.

Conversely, in analytical geometry, surfaces in space are represented by functions of two variables, so that there is between such surfaces and functions of two variables a reciprocal relationship.

For example, there corresponds to the function

$$u = \sqrt{(1 - x^2 - y^2)}$$

the hemi-sphere above the x, y plane with unit radius and centre at the origin, to the function

$$u = x^2 + y^2,$$

the so-called paraboloid of revolution, obtained by rotating the parabola $w=x$ about the u-axis, to the functions

$$u = x^2 - y^2, \text{ and } u = xy$$

the so-called hyperbolic paraboloids. The linear function

$$u = ax + by + c$$

has for its graph a plane in space.

If in the function $u = f(x, y)$ one of the independent variables, say y, does not occur, so that u depends on x only, say $u = g(x)$, the function is represented in the xyu-space by a cylindrical surface, obtained by erecting the perpendiculars to the ux-plane at the points of the curve $u=g(x)$.

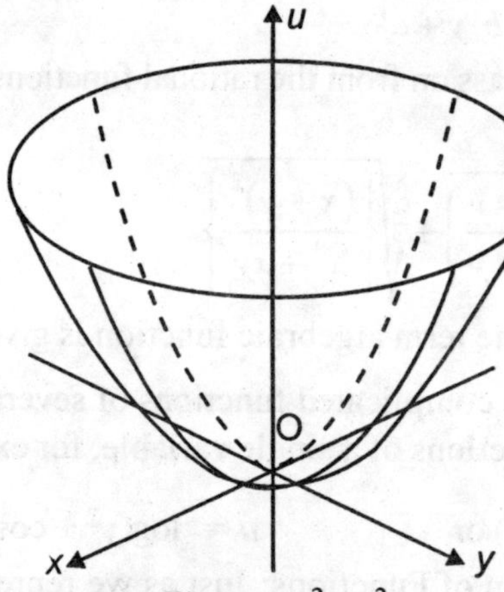

Fig. $u = x^2 + y^2$

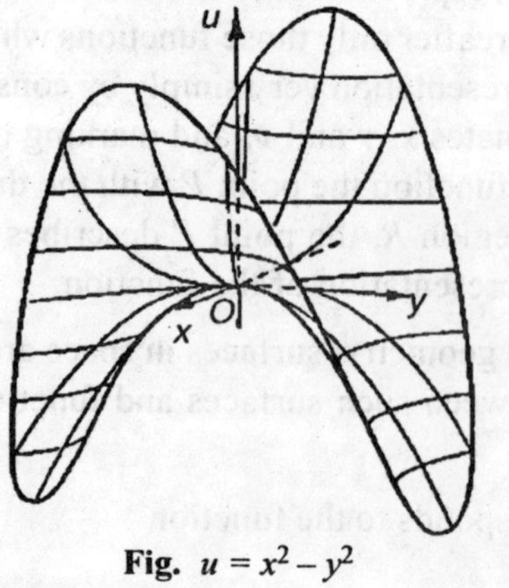

Fig. $u = x^2 - y^2$

However, this representation by means of rectangular co-ordinates has two disadvantages. Firstly, intuition fails us whenever we have to deal with three or more independent variables. Secondly, even in the case of two independent variables, it is often more convenient to confine the discussion to the *xy*-plane, since in the plane we can sketch and construct geometrically without difficulty. From this point of view, another geometrical representation of the function by means of contour lines is to be preferred. In the *xy*-plane, we take all the points for which $u = f(x, y)$ has a constant value, say $u = k$. These points will usually lie on a curve or curves, the so-called contour line for the given constant value of the function. We can also obtain these curves by cutting the surface $u = f(x, y)$ by the plane u = k parallel to the *xy*-plane and projecting the curves of intersection perpendicularly onto the *xy*-plane. The system of these contour lines, marked with the corresponding values $k_1, k_2, \ldots$ of the height k, gives us a representation of the function. As a rule, k is assigned values in arithmetic progression, say $k = n\,h$, where $n = 1, 2, \ldots$ The distance between the contour lines then gives us a measure of the steepness of the surface $u=f(x,y)$; in fact, between every two neighbouring lines, the value of the function changes by the same amount. Where the contour lines are close together, the function rises or falls steeply, where they are far apart, the surface is flattish. This is the principle on which contour maps such as those of the Ordnance Surrey and the U.S. Geological Survey are constructed.

In this method, the linear function $u = ax + by + c$ *is r*epresented by a system of parallel straight lines $ax + by + c = k$. The function $u = x + y$ is represented by a system of concentric circles. The function $u = x - y$, the surface of which has a saddle point at the origin is represented by the system of hyperbolas.

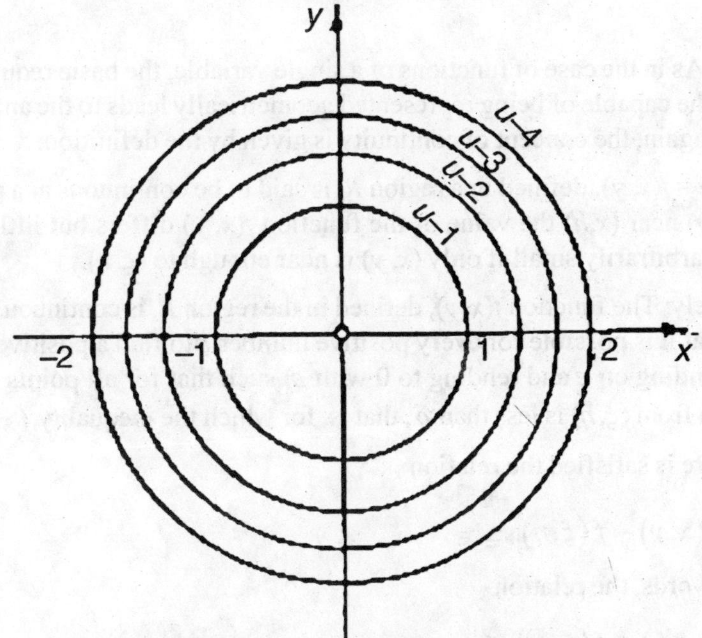

Fig. Content lines of $u = x^2 + y^2$

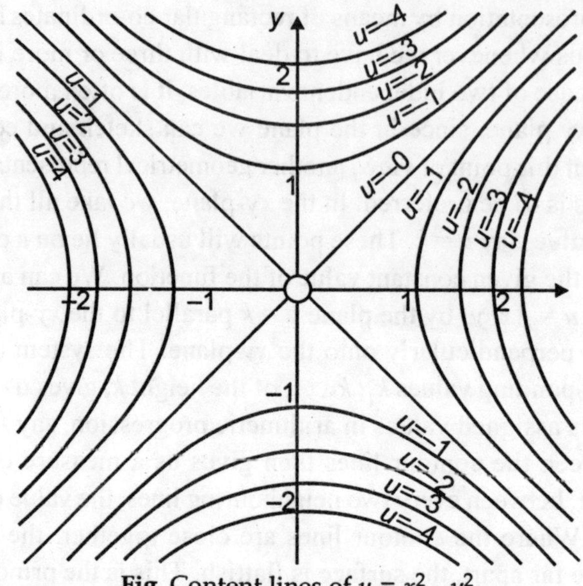

Fig. Content lines of $u = x^2 - y^2$

The method of representing the function $u = f(x,y)$ by contour lines has the advantage of it being capable of extension to functions of three independent variables. Instead of contour lines, we then have level surfaces $f(x,y,z) = k$, where k is a constant to which we can assign any suitable sequence of values. For example, the level surfaces for the function $u = x + y + z$ are concentric spheres about the origin of the co-ordinate system.

Continuity

Definition:. As in the case of functions of a single variable, the basic requirement, by which functions should he capable of being represented geometrically leads to the analytic condition of continuity. Here again, the concept of continuity is given by the definition:

A function $u = f(x, y)$, defined in a region R, is said to be continuous at a point (ξ,η) of R, if for all points (x,y) near (x,h) the value of the function $f(x, y)$ differs but little from $f(\xi,\eta)$, the difference being arbitrarily small if only (x, y) is near enough to (ξ,η).

More precisely: The function $f(x, y)$, defined in the region R, is continuous at the point (ξ,η) of R, provided that it is possible for every positive number ε to find a positive distance $\delta = \delta(\varepsilon)$ (in general, depending on ε and tending to 0 with ε) such that for all points of the region, the distance of which from (ξ,η) is less than δ, that is, for which the inequality $(x-\xi)^2 + (y-\eta)^2 + \leqq \delta^2$ holds, there is satisfied the relation

$$\left|f(x,y) - f(\xi,\eta)\right| \leqq \in$$

or, in other words, the relation

$$\left|f(\xi+h,\eta+k) - f(\xi,\eta)\right| \leqq \in$$

is to hold for all pairs of values (h, k) such that $h + k \rangle \delta$ and $(\xi + h, \eta + k)$ belongs to the region R.

If a function is continuous at every point of a region B, we say that it is continuous in R.

In the definition of continuity, we can replace the distance condition $h+k \rangle \delta \rangle$ by the equivalent condition:

There shall correspond to every $\varepsilon > 0$ two positive numbers δ_1 and δ_2 such that

$$|f(\xi+h,\eta+k)-f(\xi,\eta)| \leqq \in$$

whenever $|h| \leqq \delta_1$ and $|k| \leqq \delta_2$

The two conditions are equivalent. In fact, if the original condition is fulfilled, so is the second one, if we take $\delta_1 = \delta_2 = \delta/2$; and conversely, if the second condition is fulfilled, so is the first, if we take for δ the smaller one of the two numbers δ_1 and δ_2.

The following facts are almost obvious:

The sum, difference and product of continuous functions are also continuous. The quotient of continuous functions is continuous except when the denominator vanishes. Continuous functions of continuous functions are themselves continuous (10.4.1). In particular, all polynomials are continuous and so are all rational fractional functions except when the denominator vanishes.

Another obvious fact, which, however, is worth stating, is:

If a function $f(x, y)$ is continuous in a region R and differs from zero at an interior point P of the region, it is possible to mark off about P a neighbourhood, say a circle, belonging entirely to R, in which $f(x,y)$ is nowhere equal to zero. In fact, if the value of the function at P is a, we can mark off about P a circle so small that the value of the function within the circle differs from a by less than $a/2$, and therefore it is certainly not zero.

Examples of Discontinuities: In the case of functions of one variable, we encounter three kinds of discontinuities: Infinite and jump discontinuities, and discontinuities at which no limit is approached from one or both sides. With functions of two or more variables, no such simple classification is possible. In particular, the situation is made more complicated by the fact that discontinuities may occur not merely at isolated points, but also along entire curves.

Thus, for the function $u = 1/(x - y)$, the line $x = y$ is a line of infinite discontinuity. As we approach the line from one side or the other, the values of u increase numerically beyond all bounds through positive or negative values. The function $u = 1/(x - y)$ has the same line of discontinuity, but tends to $+ \rangle$ as we approach the line from either side. The function $u = 1/(x + y)$ has the single point of discontinuity $x = 0, y = 0$. The function $u = \sin \dfrac{1}{\sqrt{(x^2+y^2)}}$ tends to no limit as we approach the origin; the surface which it represents is obtained by rotating the graph of the function $u = \sin 1/x$ about the u-axis.

Another instructive example of a discontinuous function is given by the rational function $u = 2xy/(x + y)$. In the first instance, the function is undefined at $x = 0, y = 0$, and we supplement the definition by assuming that $u(0,0) = 0$. This function has a peculiar type of discontinuity at the origin. If we put $x = 0$, i.e., if we move along the y-axis, the function becomes $u(0,y) = 0$, which has the constant value 0 for all values of y. Along the x-axis, we likewise have $u(x, 0) = 0$. Thus, at the origin, the function $u(x,y)$ is continuous in x, if we keep y at the constant value 0, and is continuous in y, if we keep x at the constant value 0. Nevertheless, the function is discontinuous when considered as a function of the two variables x and y. In fact, at every point of the line y=x, we find that $u = 1$, so that arbitrarily near to the origin we can find points at which u assumes the value 1. The function is therefore discontinuous at the origin and cannot be defined at the origin in such a way as to make it continuous.

More generally, on the straight line $y = x \tan \alpha$, inclined at the angle a to the x-axis, we have u = 2tana/(1+tana) = 2sinacosa = sin2a. The surface corresponding to the $u = 2\,xy/(x+y)$ is therefore formed by rotating a straight line at right angle to the u-axis about that axis until it coincides with the x-axis and simultaneously raising or lowering it so that the height $\sin 2\alpha$ is associated with the angle a. As a increases to 45, the straight line rises to the height 1, and consequently falls to the level of the y-axis and below to the depth 1, thereafter rising again to the level of the x-axis. The surface enveloped by the moving straight line is known as the cylindroid and is important in mechanics.

The above example shows that a function can he continuous in x for every fixed value of y and continuous in y for every fixed value of x and yet be discontinuous when considered as a function of the two variables. The essential point in this definition of continuity is that the value of the function at a point P must be arbitrarily close to the value of the function at a point Q, provided only that Q is near enough to P; it is not permissible to restrict the position of Q relative to P in any other way.

Multiple and Repeated Integrals

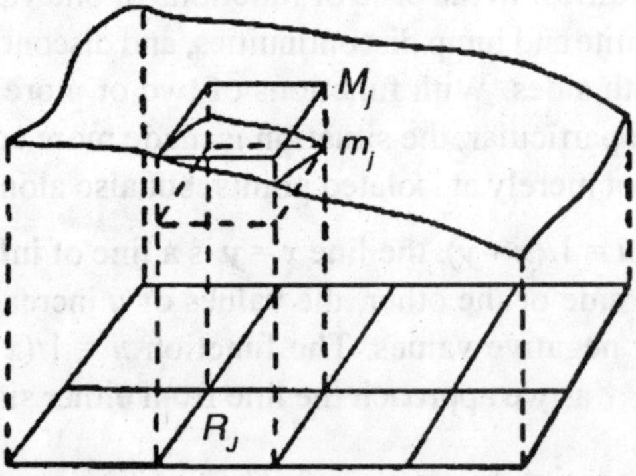

Multiple Integrals: Consider a function $u = f(x, y)$ which is defined and continuous in the rectangle R(*a* ýÿ *x* ýÿ *b*, *c* ýÿ *y* ýÿ *d*) and which assumes only positive values. We wish to assign a volume to the portion of three-dimensional space bounded by the rectangle R, the surface $u =$

$f(x, y, z)$, and the four planes $x = a$, $x = b$, $y = c$, $y = d$ perpendicular to the xy-plane. Moreover, the volume should be defined so as to satisfy certain elementary conditions:

1. If the three-dimensional region is a prism, i.e., if the function u is a constant k, the volume should be the product of the base by the height, $V=(b-a)(d-c)k$;
2. If we divide the rectangle R into smaller rectangles R_1 and R_2 by drawing straight lines, then the volume over R_1 should be equal to the volume over R_1 plus the volume over R_2;
3. If the three-dimensional region R_1 completely includes R_2, the volume of R_1 should be at least as large as that of R_2.

These considerations lead us to a method of defining V which is an immediate extension of the method of defining area in 2.1.1. By constructing lines parallel to the sides, we subdivide the rectangle R into smaller rectangles $R_1, R_2, \ldots, R_n$, the areas of which will be denoted by DR_1, DR_2, ... , DR_n. In each rectangle R_j, the fonction has a least value m_j and a larger value M_j, whence a prism with base R and height M_j completely includes the portion of the region over R_j, while this portion of the region contains the prism with base R_j and height m_j. Hence, we see that the vohume of the portion in question lies between m_jR_j and M_jR_j. Thus, the total volume V should be such that $\sum_{i=1}^{n} m_i \Delta R_i \leq V \leq \sum_{i=1}^{n} M_i \Delta R_i$

Now let the number n of rectangles increase beyond all bounds in such a way that the length of the longest diagonal tends to zero. Intuition leads us to expect that both the sums $\sum m_i \Delta R_i$ and $\sum M_i \Delta R_i$ will converge and tend to the same limit, whence we call this limit the volume V.

The reader will have observed that we have carried out an immediate generalization of the discussion. We call the common limit of the sums $\sum m_i \Delta R_i$ and $\sum M_i \Delta R_i$ the integral of the function $n = f(x, y)$ over the rectangle R and denote it by the symbol $\iint_R f(x,y)dr$

It is at once clear that, if we choose in each rectangle R_j a point (ξ_j, η_j) and find the corresponding value of the function $f(\xi_j, \eta_j)$, then there must hold the limiting relation

$$\lim_{n\to\infty} \sum f(\xi_i,\eta_i)\Delta R_i = \iint_R f(x,y)dr \, ;$$

in fact, the sum $\sum f(\xi_i,\eta_i)\Delta R_i$ lies between $\sum m_i \Delta R_i$ and $\sum M_i \Delta R_i$, both of which approach the integral as a limit.

As a particular method of subdividing r into smaller rectangles, we may subdivide the side $a \} x \} b$ into n intervals of length $\Delta x = (b - a)/n$ and the side $c \} y \} d$ into m intervals of length $\Delta y = (d - c)/m$, and then draw parallels to the axes through the points of division thus marked. The area of each rectangle R_j is then $\Delta R_j = \Delta x \Delta y$. Choosing a point (ξ_j, η_j) arbitrarily in each rectangle R_j, we form the sum

$$\sum f(\xi_i,\eta_i)\Delta R_i = \sum_i f(\xi_i,\eta_{i'})\Delta x\Delta y$$

As both n and m increase without limit, this sum approaches the integral as a limit. This type of subdivision suggests a second notation for the integral, which has been in common use since the time of Leibnitz, namely $\iint_R f(x,y)\,dxdy$.

The proof that such a limit exists if $u = f(x, y)$ is continuous can be carried out as in the A2.1). However. we shall assume without proof the even stronger statement:

The function $f(x, y)$ is continuous except along a finite number of smooth curves (curves with continuous derivatives) $y = f(x)$ or $x = \phi(y)$ along which $f(x,y)$ has jump discontinuities, then there exists double integral $\iint_R f(x,y)\,dxdy$.

We leave its proof to Volume II. It depends essentially on the fact that, as the number of rectangles increases, the total area of the rectangles having points in common with the curves of discontinmty tends to zero. Thus, even though M_j and m_j may differ considerably for such rectangles, they give rise to a little difference between the sums $\sum m_i\,\Delta R_i$ and $\sum M_i\,\Delta R_{i'}$.

With this assumption, we can find the area under surfaces $u = f(x, y)$ for which (x, y) ranges over quite complicated regions R. In fact, let the region R be bounded by a finite number of curves $x = \phi(y)$ or $y = \psi(x)$ with continuous derivatives and $f(x, y)$ be continuous in R. We enclose R in a rectangle R' and assign the points of R', which $\iint_R f(x,y)\,dr$ taken over the region R', as the volume under the surface $u = f(x, y)$, where (x, y) is in R. This integral is usually denoted by $\iint_R f(x,y)\,dxdy$.

Certain simple, but important theorems relating to these double integrals follow directly from the definition. We shall simply state the theorems, as the reader will be able to prove them without any trouble.

If $f(x, y)$ and $g(x, y)$ are integrable over a rectangle, then so are fg and cf, where c is a constant:

$$\iint_R \{f(x,y)\pm g(x,y)\}\,dr = \iint_R f(x,y)\,dr \pm \iint_R g(x,y)\,dr$$

If $f(x, y) \succ g(x, y)$ in R, then $\iint_R f(x,y)\,dr \geq \iint_R g(x,y)\,dr$

If $I_j R$ is the sum of two regions R_1 and R_2, then

$$\iint_R f(x,y)\,dr = \iint_{R_1} f(x,y)\,dr + \iint_{R_1} g(x,y)\,dr$$

Reduction of Double Integrals to Repeated Single Integrals

We now have a definition of the double integral with its interpretation as a volume and with the many possibilities of usefulness which our experience with the single integral suggests; however, as yet, we do not possess a method for evaluating such integrals. In this section, we shall see how the calculation of a double integral can be reduced to that of two single integrals.

Let $u = f(x, y)$ be a function which is defined and continuous in a rectangle R, $a \} x \} b, c \} y \} d$. If we fix upon any value x_0 in the interval $a \} x \} b$, the function $f(x_0, y)$ is a continuous function of the remaining variable y, whence there exists the integral $\int_c^d f(x_0, y)dy$ and it can be evaluated by the methods. This integral has a definite value for each value of x_0 which we may select; in other words, the integral is a function $\phi(x_0)$ of the quantity x_0:

$$\int_c^d f(x_0, y)dy = \phi(x)$$

For example, let $u = f(x, y) = xy$, $0\ x\ 1, 0\ y\ 3$. For each fixed x in the interval $\int_0^3 x^2 y^3 dy\, 0$ $x\ 1$, the integral can be evaluated and, in fact, is $81x/4$, i.e., it is a function of x. Or if $f(x, x) = e^{xy}$, $1\ x\ 2, 1\ y\ 4$, we have $\int^4 e^{xv} dy = \frac{1}{x}\left(e^{4x} - e^{x}\right)$

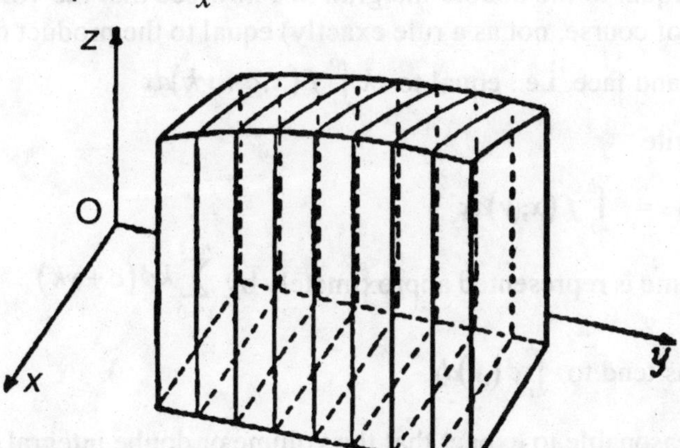

Having thus found the function $\phi(x)$, we can prove that it is continuous; this is a simple consequence of the uniform continuity of $f(x, y)$.It is therefore possible to integrate $\phi(x)$ between the limits a and b, thus obtaining the repeated integral

$$\int_a^b \phi(x)dx = \int_a^b\left(\int_c^d f(x,y)dy\right)dx$$

By reversing the order of the process, first calculating the function of y defined by $\int_a^b f(x, y)dx$ and then integrating from c to d, *we* obtain the other repeated integral $\int_c^b\left(\int_a^b f(x,y)dx\right)dy$

These integrals, as we have seen, are obtained by a double application of the ordinaly simple integration which we have studied before. Their importance lies in the following fact:

For continuous functions $f(x, y)$ and for functions $f(x, y)$ with at most jump discontinuities on a finite number of smooth curves, the repeated integrals are equal to the double integral:

$$\iint_R f(x,y)dr = \int_a^b\left(\int_c^d f(x,y)\,dy\right)dx$$
$$= \int_c^d\left(\int_a^b f(x,y)\,dx\right)dy$$

We shall content ourselves with an intuitive discussion of the case where $f(x,y)$ is continuous. In our original discussion of the double integral regarded as the volume lying above the and below the surface

$u = f(x, y)$, we obtained this volume by subdividing the solid into vertical colunms and then letting the diagonals of the bases of these columns approach zero. Instead of this, we can divide the solid into slices of thickness $k=(d-c)/n$ by drawing the lines $y = c + nk$ ($n = 0, 1, \ldots, n$) parallel to the x-axis and then constructing a plane perpendicular to the xy-plane through each line.

These planes cut the solid into n slices which grow thinner as n increases and the total volume of which is equal to the double integral. We now see that the volume of each slice is approximately (but, of course, not as a rule exactly) equal to the product of the thickness k by the area of the left-hand face, i.e., equal to $k\int_a^b f(x, c+\nu k)dx$

Hence, if we write

$$\phi(y) = \int_a^b f(x,y)dx$$

the desired volume is represented approximately by $\sum_{\nu=0}^{n-1} k\phi(c+\nu k)$

As n, these sums tend to $\int_a^b \phi(y)dy$

It is therefore reasonable to expect that the volume or doube integral is exactly equal to

$$\int_a^b \phi(y)dy = \int_c^d\left(\int_a^b f(x,y)\,dx\right)dy,$$

which is the statement made ahove. A similar discussioa makes it equally plausible that the statement

$$\int_a^b\left(\int_c^d f(x,y)\,dy\right)dx = \iint_R f(x,y)\,dr$$ is also true.

MULTIPLE INTEGRALS

Integrating functions of several variables

Recall that we think of the integral in two different ways. In one way we interpret it as the area under the graph $y = f(x)$, while the fundamental theorem of the calculus enables us to compute this using the process of "anti-differentiation" — undoing the differentiation process.

We think of the area as

$\sum f(x_i)\, dx_i = \int f(x)\, dx,$

where the first sum is thought of as a limiting case, adding up the areas of a number of rectangles each of height $f(x_i)$, and width dx_i. This leads to the natural generalisation to several variables: we think of the function $z = f(x, y)$ as representing the height of f at the point (x, y) in the plane, and interpret the integral as the sum of the volumes of a number of small boxes of height $z = f(x, y)$ and area $dx_i dy_j$. Thus the volume of the solid of height $z = f(x, y)$ lying above a certain region R in the plane leads to integrals of the form $\int\int_R = \sum_{i=1}^{n}\sum_{j=1}^{n} f(x_i, y_j)dx_i dy_j = \lim S_{mn}$.

We write such a double integral as $\int\int_R f(x, y)\, dA$.

Repeated Integrals and Fubini's Theorem

As might be expected from the form, in which we can sum over the elementary rectangles $dx\, dy$ in any order, the order does not matter when calculating the answer: There are two important orders — where we first keep x constant and vary y, and then vary x; and the opposite way round. This gives rise to the concept of the repeated integral, which we write as

$$\int\left(\int_R f(x,y)\,dx\right)dy \quad or \quad \int\left(\int_R f(x,y)\,dy\right)dx.$$

Our result that the order in which we add up the volume of the small boxes doesn't matter is the following, which also formally shows that we evaluate a double integral as any of the possible repeated integrals.

Theorem 9.1 (Fubini's theorem for Rectangles) Let $f(x, y)$ be continuous on the rectangular region R : *axb;cyd*. Then

$$\int\int_R f(x,y)\,dA = \int_c^d\left(\int_a^b f(x,y)\,dx\right)dy = \int_a^b\left(\int_c^d f(x,y)\,dy\right)dx$$

Note that this is something like an inverse of partial differentiation. In doing the first inner (or repeated) integral, we keep y constant, and integrate with respect to x. Then we integrate with respect to y. Of course if f is a particularly simply function, say $f(x, y) = g(x)h(y)$, then it doesn't matter which order we do the integration, since

$$\int\int_R f(x,y)\,dA = \int_a^b g(x)\,dx \int_c^d h(y)\,dy$$

We use the Fubini theorem to actually evaluate integrals, since we have no direct way of calculating a double (as opposed to a repeated) integral.

Example: Integrate $z = 4 - x - y$ over the region $0 \leq x \leq 2$ and $0 \leq y \leq 1$. Hence calculate the volume under the plane $z = 4 - x - y$ above the given region.

Solution. We calculate the integral as a repeated integral, using Fubini's theorem.

$$V = \int_{x=0}^{2}\int_{y=0}^{1}(4-x-y)\,dy\,dx = \int_{x=0}^{2}\left[\left(4y-xy-y^2/2\right)\right]_0^1 dx = \int_0^2 (4-x-1/2)\,dx$$

From our interpretation of the integral as a volume, we recognise V as volume under the plane $z = 4 - x - y$ which lies above $\{(x, y) \mid 0 \leq x \leq 2; 0 \leq y \leq 1\}$.

Exercise 9.3 Evaluate $\int_0^3 \int_0^2 \left(4 - y^2\right) dy\, dx$, and sketch the region of integration.

In fact Fubini's theorem is valid for more general regions than rectangles. Here is a pair of statements which extend its validity.

Theorem 9.4 (Fubini's theorem — Stronger Form) Let $f(x, y)$ be continuous on a region R

- If R is defined as $a \leq x \leq b; g_1(x) \leq y \leq g_2(x)$. Then

$$\iint_R f(x,y)\,dA = \int_a^b \left(\int_{g_1(x)}^{g_2(x)} f(x,y)\,dy\right)dx$$

- If R is defined as $c \leq y \leq d; h_1(y) \leq x \leq h_2(y)$. Then

$$\iint_R f(x,y)\,dA = \int_c^d \left(\int_{h_1(y)}^{h_2(y)} f(x,y)\,dx\right)dy$$

Proof. We give no proof, but the reduction to the earlier case is in principle simple; we just extend the function to be defined on a rectangle by making it zero on the extra bits. The problem with this as it stands is that the extended function is not continuous. However, the difficulty can be fixed.

This last form enables us to evaluate double integrals over more complicated regions by passing to one of the repeated integrals.

Evaluate the Integral

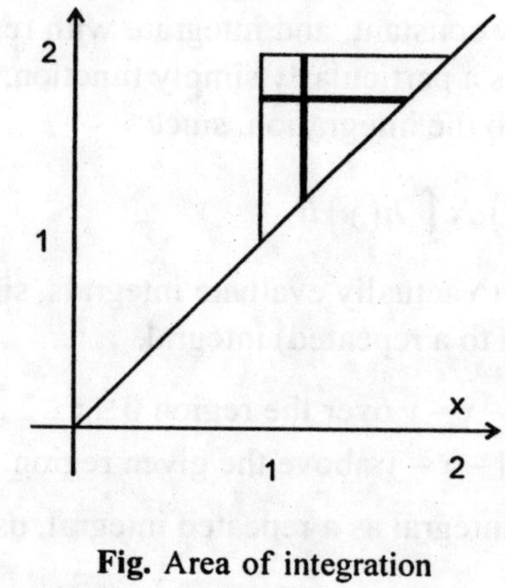

Fig. Area of integration

$\int_1^2\left(\int_x^2 \frac{y^2}{y^2}dy\right)dx$ **as it stands, and sketch the region of integration.**

Reverse the order of integration, and verify that the same answer is obtained.

Solution. **The diagram shows the area of integration.**

We first integrate in the given order.

$$\int_1^2\left(\int_x^2 \frac{y^2}{x^2}dy\right)dx = \int_1^2\left[\frac{y^3}{3x^2}\right]_x^2 dx \int_1^2\left(\frac{8}{3x^2}-\frac{x}{3}\right)$$

$$= \left[\frac{8}{3x}-\frac{x^2}{6}\right]_1^2 = \left(\frac{4}{3}-\frac{2}{3}\right)\left(\frac{8}{3}-\frac{1}{3}\right) = \frac{5}{6}$$

Reversing the order, using the diagram, gives

$$\int_1^2\left(\int_1^y \frac{y^2}{x^2}dx\right)dy = \int_1^2\left[-\frac{y^2}{x}\right]_x^y dy = \int_1^2\left(-y+y^2\right)dy$$

$$= \left[-\frac{y^2}{2}-\frac{y^3}{3}\right]_1^2 = \left(-2+\frac{8}{3}\right)-\left(-\frac{1}{2}+\frac{1}{3}\right) = \frac{5}{6}$$

Thus the two orders of integration give the same answer.

Another use for the ideas of double integration just automates a procedure you would have used anyway, simply from your knowledge of 1-variable results.

Chapter 11

Differential Equations

Integral Calculus

Integration is the reverse of differentiation. With differentiation breaking a process down to look at the instantaneous process, integration sums up the information from small time intervals to give a total result over a larger time period.

One example of this is the area under the plasma concentration time curve. Later we shall learn that this summation or integration process can be used to evaluate dosage forms, that is it can be used as a measure of performance.

Using integral calculus we can go in the reverse direction. In the section above we converted from the rate of change equation to an equation for X. We can also go further and get an area under the curve, which is a further integration.

Another example is the progression from distance, to speed (the rate of change of distance), to acceleration (the rate of change of speed).

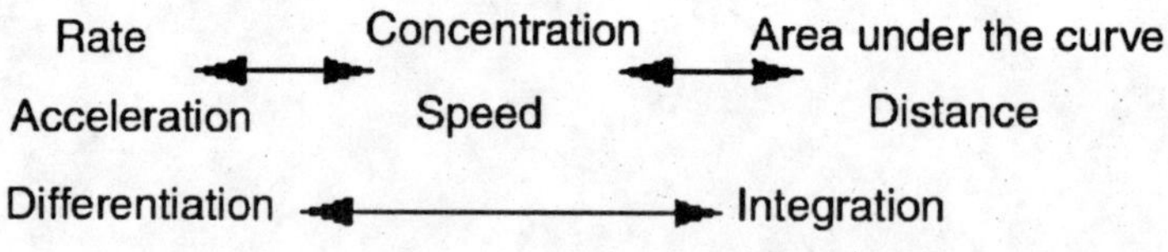

Fig: Relationship between Rate and Integral

Rate processes in the field of pharmacokinetics are usually limited to first order, zero order and occasionally Michaelis-Menten kinetics. Linear pharmacokinetic systems consist of first order disposition processes and bolus doses, first order or zero order absorption rate processes. These rate processes can be described mathematically.

First Order Equation $\dfrac{dX_1}{dt} = -k_1 \bullet X_1$

Each first order rate process ("arrow") is described by a first order rate constant (k_1) and the amount or concentration remaining to be transferred (X_1).

Zero Order Equation $\frac{dX_1}{dt} = -k_1$

Zero order rate processes are described by the rate constant alone. Amount or concentration to the zero power is 1.

Michaelis Menten Equation $\frac{dX_1}{dt} = \frac{Vm \bullet X_1}{Km + X_1}$

The Michaelis Menten process is somewhat more complicated with a maximum rate (velocity, Vm) and a Michaelis constant (Km) and the amount or concentration remaining.

The full differential equation for any component of a pharmacokinetic model can be constructed by adding an equation segment for each arrow in the pharmacokinetic model. The rules for each segment:

1. Direction of the arrow
   - If the arrow goes into the component the equation segment is positive
   - If the arrow leaves the component the equation segment is negative.
2. Type of rate process
   - If the rate process is first order multiply the (first order) rate constant by the amount or concentration of drug in the component at the tail of the arrow.
   - If the process is zero order just enter the rate constant.
   - For a Michaelis Menten processes include the amount or concentration of drug in the component at the tail of the arrow in the Michaelis-Menten equation.

An example

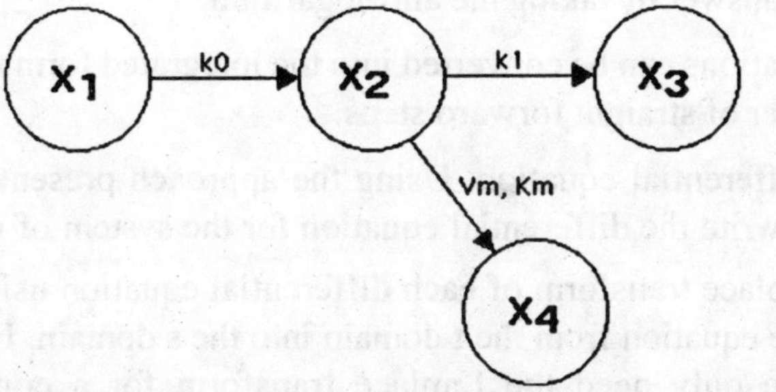

Fig. An example pharmacokinetic model with zero order, first order and Michaelis Menten processes. In Figure the rate process from one to two is zero order. The process from two to three is first order and the process from two to four follows Michaelis-Menten kinetics. We can now systematically write the differential equations for each component of the model.

- Component 1 There is one arrow leading out of component one so the rate process if negative. This process is zero order so we just write the rate constant. The equation for this component is: $\frac{dX_1}{dt} = -k0$

- Component 2 This component is more of a challenge. There are three arrows connected with this component. One arrow leads to the compartment so this equation segment is positive. The other two arrows lead away from the component and these equation segments are negative. Starting with the zero order process provides is a positive k0 to the differential equation. The first order process is developed as the rate constant multiplied by the amount remaining in component 2. This is negative and is $-k_1 \times X_2$. The final segment is the Michaelis Menten process from component 2 to component 4. This is also negative and is described as $-Vm \times X_2/(Km + X_2)$. The total differential equation for this component is: $\frac{dX_2}{dt} = -k0 - k_1 \bullet X_2 - \frac{Vm \bullet X_2}{Km + X_2}$
- Component 3 OK; downhill from here. Now we have one arrow going to this component from component 2, thus the equation segment is positive. The rate process is first order so we multiply the rate constant by the amount remaining in the component at the tail of the arrow, component 2. The equation is: $\frac{dX_3}{dt} = K_1 \bullet X_2$
- Component 4 The fourth component is also described with one equation segment. The arrow leads to this component so the segment is positive. The rate process is a Michaelis Menten process. The equation for this component is: $\frac{dX_4}{dt} = \frac{Vm \bullet X_2}{Km + X_2}$

Integrating Differential Equations using Laplace Tranforms

Laplace transforms are a convenient method of converting differential equations into integrated equations, that is, integrating the differential equation. It is similar to the use of logarithms to multiple or divide numbers. To multiple two numbers we convert each number into their respective logarithm and add. The sum is converted into the 'anti'-logarithm and the product of the original numbers is the result. For division, the logarithm of the two numbers are subtracted before obtaining the answer by taking the anti-logarithm.

Differential equations can be converted into the integrated form using Laplace transforms by following a number of straight forward steps.

1. Write the differential equation. Using the approach presented on the previous page you need to write the differential equation for the system of interest.
2. Take the Laplace transform of each differential equation using a few transforms. We transform the equation from the t domain into the s domain. For most pharmacokinetic problems we only need the Laplace transform for a constant, a variable and a differential.
3. Use some algebra to solve for the Laplace of the system component of interest. Often the Laplace of a component 'up-stream' will need to be solved first and substituted into the equation of interest.
4. Finally the 'anti'-Laplace for the component is determined from tables such as that below or by using the 'finger-print' method describe later.

In step two above, only three Laplace transforms are necessary for most of the linear of pharmacokinetic systems that you might encounter. These are:

The Laplace of a constant:

$$L(A) = \frac{A}{s}$$

The Laplace of a variable (i.e. amount in a component of the model; e.g. X_1):

$$L(A \bullet X_1) = A \bullet \bar{X}_1$$

The Laplace of the differential of a variable (e.g. dX_1/dt):

$$L\left(\frac{dX_1}{dt}\right) = s \bullet \bar{X}_1 - X_1(0)$$

Time for some examples

IV Bolus - Linear one compartment model

The IV bolus does is placed in component one at zero time. Elimination is by a single first order process described by the rate constant kel (elimination rate constant).

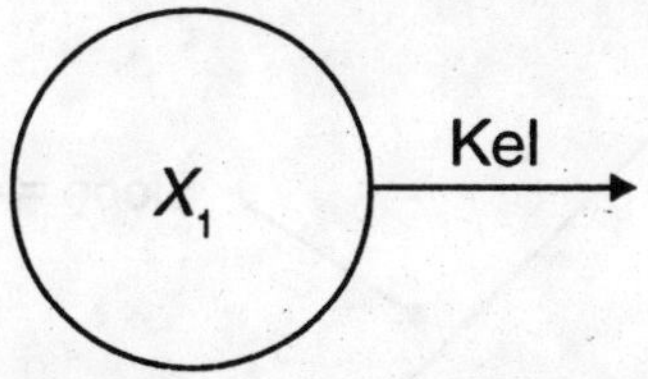

Fig. Linear one compartment model with one elimination process

1. **Write the differential equation for X_1. There is only one component.**

   $$\frac{dX_1}{dt} = -\text{kel} \bullet X_1$$

2. **Transform the equation into the Laplace form**

   $$s \bullet \bar{X}_1 - X_1(0) = -kel \bullet \bar{X}_1$$

3. **Rearranging and solving for $L(X_1)$. Substitute Dose for $X_1(0)$.**

   $$\bar{X}_1 \bullet (s + \text{kel}) = X_1(0) = \text{Dose}$$

   $$\bar{X}_1 = \frac{Dose}{(s + kel)}$$

4. **The final result can be determined from the Laplace Transform table (below - line 3 with A == Dose: a == kel).**

   $$X_1 = \text{Dose} \bullet e^{-\text{kel} \bullet t}$$

INTEGRATING DIFFERENTIAL EQUATIONS NUMERICALLY

Point-Slope methods - Euler's Method

If you don't like Laplace transforms you might prefer using a numerical method of integrating the differential equations that you might find in pharmacokinetic studies. Actually a computer program such as Boomer will do this for you but doing this once or twice using a calculator might be informative. There are a number of different types of numerical integration methods but the easiest to understand is probably the point-slope methods. Euler's method is the simplest of these point-slope type of numerical integration. If we have an initial amount or concentration (the point) and a differential equation (the slope) Euler's method could be used. As an example consider an IV Bolus dose and a linear one compartment mode. At time zero we have a Dose ($X_1(0)$) and a slope expressed as the differential equation for a one compartment linear model:

$$\frac{dX_1}{dt} = -\mathrm{kl} \bullet X_1$$

This first step of Euler's method can be illustrated graphically as shown.

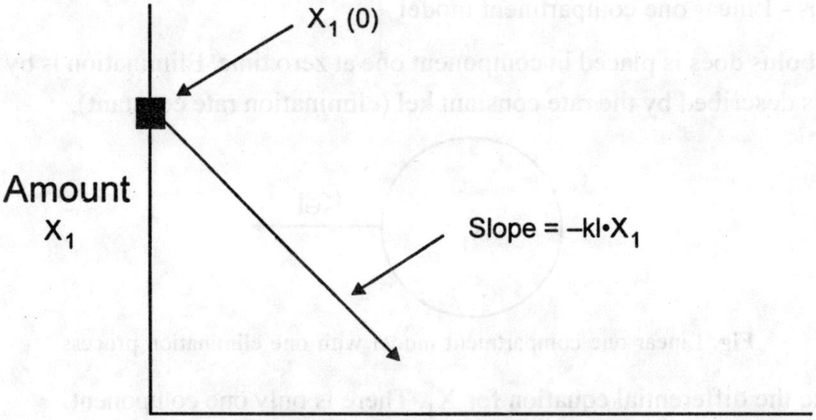

Fig. A first step using Euler's method

In practice the method involves using this initial slope for a finite (small) stepsize (in the time direction). The method is then repeated for another stepsize using the result from the previous step as the new starting point. If we start with a dose of 100 mg, k1 of 0.25 hr^{-1} and a step size of 0.1 hour the amount of drug remaining is 97.5 mg.

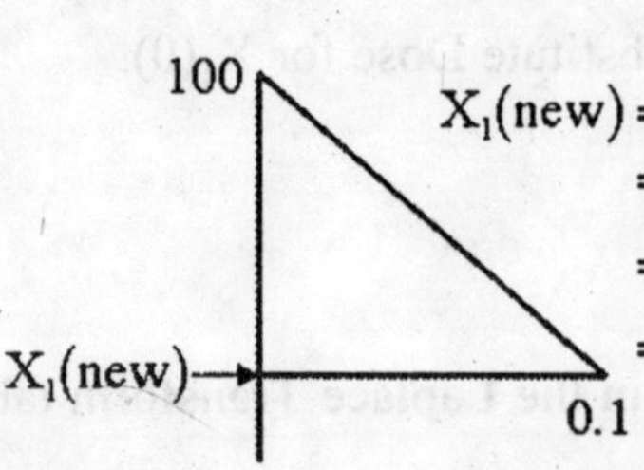

$$\begin{aligned} X_1(\text{new}) &= X_1(\text{old}) + \text{slope} \bullet \text{ss} \\ &= X_1(\text{old}) + (-\text{k1} \bullet X_1(\text{old})) \bullet \text{ss} \\ &= 100 + (-0.25 \times 100) \times 0.1 \\ &= 100 - 2.5 = 97.5 \end{aligned}$$

Fig. One step with Euler's method

If this process is continued we can estimate the amount remaining at the time or times of interest.

Using the integrated equation for this calculation gives an *exact* amount of 77.88 mg after 1 hour. This means Euler's method has an error rate of 0.32 % with this stepsize. With a stepsize of 0.05 hour (20 steps) the error is 0.16%. Even smaller steps and more steps would be needed to more accurate results. This might illustrate a limitation of Euler's method. Although the method is simple to implement very small stepsizes are needed to achieve accurate results.

Area Under the Plasma Concentration Time Curve (AUC)

The area under the plasma (serum, or blood) concentration versus time curve (AUC) has an number of important uses in toxicology, biopharmaceutics and pharmacokinetics.

ToxicologyAUC can be used as a measure of drug exposure. It is derived from drug concentration and time so it gives a measure how much - how long a drug stays in a body. A long, low concentration exposure may be as important as shorter but higher concentration. Some drugs are dosed using AUC to quantitate the maximum tolerated exposure (AUC Dosing).

BiopharmaceuticsThe AUC measured after administration of a drug product is an important parameter in the comparison of drug products. Studies can be performed whereby different drug products may be given to a panel of subject on separate equations. These bioequivalency or bioavailability studies can be analysed by comparing AUC values.

Pharmacokinetics Drug AUC values can be used to determine other pharmacokinetic parameters, such as clearance or bioavailability, F. Similar techniques can be used to calculate area under the first moment curve (AUMC) and thus mean resident times (MRT).

Calculation of AUC using the Trapezoidal Rule

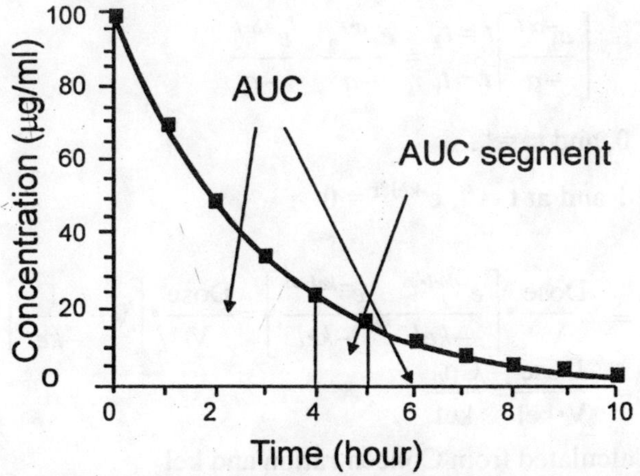

Fig. Linear Plot of Cp versus Time showing AUC and AUC segment

The area under the plasma concentration time curve (**AUC**) is very useful for calculating the relative efficiency of different drug products. It can used to calculate the total body clearance (CL) and the apparent volume of distribution.

If we have a smooth line for concentration versus time or an equation for Cp versus time from a pharmacokinetic model we could slice the area into vertical segments. Each segment would be very thin, Ät and in extreme dt, in width. The total AUC is calculated by adding these segments together. In calculus this would be the integral. Each very narrow segment has an area = Cp*dt. Thus the total area (AUC) is given by Equation:

$$AUC = \int_{t=0}^{t=\infty} Cp_t \bullet dt$$

Equation. Total AUC calculated from Very Narrow Segment

The integrated equation for plasma concentration as a function of time is:

$$Cp_t = Cp_0 \bullet e^{-kel \bullet t} = \frac{\text{Dose}}{\text{V}} \bullet e^{-kel \bullet t}$$

Equation. Cp versus Time after an IV Bolus dose

This is essentially the same as the integrated equation we derived for amount of drug remaining using Laplace transforms. The difference here is the use of V, apparent volume of distribution, to convert amount into concentration. We can substitute this equation for Cp_t into Equation to derive an analytical equation for AUC:

$$AUC = \int_{t=0}^{t=\infty} \frac{\text{Dose}}{\text{V}} \bullet e^{-kel \bullet t} \bullet dt = \frac{\text{Dose}}{\text{V}} \bullet e \int_{t=0}^{t=\infty} e^{-kel \bullet t}$$

Equation. AUC calculated as the integral of Cp versus time

From Math Table we would find that:

$$\int_{t=t_1}^{t=t_2} e^{-a \bullet t} \bullet dt = \left[\frac{e^{-a \bullet t}}{-a} \right]_{t=t_1}^{t=t_2} = \frac{e^{-a \bullet t}{}_2}{-a} - \frac{e^{-a \bullet t}}{-a}$$

With a = kel, $t_1 = 0$, and t_2 = ",

At t = 0, $e^{-kel*t} = 1$ and at t = ", $e^{-kel*t} = 0$

Therefore:

$$AUC_{0-\infty} = \frac{\text{Dose}}{\text{V}} \bullet \left[\frac{e^{-kel \bullet \infty}}{-kel} - \frac{e^{-kel \bullet \infty}}{-kel} \right] = \frac{\text{Dose}}{\text{V}} \bullet \left[0 - \frac{1}{-kel} \right]$$

$$AUC_{0-\infty} = \frac{\text{Dose}}{\text{V} \bullet \text{kel}} = \frac{\text{Cp}_0}{\text{kel}}$$

Equation. AUC Calculated from Concentration and kel

This is analytical integration (exact solution, given exact values for V and kel)

Note: For t = 0 to ∞ , AUC = Cp_0/kel. With t = t to ∞ , AUC is calculated as Cp_t/kel. We will use this result further down on this page

That is:

$$AUC_{t-\infty} = \frac{\text{Dose}}{\text{V}} \bullet \left[\frac{e^{-kel\bullet\infty}}{-kel} - \frac{e^{-kel\bullet t}}{-kel}\right] = \frac{\text{Dose}}{\text{V}} \bullet \left[0 - \frac{e^{-kel\bullet t}}{-kel}\right]$$

$$AUC_{t-\infty} = \frac{\text{Dose}\bullet e^{-kel\bullet t}}{\text{V}\bullet \text{kel}} = \frac{\text{Cp}_t}{\text{kel}}$$

Rearranging Equation to solve for V gives:

V = Dose/(AUC * kel)

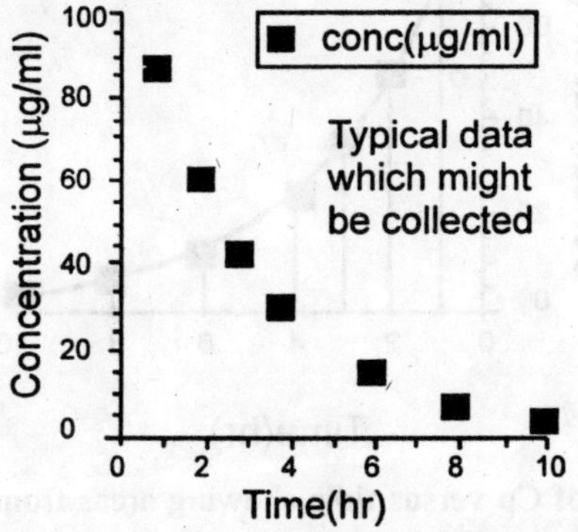

Fig. Linear Plot of Cp versus Time showing Typical Data Points

We can calculate the AUC of each segment if we consider the segments to be trapezoids.

The area of each segment can be calculated by multiplying the average concentration by the segment width. For the segment from Cp_2 to Cp_3:

for example: example : $AUC_{2-3} = \frac{Cp_2 + Cp_3}{2} \bullet (t_3 - t_2)$

This segment is illustrated in figure below.

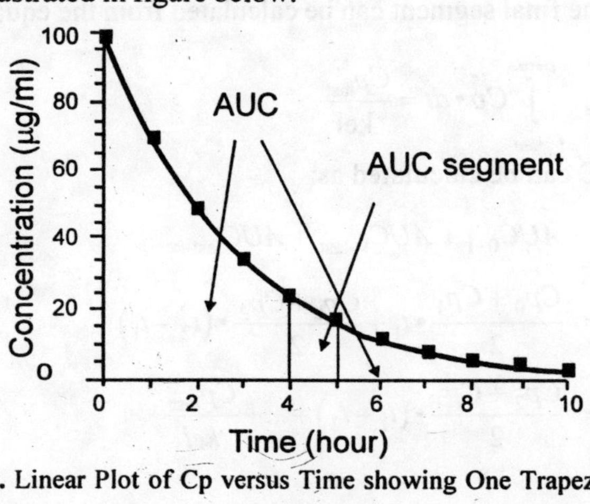

Fig. Linear Plot of Cp versus Time showing One Trapezoid

The area from the first to last data point can then be calculated by adding the areas together.

$$AUC_{1-n} = \sum\left\{\frac{Cp_1 + Cp_2}{2}\bullet(t_2 - t_1)\right\} + \left\{\frac{Cp_2 + Cp_3}{2}\bullet(t_3 - t_2)\right\} + \ldots$$

Note: Summation of data point information (non-calculus)

This gives:

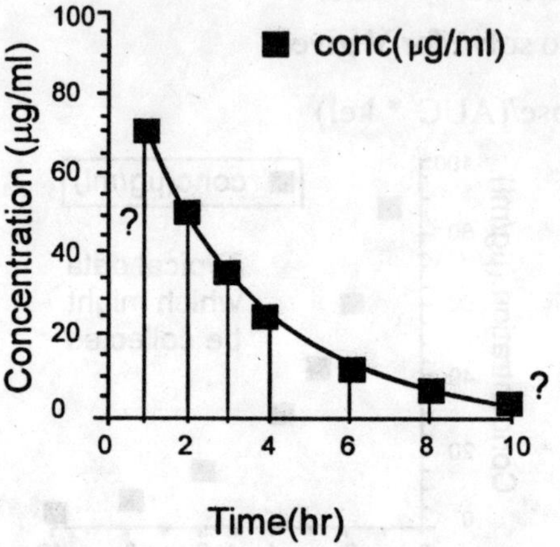

Fig. Linear plot of Cp versus time showing areas from data 1 to data n

To finish this calculation we have two more areas to consider. The first and the last segments.

After a rapid IV bolus, the first segment can be calculated after determining the zero plasma concentration Cp^0 by extrapolation.

Thus

$$AUC_{0-1} = \left\{\frac{Cp_0 + Cp_1}{2}\bullet(t_1)\right\}$$

If we assume that the last data points follow a single exponential decline (a straight line on semi-log graph paper) the final segment can be calculated from the equation above from t_{last} to infinity:

$$AUC_{t_{last}-\infty} = \int_{t=t_{last}}^{t=\infty} Cp \bullet dt = \frac{Cp_{\text{last}}}{\text{kel}}$$

Thus the total AUC can be calculated as:

$$AUC_{0-\infty} = AUC_{0-1} + AUC_{1-\text{last}} + AUC_{\text{last}-\infty}$$

$$= \frac{Cp_0 + Cp_1}{2}\bullet t_1 + \frac{Cp_1 + Cp_2}{2}\bullet(t_2 - t_1)$$

$$= \frac{Cp_2 + Cp_3}{2}\bullet(t_1 - t_2) + \ldots + \frac{Cp_{\text{last}}}{kel}$$